THE WHETSTONE OF WITTE

ROBERT RECORDE

THE WHETSTONE OF WITTE

A FACSIMILE OF THE
SOLE EDITION
IMPRINTED AT LONDON BY
JOHN KINGSTON
1557

RENASCENT BOOKS

THE WHETSTONE OF WITTE
First imprinted by Iohn Kyngstone in 1557

This facsimile edition published by
TGR Renascent Books
27 Springdale Court
Mickleover, Derby DE3 9SW
United Kingdom
2010

Paperback edition first published 2013

ISBN 978-1-4825893-0-6
www.renascentbooks.co.uk

Origination, layout and typesetting by
Gordon Roberts
Transcript and proofing by
Elizabeth Roberts

Printed and bound by
CreateSpace, Charleston, South Carolina, U.S.A.

for
HENRY

INTRODUCTION

This book is a facsimile of Robert Recorde's *The Whetstone of Witte*, originally printed in London by John Kingston in 1557. Like most of Recorde's books, it is written in the form of a dialogue between a master and a rather clever scholar. Robert Recorde was a consummate Renaissance man; a physician at the courts of Henry VIII, Edward VI and Mary, a very learned scholar, an able mathematician and a teacher of outstanding ability. He wrote a progression of justly famed mathematical textbooks with the following titles: *The Grounde of Artes, The Pathway to Knowledge, The Gate of Knowledge, The Castle of Knowledge, The Treasure of Knowledge* and *The Whetstone of Witte*. Of these six, *The Gate* and *The Treasure* are no longer extant. The last book in the series (the one you now hold) is partly an additament to Recorde's treatise on arithmetic, begun in *The Grounde* and here continued as *The Seconde Parte of Arithmetike* and *The Extraction of Rootes*. But perhaps of more importance for the history of mathematics, the book also contains a detailed treatment of *The Arte of Cossike Nombers*. As such it can be seen historically as the first book on algebra written in the English language. It also contains a treatise on *The Arte of Surde Nombers*. While *surd* remains in use in modern-day mathematics (meaning irrational numbers), *cossike* (or *cossic*) is now completely archaic and so perhaps requires explanation. The word *cosa* is Latin for 'thing' and was used for the unknown in early European texts on algebra. Where today we refer to someone who studies algebra as an algebraist, so in earlier times were they called cossists and for many years algebra was known as the cossic art.

Scholarly study of Recorde's treatise has shown that he was much influenced by the works of German algebraists, in particular by the book *Coss*. This was published in 1525 by Christoff Rudolff and is generally acknowledged to be the first German algebra text. Recorde adopted Rudolff's algebraic symbols, including those for the square, cube and fourth roots, in their entirety. Like many contemporary German mathematicians, Recorde also adopted Rudolff's terminol-

Introduction

ogy for powers, for example *zenszensdezens* (the eighth power), which he anglicised so that in *The Whetstone* it is rendered *zenzizenzizenzike*. This particular terminology is discussed in more detail below. Since, so far as is known, Recorde did not read German, he may also have made use of Michael Stifel's *Arithmetica Integra*, published in Latin in 1544, in which Stifel utilised the symbols and terminology of his German compatriot Rudolff.

BRIEF BIOGRAPHY

Robert Recorde was born between 1510 and 1512 in Tenby, Pembrokeshire. He was the second son of Thomas Recorde and Rose Jones. Practically nothing is known of his childhood and the first thing in his life about which we can be certain is his entrance into Oxford University in about 1525. It is not known what he studied but we may assume it was the usual (for the time) *trivium* of grammar, logic and rhetoric, followed by the *quadrivium* of arithmetic, geometry, music and astronomy. He graduated with a B.A. in 1531 and was elected a Fellow of All Souls College in the same year. He may have taught at Oxford for a few years but the evidence for this is scanty. At some time he moved from Oxford to Cambridge, where he studied for an M.D. and graduated in 1545 at the age of 35. He then moved to London, where for a few years he practised medicine. In later years he was always to describe himself as 'physician'. A defining moment in his life occurred in 1549 when he was appointed Controller of the Bristol Mint. It was during his time there that he made a very powerful and ruthless enemy. Sir William Herbert was sent by Edward VI to help suppress a revolt by John Dudley, Earl of Warwick, in the west country. Herbert demanded that Recorde divert funds from the mint to pay and support his army, but Recorde refused on the grounds that the order did not come from the king. Herbert countered and accused Recorde of treason. He was lucky to incur the mild penalty of confinement to court for 60 days. However, apparently all was later forgiven because in 1551 he was appointed general surveyor of Mines and Monies in Ireland. He was placed in charge of the Wexford silver mines and also became the technical supervisor of the Dublin mint. In the meantime, Sir William Herbert was created Earl of Pembroke for his services to the crown during the rebellion, and there was continued animosity between him and Recorde. Although the silver mines at Wexford had great potential, the enterprise was largely unsuccessful, mainly due to a lack of royal investment and the imperfect state of mining technology. The mines closed in 1553 and Recorde was re-called to England. Upon the accession to the throne of Mary, the daughter of Henry VIII, Recorde's old enemy the Earl of Pembroke was made a privy councillor for his support of Mary's claim to the throne. For some strange reason, Recorde chose the moment when Pembroke was strongest to try and get his revenge, charging him with misconduct in gaining his court positions. The allegation was probably true, but Pembroke was in favour with the monarchy and so had almost perfect immunity. He responded by suing Recorde for libel. There was a hearing in January 1557 and Recorde was ordered to pay the huge sum of £1000 compensation. He either could not or would not pay and so was sentenced to imprisonment in the King's Bench Prison in Southwark, for debt. Whilst in prison he made his will, leaving small sums of money to various people, including £20 to his mother. The date of his death is not known with any certainty, but is generally supposed to have been in the later part of 1558, only a short time after making his will.

The notes which follow are provided as a guide to reading, un-derstanding and enjoying this facsimile edition of Recorde's last, and perhaps greatest, mathematical work.

PAGINATION
Pages are not numbered consecutively as in modern practice. How-ever, most right-hand pages do bear the printer's signature numbers quite prominently at the bottom. This system of numbering was used by the printer to ensure the correct gathering and folding of pages prior to sewing and binding. The signatures of the front mat-ter are signed with the lowercase letters a to b, the main text uses the uppercase letters A to Z, then Aa to Rr. Intermediate pages in the signatures are numbered in Roman numerals – i, ii, and iii. Since

Introduction

only right-hand pages carry a signature number, when referencing pages it is usual to label right-hand pages as recto and left-hand pages as verso; thus A.i.r, A.i.v, A.ii.r, A.ii.v, etc. Those recto pages which do not have a signature number are simply the other half of a folded leaf, and so for referencing it is necessary to extrapolate the numbering forward to the start of the next signature. However, for the benefit of modern day readers who expect consistent pagination, modern page numbering is applied over a line drawn above Recorde's text.

SPELLINGS AND PUNCTUATION

Dictionaries and standard spellings did not exist when *The Whetstone of Witte* was first printed, and many words are not even spelt phonetically. Therefore many spellings in the book appear peculiar to the modern reader, but a little practice at reading Early Modern English soon renders the text intelligible. Many familiar words look strange simply because, unlike modern spellings, they end with the silent letter e and the last consonant might or might not be doubled, hence *mane* or *manne* (man), and *rune* or *runne* (run). The letter y is often used in place of i, for example *fynde* (find) or *fyrste* (first). Early printing conventions were to use the terminal letter s at the end of words, as today, but the long form everywhere else, for example *poſſeſs* (possess). The letters u and v were not considered to be two distinct letters, but different forms of the same letter. Typographically, v was often used at the start of words and u elsewhere, hence *vnmoued* (unmoved) or *vnloued* (unloved). But conversely, the letter v was often used where today we would expect the letter u, as in, for example, *thervnto* (thereunto). Neither were the letters i and j considered distinct, so that the word judge would be spelt *iudge*. In short, and for example, expect to read modern spellings such as sum, divisor and just, as *ſomme, diuiſor* and *ivst*.

Punctuation appears eccentric to modern eyes – for example, commas are sometimes present when (by modern standards), their use is not required. Conversely they are often missing where present-day usage would insert them. Full stops and the colon are used arbitrarily and interchangeably, and the ends of sentences are sometimes

Introduction

not terminated by any punctuation at all. The first word of a following sentence might or might not be capitalised.

CONTRACTIONS

Often the words 'the' and 'that' are contracted to y^e and y^t respectively, with the small letters e and t placed directly above the y. Of course, these contractions should be read in the text with their full pronunciation – 'the' and 'that'. Note that y^e (the) is not to be confused with *ye* (you).

DIACRITICAL MARKS

Diacritical marks have been used to abbreviate printed words ever since Gutenberg, and early English printers adopted the same conventions that Gutenberg did for Latin texts (which he copied, in turn, from the handwritten texts of medieval scribes). Diacritical marks are used on many pages of *The Whetstone of Witte* to indicate the omission of the consonant m or n where this follows a vowel. The missing letter is indicated by placing the mark (a bar) over the vowel. Instances are *exāple* (example), *scā* (scan), *takē* (taken), *quotiēt* (quotient), *ī* (in), *frō* (from), *nōber* for *nomber* (number) and *chaūce* for *chaunce* (chance). Words abbreviated with diacritical marks should also be read, of course, with their full pronunciation.

TYPOGRAPHICAL FEATURES

All the contractions and abbreviations found in the pages of *The Whetstone of Witte* are compositors tricks to help in the justification of entire paragraphs – something that was considerably easier in the days before orthography and standardised spellings. Justification of paragraphs was not merely a cosmetic feature (as it is today). Early printers would be laying out discrete pieces of movable metal type into a square wooden frame and if the frame was not completely filled, the types would move under the action of the press and perhaps even fall out. In other words, each line of each paragraph had to extend fully from left to right, with the letters 'jammed' in the frame or the page would be unprintable.

One way for Renaissance printers to do this might have been by inserting blank spaces of suitable lengths between the words on each

Introduction

line, but this is not a satisfactory solution. The result is usually 'rivers' of blank space flowing down the page, which seriously interrupt reading and which was recognised as a problem from the very earliest days of printing. Hence the use of aggressive hyphenation, contracted words, diacritical marks and variant spellings of the same word, like hed (which has three letters), head (which has four), or hedde (which has five), all very useful when striving to obtain justification and spelling is not a problem. The compositor would use any or all of these tricks at will in order to obtain a solid block of text on each page.

It was also an early printing convention to enclose numbers within dots, or full stops, ostensibly to distinguish counting numbers from the cardinal numbers used to represent words like first, second, third, etc. The compositors of *The Whetstone of Witte* have made much use of dots around numbers, to the point of possibly irritating readers by their overuse. Conversely, they show no compunction about omitting the dots whenever the need to justify a line of text demands it. It might easily be thought that the use or omission of dots surrounding numbers and algebraic symbols within these pages is quite arbitrary, done more for typographical layout reasons than for any increase in clarity or meaning.

FAULTS
There are no errata in *The Whetstone of Witte*. However, the attentive reader will find some faults and errors in its pages and, of course, no attempt has been made to indicate or remedy these in this facsimile reprint. Sometimes spelling mistakes occur despite the relaxed spelling conventions of the time, although these should more properly be regarded as typographical errors rather than mistakes in spelling as we understand the term today. Examples are 'sholar' (scholar) and 'rooe' (roote - root), where the letter c and the letter t respectively have been omitted and the words as printed are obviously wrong. Occasionally letters in a word are accidently transposed and so the meaning can sometimes be a little puzzling. Another example of what appears at first sight to be an odd word is 'likemaies' (likeways

Introduction

– likewise), where the w has obviously been printed upside down. A good example of a peculiar spelling in these pages (to modern eyes) is 'columpne' for column, as in a tabular column. Another strange error, if such it may be called, is the decorated initial capital N which appears at the start of the section on *The Extraction of Rootes*. This has been peculiarly printed as though looking at it through a mirror. The engraving appears again, uncorrected, at the beginning of *The Arte of Cossike Nombers* and also at the beginning of *The Arte of Surde Nombers*.

More seriously from a mathematical point of view, the original typesetters have sometimes accidently transposed the digits of a number. An example here occurs in the table on Sig. F.iii.r, where the scholar sets forth a table showing all the factors of the number 5460 in pairs. The pairing of 52 with 150 is wrong – the pairing of course should be 52 with 105. An example of a genuine mathematical mistake (rather than a typesetting error) occurs on Sig. N.iii.r where the scholar, preparatory to extracting a square root, sums the numbers 5625, 9216 and 15129 to get 29960, whereas the correct sum should be 29970. He then works with this error in the text, finally arriving at a square root of 173, remainder 31. Although this particular error has no effect on the solution the scholar is seeking, the remainder is actually greater by 10, that is 41.

When these and other typographical and/or mathematical errors are encountered in this book, remember that they occur in the original printing and are faithfully reproduced in this facsimile reprint. But before one becomes too smug about detecting errors with the benefit of hindsight and an electronic calculator, it would perhaps be salutary to read Recorde's rhyme to the 'curious scanner' (interested reader) on the page following the preface, concerning the redress of errors.

THE TITLE

Many people have been puzzled by the title of this book. *The Whetstone of Witte* is in fact rather a clever pun, but one which is incomprehensible today and only amusing when it is explained or when

Introduction

one is well versed in Latin. It has already been remarked that *Cosa* is the Latin word for 'thing', that is, an algebraic unknown. However, a similar word *Cos* is Latin for 'whetstone', a stone used for sharpening axes and other edge tools. Hence Recorde's pun – his book on algebra is one on which to sharpen one's mathematical skills, or as he would have it, one's wit.

THE EPISTLE DEDICATORY
Recorde dedicates his book to the Muscovy Company, a body of English merchants trading with Russia. In 1553 Sir Hugh Willoughby and Richard Chancellor set sail to seek out a Northeast Passage to China and the East Indies. Willoughby's ship was lost, but Chancellor reached Archangel on the White Sea and established trade links with Moscow. As a result the Muscovy Company was formed in 1555 by the navigator and explorer Sebastian Cabot and various London merchant adventurers, who were granted a monopoly of Anglo-Russian trade. It was the first English joint-stock company in which the capital remained regularly in use instead of being repaid after every voyage. Recorde became an advisor to the company on matters pertaining to mathematics and navigation. In his dedication Recorde promises that, if *The Whetstone of Witte* is well received (which he doubts not), he will set forth a book on navigation for the company in which he will show how, without great difficulty, ships may sail by a northeast passage to the North East Indies. Alas, Record (who so far as is known never left the shores of England) died before he could keep his promise.

POWERS OF NUMBERS
As mentioned above, German writers on algebra used the word *zensus* for a squared number. They in turn derived it from the medieval Italian word *censo*, which is a close relative of the Latin *census*. The Italians used *censo* to translate the Arabic word *mál*, literally 'possessions; property' which was the usual word in that language for the square of a number. This came about because the Arabs, like most mathematicians of those and earlier times, thought of a square number as a depiction of an area, especially of land, hence property.

Introduction

So, for a time, *censo* in Italian and *zensus* in German was the word for a squared number. Record adopts the word for his treatise on algebra, and the first mention of it occurs on Sig H.ii.v, where it is spelt *zenzike*.

Up to this point Recorde happily refers to squares (the second power of a number) and cubes (the third power), which can be easily visualised as geometrical entities. However, in his time there was no easy way of denoting the higher powers of numbers, which was a great hindrance to him and to effective mathematics in general. So like his predecessors he represents a fourth power by the square of a square or, by extension of the German terminology, a *zenzizenzike*. This is nothing more than a condensed form of the Italian *censo di censo* used by Leonardo of Pisa (Fibonacci) in his famous book *Liber Abaci* of 1202. A sixth power is a *zenicube* or a square of a cube and again by obvious extension the eighth power, a square of squared squares, is a *zenzizenzizenzike*. Although none of the *zenzi-* words survive in the English language except as historical curiosities, *zenzizenzizenzike* is often held up as a weird word best known for being an example of a weird word. It also features in 'trivial pursuit' type quizzes as being the word with the most z's in the English language.

THE EQUALS SIGN

It has to be admitted that *The Whetstone of Witte*, while justly famed as a pedagogical work, has little that is original in algebraic techniques. However it does have one great claim to originality. This is the creation of the modern mathematical symbol for equality. On Sig. Ff.i.v, where Recorde is explaining to the scholar the work of equations, he utters the now famous phrase 'And to avoide the tediouse repetition of these woords : is equalle to : I will sette as I do often in woorke use, a paire of parallels, or Gemowe lines of one lengthe, thus : = because noe 2 thyngs, can be moare equalle'. In printed books before Recorde equality was usually expressed rhetorically, often by the Latin word *æquales* or its abbreviated form *aeq*.

Today it is difficult to think of any other symbol for equality being used, but Recorde's parallel lines had several competitors and nearly perished in a fierce struggle for existence. This is evidenced by

Introduction

the fact that over sixty years were to elapse before the = sign again found its way into print. Readers of this book will quickly notice that Recorde's equals sign (and the signs for plus and minus, which although not original to him appear here for the first time in an English book) are much longer than those we are accustomed to today.

RECORDE'S TROUBLESOME CONDITION

Readers studying the 16th century mathematics of Robert Record in this book will perforce come across a number of passages which, at first sight, appear strange and somewhat perplexing. For example, on Sig. C.ii.r he tells the scholar that he will omit some explanations concerning proportions until another time 'consideryng ... the troublesome condition of my unquiete estate'. Again, on Sig. Q.ii.v, he tells the scholar that '... because I have litle leiser, to spende moche tyme with you (save that zeale to your furtheraunce doeth make me partly to forgette my owne businesse) therefore will I leave this question to your self, to be aunswered at your laisure'. And yet again, on Sig. Mn.iiii.v, after saying that he could provide the scholar with a demonstration which depends on the 38th theorem of the patthewaie (i.e. a theorem in his book on geometry, *The Pathway to Knowledge*) he declines to do so by saying 'but haste of other businesse, maketh me to omit the demonstration at this tyme'.

Reading such passages makes one suspect that we have here an author whose troubles, whatever they might be, weigh so heavily on his mind as he writes that he cannot avoid them breaking into his train of thought, and so via his pen inadvertently appearing on the printed page. This suspicion is reinforced when we come to the penultimate page in the book. Engaged in a discussion of universal roots with the scholar, Recorde (in the guise of the master) abruptly interrupts their talk thus:

Master. You say truth. But harke, what meaneth that hastie knocking at the door?
Scholar. It is a messenger.
Master. What is the message? Tel me in mine eare.
Yea, sir, is that the matter? Then there is no remedie, but that I must neglect all studies and teaching, for to withstande these

Introduction

daungers. My fortune is not so good, to have quiet tyme to teache.

And so on for a few paragraphs more, until the scholar says, 'My harte is so oppressed with pensiveness, by this sudaine unquietnesse, that I cannot express my grief. But I will praie, with all theim that love honeste knowledge, that God of his mercie will sone ende your troubles and graunte you such reste as your travel doth merite. And all that love learning say thereto, Amen.'

At which Recorde concludes his book by having the master reply, 'Amen, and Amen'. These were the last words he ever wrote. This is not the place to offer a full biography of Recorde, but in order to understand these somewhat cryptic passages, placed as they are in the unlikely context of a mathematical textbook, it is necessary to know something of the troubles besetting him at this time.

In January 1549 Recorde was appointed comptroller of the Bristol Mint. His refusal later that year to divert monies without the King's authority, to troops suppressing a rebellion in the southwest, led to an accusation of treason by William Herbert, Earl of Pembroke. Recorde was punished by confinement at court for 60 days, and although he survived the clash relatively unscathed, he had made a powerful enemy. By 1551 he had returned to favour at court, and he became surveyor of the mines and monies in Ireland, but the silver mines failed to show a profit and he was dismissed. In 1556 Recorde accused Pembroke of malfeasance as commissioner of the mints, which was a very reckless accusation to level against a favoured member of the nobility married to Anne Parr, sister of Catherine Parr, sixth wife of Henry VIII. Pembroke sued for defamation and Recorde was ordered to pay the enormous sum of £1000 in compensation, which he was unable to do. It is probable that at the same time he was also under investigation by the privy council over his stewardship of the Irish mines. Hence we see reasons for the anxiety which weighed so heavily on his shoulders while he struggled to finish *The Whetstone*, and which he found so oppressive that he was unable to keep his worries out of his writing. His fears proved all too real. The knock did come at his door and he was committed

Introduction

to the King's Bench prison, but whether for debt to Pembroke, for irregularities in the management of the mines, or perhaps for both, is unclear. Within a year 'goal fever' – probably cholera, typhoid or dysentery – took his life. Such was the cruel fate which befell this learned and kindly spirit. Think on this as you read *The Whetstone*, his final book.

SOURCES
Readers wanting to know more about Robert Recorde and his famous series of mathematical books should consult the following:

For an easily accessible biography visit the MacTutor History of Mathematics, 'Robert Recorde', [online] http://www-history.mcs.st-andrews.ac.uk/Biographies/Recorde.html.

Written sources are:

Stephen Johnston, 'Recorde, Robert (c1512–1558)' *Oxford Dictionary of National Biography*, Oxford University Press, 2004.

Howell Lloyd, 'Famous in the Field of Number and Measure: Robert Record, Renaissance Mathematician', *Welsh History Review*, Vol. 2 (2000), pp. 254-282.

William Barr, 'A World View of Robert Recorde: A Brief Study of Tudor Cosmology, *Albion: A Quarterly Journal Concerned with British Studies*, Vol. 1, No. 1 (1969), pp. 1-9.

Joy B. Easton, 'The Early Editions of Robert Recorde's Ground of Artes', *Isis*, Vol. 58, No. 1 (Winter 1967), pp. 515-532.

Joy B. Easton, 'On the date of Robert Recorde's birth', *Isis*, Vol. 57, No. 1 (Spring 1966), p. 121.

Margaret E. Baron, 'A Note on Robert Recorde and the Dienes Blocks', *The Mathematical Gazette*, Vol. 50, No 374 (Dec 1966), pp. 363-369.

Louise Diehl Patterson, 'Recorde's Cosmography, 1556', *Isis*, Vol. 42, No. 3 (Oct 1951), pp. 208-218.

E.R. Sleight, 'Early English Arithmetics', *National Mathematics Magazine*, Vol. 16, No. 4 (Jan 1942), pp. 198-215 and Vol. 16 No. 5 (Feb 1942), pp. 243-251.

Introduction

Francis R. Johnson & Stanford V. Larkey, 'Robert Recorde's Mathematical Teaching and the Anti-Aristotelian Movement', *The Huntingdon Library Bulletin,* No. 7 (Apr 1935), pp. 59-87.

David Eugene Smith & Frances Marguerite Clarke, 'New Light on Robert Recorde', *Isis,* Vol. 8, No. 1 (Feb 1926), pp. 50-70.

David Eugene Smith, 'New Information Respecting Robert Recorde', *The American Mathematical Monthly,* Vol. 28, No. 8/9 (Aug–Sep 1921), pp. 296-300.

Frank V. Morley, 'Finis Coronat Opus', *The Scientific Monthly,* Vol. 10, No. 3 (Mar 1920), pp. 306-308.

BIBLIOGRAPHY

Williams, Jack (2011) *Robert Recorde. Tudor Polymath, Expositor and Practitioner of Computation*, London, Springer-Verlag.

Roberts, Gareth and Smith, Fenny (Eds.) (2012) *Robert Recorde. The Life and Times of a Tudor Mathematician*, Cardiff, University of Wales Press.

HERE BEGINS
THE WHETSTONE OF WITTE

The whetstone of witte,

whiche is the seconde parte of
Arithmetike : containyng thextrac‑
tion of Rootes : The *Cossike* practise,
with the rule of *Equation* : and
the woorkes of *Surde*
Nombers.

Though many stones doe beare greate price,
The whetstone *is for exersice*
As neadefull, and in woorke as straunge:
Dulle thinges and harde it will so chaunge,
And make them sharpe, to right good vse:
All artesmen knowe, thei can not chuse,
But vse his helpe: yet as men see,
Noe sharpenesse semeth in it to bee.
 The grounde of artes *did brede this stone:*
His vse is greate, and moare then one.
Here if you list your wittes to whette,
Moche sharpenesse therby shall you gette.
Dull wittes hereby doe greatly mende,
Sharpe wittes are fined to their fulle ende.
Now proue, and praise, as you doe finde,
And to your self be not vnkinde.

❡ These Bookes are to bee solde, at
the Weste doore of Poules,
By Jhon Kyngstone.

❡ To the right worshipfull, the go=
uerners, Consulles, and the reste of the com=
panie of venturers into Moſcouia, Robert Re=
corde Phiſitian, wiſſheth healthe with
continualle increaſe of commodi=
tie, by their worthie and
famous trauell.

 Wil not, nother ought I
ſo euilly to iudge of my
countrie, that learnyng
here can haue no liber=
tie : but by aide of frende=
ſhippe, or ſtrength of po=
wer. For as Englande
did neuer wante learned
wittes, ſo at this tyme I doubt not, but there
be a great multitude, that deſirouſly embrace
all kindes of knowledge, and frendely are af=
fected toward the furtherance of it. And ther=
fore I dare ſaie, thei can not malice me, whi=
che am ſo willyng to helpe the ignoraunte, ac=
cordyng to my gifte and ſimple talēte. Wher=
by alſo this moche praiſe I maie iuſtly craue,
to haue the commendation and rewarde of a
ſolliciter in this cauſe. For though my trauell
can not moche profite them, that be well lear=
ned, yet doeth it excite the beſte learned, to re=
member their duetie to their countrie : and to
be aſhamed, that thei hauyng ſo greate habi=
litie, ſhall be founde moare ſlacke to aide their
coūtrie, then he that hath ſmaller knowledge,

a.ii.   and

The Epiſtle

and leſſe occaſion otherwaies. Accoꝛdyng as
men haue receiued, ſo are thei bounde to yeld.
Theſe excellente giftes are not lente vnto mē,
to be hidden. And there are a great multitude
that thurſt, and long moche foꝛ ſoche aide. Foꝛ
bothe theſe cauſes J ſaie, that naturalle bōde
to our countrie doeth chalenge it : and foꝛ that
the honeſte deſires of ſo many good natures
ſo moche requireth it, J exhoꝛte them that be
beſte hable, to take from me this chargeable
wooꝛke, and to further their countrie men, as
equitie would. And in the meane ceaſon while
J ſee them ſo ſlacke, let them not bee offended
with me, foꝛ pꝛeuentyng them. Foꝛ better it is
that a ſimple Coke doe pꝛepare thy bꝛekefaſt,
then that thou ſhouldeſt goe a hungered to
bedde. Yea better it is to haue ſome groſſe re=
paſte, then to ſterue foꝛ honger. And the com=
mon ſoꝛte will kinde ſmalle faulte of wante,
as long as thei ſee any man ſerue their expec=
tation. So that foꝛ this cauſe alſo, that my
paines foꝛ a time, doeth excuſe other finer wit=
tes, thei ought to render me ſome thankes a=
gain. But if thei ſtaie foꝛ feare of tauntes, and
barkyng of curſes, their coꝛage is ſmalle. Jf
thei miſdoubte the gratefull acceptation of
their ſtudies, thei doe iniurie to their countrie.
Foꝛ whoe cā doubt but ſo ciuile a coūtrie, will
thankefully receiue, and moſte frendly recom=
penſe the trauelle, of ſoche as ſtudie foꝛ their
                                                    benifite,

Dedicatorie.

benifite, and serueth their necessarie commo=
dities. This perswasion maketh me so bolde,
that I can not thinke it neadefull, to seke any
protector, for this or any like woozke. Sith
euery good man will offer hymself, to defende
that, wherby his natiue countrie is benifited.
Excepte at some tyme, by excitation of the fu=
ries, some naughtie natures doe practice their
fraude, to berefte the realme of some singu=
lare commoditie. But as I feare no soche, so
at this tyme I seke no soche aide against thē.
Yet for testefiyng of frendeshippe, and grate=
fulle remembraunce, I could doe noe lesse, but
sende this Booke to soche as I thought, not
onely to deserue it, but also would gladly re=
ceiue it. And if I maie perceiue, that you doe
accepte it (as I doubte not) with as good a
wille, as I dooe sende it, I will for your plea=
sure, to your coumforte, and for your commo=
ditie, shortly set forthe soche a booke of Naui=
gation, as I dare saie, shall partly satisfie and
contente, not onely your expectation, but also
the desire of a greate nomber beside. Where=
in I will not forgett specially to touche, bothe
the olde attempte for the Northlie Nauigati=
ons, and the later good aduenture, with the
fortunate successe in discoueryng that voiage,
whiche noe men before you durste attempte,
sith the tyme of Kyng Alurede his reigne. I
meane by the space of. 700. yere. Nother euer

　　　　　　　　　　　a.iii.　　any

The Epiſtle

any befoze that tyme, had paſſed that voiage,
excepte onely Ohthere, that dwelte in Hal=
golande : whoe repozted that iozney to the no=
ble Kyng Alurede : As it doeth yet remaine in
aunciente recozde of the olde Saxon tongue.
So that if you continue with cozage, as you
haue well begon, you ſhall not onely winne
greate riches to your ſelues, and bzyng won=
derfull commodities to your coütrie. But you
ſhall purchaſe therewith immoztall fame, and
be pzaiſed foz euer, as reaſon would : foz ope=
nyng that paſſage, that ſhall pzofite ſo many.
In that Boke alſo I will ſhewe certain mea=
nes, how without greate difficultie, you maie
ſaile to the Pozthe Eaſte Indies. And ſo to
Camul, Chinchital, and Baloz, whiche bee
coütries of greate commodities. As foz Cha=
tai lieth ſo farre within the lande, toward the
Southe Indian ſeas, that the iozneie is not
to be attempted, vntill you be better acquain=
ted with theſe countries, that you muſt firſt ar=
riue at. But theſe thynges come in this place
vntimely. I pzaie you accepte frendely in the
meane ceaſon this Booke, whiche will bee a
greate aide to the well vnderſtandyng of the
reſte that is behinde. And as I ſhall vnder=
ſtande your deſire, ſo will I haſte the other.
God pzoſpere well your endeuoure, and ſende
you ſoche good ſucceſſe, as ſo wozthie aduen=
ture doeth deſerue : Whiche I doubte not will,
                                        inſue,

Dedicatorie.

inſue, if cankered malice of ſome ſpitefull ſto=
mackes doe not pꝛeuaile, as thei can not ceaſe
to pꝛactice, to hinder your commoditie, and
deface your trauel. But as it is euer ſeen, and
therfoꝛe commonly knowen, that enuie doeth
ſtill repine at gloꝛie, ſo ought all honeſte har=
tes, to pꝛoſecute their good attemptes, and
contempne the ballynge of dogged
curres. So fare you well. And
loue hym againe, that de=
lighteth and ſtudieth
to farther your
cōmoditie.

At London the .xii. daie of
Nouember .1557.

THE PREFACE
to the gentle Reader.

Lthough nomber be in=
finite in increaſyng : ſo that
there is not in all the woʒlde,
any thing that can excede the
quantitie of it : Nother the
graſſe on the ground, nother
the droppes of water in the
ſea, no not the ſmall graines
of Sande through the whole
maſſe of the yearth : yet maie it ſeme by good reaſon,
that noe man is ſo experte in *Arithmetike*, that can nõ=
ber the commodities of it. Wherefoʒe J maie truely
ſaie, that if any imperfection bee in nomber, it is bi=
cauſe that nomber, can ſcarſely nomber, the commo=
dities of it ſelf. Foʒ the moare that any experte man,
doeth weigh in his mynde the benifites of it, the moʒe
of them ſhall he ſee to remain behinde. And ſo ſhall he
well perceiue, that as nomber is infinite, ſo are the
commodities of it as infinite. And if any thyng doe oʒ
maie exceade the whole woʒlde, it is nomber, whiche
ſo farre ſurmounteth the meaſure of the woʒlde, that
if there were infinite woʒldes, it would at the full cõ=
pʒehend them all. This nomber alſo hath other pʒe=
rogatiues, aboue all naturalle thynges, foʒ neither is
there certaintie in any thyng without it, nother good
agremente where it wanteth. Whereof no man can
doubte, that hath been accuſtomed in the Bookes of
Plato, Ariſtotell, and other aunciente Philoſophers,
where he ſhall ſee, how thei ſearche all ſecrete know=
ledge and hid miſteries, by the aide of nomber. Foʒ
not onely the conſtitution of the whole woʒlde, dooe
thei referre to nomber, but alſo the compoſition of
                                    b.i.   man,

The excel-
lencie of
nomber.

manne, yea and the verie substaunce of the soule. Of whiche thei professe to knowe no moare, then thei cã by the benifite of nomber attaine. Furthermore, for knowledge and certaintie in any other thynge, that mannes witte can reche vnto, there is noe possibilitie without nomber. It is confessed amongeste all men, that knowe what learnyng meaneth, that beside the Mathematicalle artes, there is noe vnfallible knowe= ledge, excepte it bee borowed of them. And emongeste them, it is sufficiently knowen, and well declared by *Nicomachus,* and diuerse other wziters, that *Arithme= tike* is the fountaine of all the other, and their ground and bonde, as he calleth it. If any man will saie, that Diuinitie, Lawe, and Physike, maie be had without it : oz that thei take litle aide therby. Although I haue

Diuinitie. befoze this tyme aunswered thereto, yet now I saie again : that in Diuinitie there are greate hidde secre= tes in nombers. So that diuerse excellente Diuines, haue wzitten whole Bookes of the misteries of nom= bers. And some of their Bookes intituled : *The Diui= nitie of Nombers.* But what Chzisten manne is igno= raunte, that betwene *Trinitie* and *vnitie,* doeth consiste the full grounde of al Diuinitie? Wherefoze I neade not to allege the other hollie and sacred Nombers. Saue that .7. will not permitte me to passe it with si= lence. In whiche is contained, not onely the secretes of the creation of all thynges : and the consummation of the whole wozlde againe, with the state of eterni= tie : But also by it is the Sabbothes reste, and therby the full life and conuersation of godlie persones, re=

Lawe. pzesented and insinuate. In Lawe twoe kyndes of Iustice are the somme of the studie : *Iustice Distribu= tiue,* and *Iustice Commutatiue,* whiche termes I vse, as beste knowen in that arte : But what is any of thē bothe without Nomber? I haue said in an other place (as I learned of that noble Philosopher *Aristotell)* that

THE PREFACE.

that if the knowledge and diſtinction, of *Geometricalle*
and *Arithmeticall* proportion bee not well obſerued,
there can noe Juſtice well bee executed. And how of=
ten the miniſters of the Lawe vſe aide of Nomber. I
neade not repete, bicauſe none but madde men doubt
of it. And as for Phyſike, without knowledge and   *Phyſike.*
aide of nomber is nothynge. Wee ſee that nature in
generation, bothe of manne and beaſtes, yea and of al
thynges els doeth obſerue nomber exactly. As well in
the tyme of formation, as in the monethes of quicke=
nyng, and of birthe. The miſteries of the ſeuenth and
nineth monethes are ſufficiente teſtimonies therein.
Beſide that from the fourthe monethe til the ſeuenth
many thynges bee permitted, that els bee not conue=
niente. For the vſe of the pulſe, and for criticalle da=
yes, beſide the proportion in degrees in ſimple medi=
cines, and mixture of compounde medicines, and o=
ther infinite maters, what nomber can doe, and what
aide it giueth, onely the ignoraunte doe doubte.

But where can there bee any better teſtimonie for   *Aſtronomie.*
Nomber, then that the celeſtialle bodies doe kepe an
vnfallible nomber, in all their wonderfulle motions?
By means whereof, mannes witte is habled to at=
taine the knowledge of them. As by the *Arithmeticalle*
tables, of their motions it is eaſily knowen. There=
fore and for that we ſee the yere, and all the diſtincti=
on of times, beſide the common vſe of trafike betwene
menne, to depende of nomber, wee muſte neades not
onely confeſſe it to bee, as it were the onely ſtate of
all natures woorkes, and of all ciuilitie : but we muſt
alſo honoure and reuerence it, as often as wee duely
remember the excellencie and benefite of it. Was not
Nomber, thinke you, wonderfullie honoured, when
noe name was thought moare meter for God, then
the name of Nomber? I meane .1. and .3. the name of
the Trinitie. But to come to moare familiare ma=

THE PREFACE.

ters, J will ſaie, as *Plato* ſaieth in his Booke De ſum=
mo bono. *Take awaie Arithmetike, with meaſure and weigh=*
tes, from all other artes, and the reſte that remaineth is but
baſe, and of noe eſtimation. Where although *Plato* dooe
name three thinges in appearaunce, that is Nomber,
Meaſure, and Weighte. What are Meaſure and
Weighte, but nomber applied to ſeueralle vſes? For
Meaſure is but the nombryng of the partes of lengthe,
bredthe, oʒ depthe. And ſo *weighte* (as here it is taken)
is the nomberyng of the heuineſſe of any thyng. So
that if nomber were withdrawen, no manne could
either meaſure, oʒ weigh any quantitie. And therfoʒe
it muſt followe : that nomber onely maketh all artes
perfecte, and woʒthie eſtimation : ſeyng that without
it, all artes are but baſe, and without commendation.
This maie ſuffice foʒ the iuſte cõmendation of *Arith=*
metike. But yet one commoditie moare, whiche all
menne that ſtudie that arte, doe fele, J can not omitte.
That is the filyng, ſharpenyng, and quickenyng of
the witte, that by pʒactice of *Arithmetike* doeth inſue.
It teacheth menne and accuſtometh them, ſo certain=
ly to remember thynges paſte : So circumſpectly to
conſider thynges pʒeſente : And ſo pʒouidently to foʒ=
ſee thynges that followe : that it maie truelie bee cal=
led the *File of witte.* Yea it maie aptly bee named the
Scholehouſe of reaſon. The like iudgemente had *Plato* of
it, as appeareth by his wooʒdes in the ſeuenth booke
De republica. Where he ſaieth thus : *Thei that be apte of*
nature to Arithmetike, bee readie and quicke to attaine all
kindes of learnyng. And thei that bee dulle witted, and yet bee
inſtructed and exerciſed in it, though thei gette nothyng els, yet
this ſhall thei all obtain that thei ſhall bee moare ſharpe wit=
ted, then thei were before. What a benefite that onely
thynge is, to haue the witte whetted and ſharpened, J
neade not trauell to declare, ſith all menne confeſſe it
to be as greate as maie be. Excepte any witleſſe per=
ſone

Meaſure.
Weighte.

THE PREFACE.

ſone thinke he maie bee to wiſe. But he that moſte feareth that, is leaſte in daunger of it. Wherefoꝛe to conclude, I ſee moare menne to acknowledge the be= nifite of nomber, then I can eſpie willyng to ſtudie, to attaine the benifites of it. Many pꝛaiſe it, but fewe dooe greatelp pꝛactiſe it : onleſſe it bee foꝛ the vulgare pꝛactice, concernyng Merchaundes trade. Wherein the deſire and hope of gain, maketh many willyng to ſuſtaine ſome trauell. Foꝛ aide of whom, I did ſette foꝛth the firſte parte of *Arithmetike*. But if thei knewe how farre this ſeconde parte, dooeth excell the firſte parte, thei would not accoumpte anp tyme loſte, that were imploied in it. Pea thei would not thinke any tyme well beſtowed, till thei had gotten ſoche habili= tie by it, that it might be their aide in al other ſtudies. And if *Plato* doe require *Arithmetike*, as a ſpecialle and a neceſſarie qualitie in hym, whom he would admitte as a citezein in his politike toune : How maie wee thinke of our ſelues, that deſire to gouerne other, and pet can ſcante ſkille of common nomber? So farre are many, pea moſte parte of vs from cunnyng in nom= ber. *Plato* thinketh noe manne hable to bee a good ca= pitaine, excepte he bee ſkilfulle in this arte : And wee accoumpte it noe parte of thoſe qualities, that bee re= quired in any ſoche manne. Howbeit foꝛ the better trialle thereof, I haue in this Booke framed ſome of the queſtions in ſoche ſoꝛte, as thei maie appꝛoue the vſe of this arte, not onely good foꝛ capitaines, but al= ſo moſte neceſſarie foꝛ theim. So that without it, thei can not Marſhall their battaile, nother vewe their e= nemies campe oꝛ foꝛte. And if I ſhall ſaie as I thinke, without it a capitaine is noe capitaine. In this booke what I haue wꝛitten, foꝛ the aide of all menne, and namelp ſoche of my countrie menne, that vnderſtand nothyng but Engliſhe, I neade not to repete particu= larelp, but remitte them to the booke it ſelf, to ſee it at

b.iii.   large

THE PREFACE.

large. Onely this maie I saie : that as I haue doen in
other artes, so in this I am the first venturer, in these
darke maters. Wherfore I trust thei that be learned,
and happen to reade this worke, wil beare the moare
with me, if thei finde any thyng, that thei doe mislike :
Wherein if thei will vse this curtesie, either by wri=
tynge to admonishe me thereof, either theim selfes to
sette forthe a moare perfecter woorke, I will thynke
them praise worthie. But if any manne will be so ha=
stie, other to blame that, whiche he is not hable to a=
mende, or to condempne that, whiche he did neuer vn=
derstande : As some ofte tymes doe of a fonde curiosi=
tie, I will wisshe hym a better witte, and moare mo=
destie. And to preuente all soche seuere Judges, I
thought it good to admonishe you before, that by oc=
casion of trouble vpon trouble, I was hindered from
accomplishyng this worke, as I did intende. But yet
is here moare, then any manne might well looke for
at my handes, if thei did knowe and consider myne
estate. And this moche moare I saie : that if I maie
perceiue, that this Booke bee as well receiued, as the
firste parte was, I will striue moche, to stele from my
troubles so moche tyme, as to set out the reste of
this arte, moare completely in Englishe,
then euer I sawe it in any toungue,
hetherto doen : trust thereto ad=
suredly. And wisshe hym
good, that traueleth
for thy benifite.

Of the rule of Cofe.

One thyng is nothyng, the prouerbe is,
Whiche in fome cafes doeth not miffe.
Yet here by woorkyng with one thyng,
Soche knowledge doeth from one roote fpryng,
That one thyng maie with right good fkille,
Compare with all thyng : And you will
The practice learne, you fhall fone fee,
What thynges by one thyng knowen maie bee.

To the curioufe fcanner.

If you ought finde, as fome men maie,
That you can mende, I fhall you praie,
To take fome paine : fo grace maie fende,
This worke to growe to perfecte ende.
    But if you mende not that you blame,
I winne the praife, and you the fhame.
Therfore be wife, and learne before,
Sith flaunder hurtes it felf mofte fore.

The seconde parte of Arithmetike,
containyng the extraction of Rootes in di-
verse kindes, with the Arte of Cossike
nombers, and of Surde nombers
also, in sondrie sortes.

The interlocutors, Master.Scholar.

The Master.

 See your desire can not
bee satisfied, neither your re-
quest stated, vntill I maie in-
stly aunswere you, that I can
teache you no more : whiche
aunswere maie state your re-
quest, although it content not
your desire.

Scholar. I beseche God of
his mercie, to withstande all suche occasion : except it
maie be more to your owne contentation and profite,
then it would be pleasaunt to the louers of learning.

Master. Yet a iuste excuse maie stande for my declaration : As if ignoraunce doe inforce me to state
my trauell.

Scholar. Your owne ignoraūce, I trust, you will
not allege : and as for the ignoraūce of other, it ought
to bee no state : sith the ignoraunte multitude doeth,
but as it was euer wonte, enuie that knoweledge,
whiche thei can not attaine, and wishe all men igno-
raunt, like vnto themself, but all gentle natures, con
tenneth suche malice : and despiseth theim as blinde
wormes, whom nature doeth plague, to stay the poi-
sone of their venemous stynge.

Master. We shall not nede to stande on this talke,
but trauell with knowledge to vanquishe ignoraūce :
And beleue that the *pricke* of knoweledge, is more of
force then the stynge of ignoraunce : yea, the *poincte* in

A.i.   Geometrie,

The seconde parte

Geometrie, and the vnitie in *Arithmetike*, though bothe be vndiuisible, doe make greater woorkes, & increase greater multitudes, then the brutishe bande of igno=raunce is hable to withstande.

Scholar. Our talke groweth well to our mater. I beseke you therfore, with that vnitie beginne, and builde on it your worke, as a forte against ignorance

Master. Vnitie is of it self vndiuisible, and yet is it in al partes of the worlde, and in euery thing. Yea, the worlde it self consisteth of vnitie, is named of vni=tie, was made by vnitie, and is preserued by vnitie, and onely ignoraunce with her broode secluded from vnitie, so that of it to repete the fulle force, would oc=cupie muche time, and make greate volumes.

Scholar. Sith vnitie is so mightie, and of suche force (as you saie) what maie be thought then of nom=ber, whiche containeth a multitude of vnities? And is nothyng els but a collection of vnities.

Master. Vnitie is the fountaine and originalle of number, yea vnitie by addition onely shall make a greater nomber, then any nombers can doe by mul=tiplication. But this is marueilouse, that no number repineth against diuisiõ, till it come to an vnitie : and then will it permit no farther diuisiõ. And therfore it is said, that vnitie doeth neither multiplie nor diuide.

And as al nombers maie be more or lesse, so the les=ser is euer a *parte or partes* of the greater.

As 5 vnto 10 is a parte, named a halfe : but vnto 7,5, is not a parte, but partes, and is called $\frac{5}{7}$. So 8 to. 24. is a parte that is $\frac{1}{3}$: but vnto. 36. it is partes, that is $\frac{2}{9}$.

Scholar. J perceiue, you call it a parte, when the numerator in the fraction (reduced to the smalleste) is an vnitie. And when the numerator is a nomber, then that fractiõ betokeneth partes of a nomber.

But J praie you, what varieties of nombers bee there principally to be considered in this arte?

Master.

Vnitie.

Nomber.

A parte.
Partes.

of Arithmetike.

Maſter. Nomber is diuided into diuerſe kindes, for ſome are *whole nombers*, and thei onely of *Euclide*, *Boetius*, and other good writers are called nombers. Other are *broken nombers*, and are commonly called *fractions*. Of theſe bothe I haue written in the firſte and ſeconde partes of *Arithmetike* : So that I mighte ſeme to curiouſe to repete any parte of it again.

But now in eche kinde of theſe, there are certaine nombers named *Abſtracte* : and other called nombers *Contracte*.

Abſtracte nombers are thoſe, whiche haue no denomination annexed vnto them. And thoſe that haue any denomination ioyned to theim, are called *Contracte* nombers.

Scholar. This I ſee to be a reaſonable diſtinctiõ, and agreable to the ſignification of the names.

For as that nomber is cõtracte, from his generall libertie of ſignification, whiche is boũde to one denomination, as in ſaiyng. 10. grotes (where. 10. is reſtrained frõ the libertie of balowyng any other thing but grotes) ſo if it had no denomination adioined, it might then ſignifie the nomber of daies, or of miles, or any like thyng, as well as of grotes. For when I ſaie. 10. and doe not limitte any denominatiõ, then is that. 10. abſtracte and ſeuered frõ all ſpecialties, and ſtandeth free to any name of thing.

But this (me thinketh vnder your correction) can not extend to broken nombres : whiche euermore carry with them their denomination : ſeyng thei conſiſte of a numerator and a denominator.

Maſter. You ſeme to ſaie well. And the like iudgemete doeth appere to be in ſome writers of this arte. But yet ſeyng that fractions maie haue all other artificiall denominations, that whole nombers maie receiue : and maie alſo bee without theim : therefore muſt wee either make a more curiouſe diſtinction of

The firſte deuiſion of nombers.

The ſeconde diuiſion of nombers.

Abſtracte.

Contracte.

whether broken nombers be contracte, or not.

A.ii.   that

The seconde parte

that name of denomination : o2 els wee must seclude
fractions, frō the necessitie of that name : o2 els third=
ly, to auoied contention, cal them nombers contracte
imp2operly.

Scholar. J assente thereto as reason would.

Yet one thyng mo2e J must demaunde of you, why
Euclide, and the other learned men, refuse to accompte
fractions emongest nombers.

Master. Bicause all nombers doe consiste of a mul=
titude of vnities : and euery p2oper fraction is lesse
then an vnitie, and therefo2e can not fractions exact=
ly be called nombers : but maie bee called rather frac=
tions of nombers.

Scholar. In deede now that J doe waie the mater
mo2e exactly, it appereth that a fractiō is not p2oper=
ly a nomber, but a connexion and conference of nom=
bers, declaryng the partes of an vnitie. Fo2 the nu=
merato2 doeth signifie one nōber, and the denomina=
to2 an other : The denominato2 declarynge into how
many partes the vnitie is diuided, and the numera=
to2 signifiyng that of those partes, not all, but so ma=
ny onely are to be takē, as the numerato2 impo2teth.

Master. Well, then to p2ocede, nombers abstracte
are considered in. 3. p2incipall varieties : That is, first
without comparison to any other nomber o2 figure.
And that nomber maie well be called *nomber absolute.*

Secondarily, some nombers bee vsed onely in re=
lation to other, and therfo2e ought to bee called *nom=
bers relatiue.*

Thirdly, many nombers are referred to some fi=
gure, that maie rise by multiplicacion of their partes
together, and that diuersly. And those nombers ther=
fo2e maie bee called *figuralle nombers.*

Scholar. If J conceiue your wo2des rightly, this
is your meanynge : that when J saie. 10. 25. 100. o2
200. &c. these nombers stand absolute from all deno=
minacion

of Arithmetike.

minacion, and clere from all relatiō and compariſon.

But when I ſaie. 6. is halfe of. 12. oʒ. 15. is triple to 5. here the numbers beepng compared together, are aptly called *nombers relatiue* : So if I ſaie, that. 16. is a *ſquare nomber*, bicauſe it is made of. 4. multiplied by. 4 then is. 16. here to be called a *figuralle nomber.*

Maſter. You take it well. Therfoʒe will I bʒiefly touche the membʒes of euery kinde.

Firſt of abſolute nombʒes, ſome are *euen nombers,* and ſome are *odde.*

Scholar. All men knowe that. And farther, that *Nombers, euen, & odde.* euen numbers are thoſe, whiche maie be diuided into e= qualle halfes : and ſo can not *odde nombers,* without a fraction.

Maſter. Of this plaine eaſie thyng, marke what foloweth : a greater doubte diſſolued. Foʒ if an odde nomber (as. 7. foʒ example) can not bee parted into. 2. equalle nombers, eche beepng halfe of. 7. then. $3 \cdot \frac{1}{2}$. whiche is commonly called the halfe of. 7. is no nōber.

Scholar. It can not be denied. And ſo (I ſee now) no fraction can bee a nomber. This greate doubte is plainly diſſolued, by a very certaine and moſte kno= wen pʒinciple.

Maſter. Now farther. Of bothe theſe kindes of *Nombers cō= pounde, and ſimple.* nombers, ſome bee *compounde,* and ſome bee *ſimple* and *vncompounde. Compounde* nombers are made by multi= plicacion of. 2. nombʒes together, and not by addiriō, though the name might ſeme to ſerue to bothe.

Scholar. So I perceiue, that 5. is no *compounde* nō= ber, although it bee made by addition of. 2. and. 3. but 6. whiche is made by multiplication of. 2. and. 3.

Likewaies. 9. is *compounde,* bicauſe that. 3. multi= plied by. 3. doeth make. 9.

And. 15. alſo is *compounde* by multiplipng. 5. and. 3. together. *One is no nomber.*

And hereby I ſe that. 1. is not to be called a nomber

foʒ

The seconde parte

fo₂ then all nōbers aboue it, muſt nedes be *compounde*, bicauſe thei conſiſt all of vnities.

Maſter. But yet by multiplication of. 1. no other nomber is *compounde*.

Scholar. By thoſe wo₂des J am taught to knowe mo₂e, and ſpeake better.

Maſter. *Euen nombers* are yet diuerſly to be conſide=red in their diuiſions. Fo₂ although the greate multi=tude of euen nombers bee *compounde*, yet. 2. is accomp=ted truely an euen nomber, o₂iginall, and *vncompoūde*. So that it maie make other nombers, & is made of no nōbers, but of vnities onely, as al odde nombers are.

Two is vn-compounde.

All other euen nombers are *compounde*, and are di=uerſly diuided, fo₂ ſome are *euen nombers euenly*, and ſome are *euen nombers oddely*, and ſome are *euen nombers* bothe *euenly and oddely. Euen nombers euenly*, are ſuche nombers as maie bee parted continually into euen halfes, till you come to an vnitie. As fo₂ example. 32. firſt is diuided into. 16. as his euen halfe : and again, 16. into. 8. as his halfe : And. 8. againe by. 4. is parted into. 2. euen partes : Then. 4. alſo by. 2. And that. 2. is diuided into. 2. vnities, as his iuſte halfes.

Euen nom-bers, euenly.

But *euen nombers vneuenly*, are ſuche nombers as maie bee diuided into. 2. equalle partes : whiche are odde nombers. As. 18. is diuided into. 9. and. 9. as his halfes, and thei are odde. So. 10. is diuided by. 5. And 30. by. 15. with a greate nomber mo₂e of ſuche ſo₂te.

Euē nombers vneuenly.

Nombers euen euenly and oddely, bee commonly called ſuche nombers, as maie bee diuided into. 2. equalle and euen halues : but befo₂e you come to an vnitie, the halfes will be odde nombers. As. 60. maie be firſt parted into. 30. and. 30. whiche are euen : And thei a=gaine diuided by. 15. whiche is odde.

Euen nom-bers, euenly and oddely.

Likewaies. 24. is diuided firſt by 12. And that 12. by 6. & laſtly. 6. is diuided by. 3. whiche is an odde nōber.

So. 28. maie bee diuided into. 2. equalle and euen halues,

of Arithmetike.

halues, that is into 14. And that. 14. into. 7. whiche is
the halfe of. 14. but is an odde nomber.

Scholar. This I perceiue well. And, as I iudge,
the diſtinction into thoſe. 3. kindes, is not onely rea=
ſonable, but alſo nedefull. And yet you ſeme to ſpeake
doubtfully, of this laſte membʒe. Bicauſe I remem=
ber not that you vſe this woʒde commōly, but where
you giue place rather to cuſtome, then to reaſon.

Maſter. Or elſ to cuſtome of the common ſoʒte of
wʒiters, rather then to the iudgemente of the moſte
aunciente wʒiters.

And ſo in this caſe *Euclide* doeth not ſeme to admitte
this thirde member. But accompteth it vnder the ſe=
conde kinde. As maie well appere in his. 9. boke, and
34. pʒopoſition, where he calleth ſuche a nomber, *euen=
ly euen,* and *euenly odde* alſo, whiche place cōferred with
the definitions in the ſame booke, doeth appʒoue in
many wiſe mennes opinions, that *Euclide* minded but
2. onely kindes of thoſe nōbers. And yet in this thing
(I thinke) he did rather appʒoue. 3. varieties by his
pʒopoſitiōs, then eſtabliſhe onely. 2. ſoʒtes by his firſt
definitions.

But herein I will ſpende no moʒe tyme. But ſaie
bʒiefly that the diſtinctiō of. 3. kindes, ſerueth to good
vſe, and eaſe in teachyng.

And now foʒ farther knowledge of nombers, ſome
are called *nombers perfecte,* & ſome are *nombers imperfect.*

Perfecte *nombers* are ſuche ones, whoſe partes ioy=
ned together, will make exactly the whole nomber.

Nombers
perfecte.

And therfoʒe are. 6. and. 28. accompted perfecte nō=
bers : bicauſe the partes of eche of theim added toge=
ther, doe make the ful and intere nomber, whoſe par=
tes thei bee. As of. 6. the halfe is. 3. the thirde parte is
2. the firſt parte is. 1. As foʒ a quarter, and fifte parte
it hath not in whole nomber. Now put together. 1. 2.
and. 3. and thei make iuſte. 6. whoſe partes thei bee.

6.

And

The seconde parte

And therfoze is. 6. a perfecte nomber.

28.

Likewaies. 28. hath foz his halfe. 14. foz his quar=
ter. 7. foz his seuenth parte. 4. and foz his fowertenth
parte. 2. and foz his. 28. parte. 1. all whiche put toge=
ther, that is. 1. 2. 4. 7. and. 14. doe make. 28. of this soɀt
there are very fewe moɀe in cōpariſō. And foz an exā=
ple, I ſett here, as many as are vnder. 6000000000.
and thei are theſe. 6. 28. 496. 8128. 130816. 2096128.
33550336. 536854528.

Nombers
imperfecte.

But now of the contrary kind, *imperfecte nombers* be
ſuche, whoſe partes added together, doe make either
moɀe oɀ leſſe, then the whole nomber it ſelf : whoſe
partes thei bee.

Nombers
ſuperfluouſe.

And if the partes make moɀe then the whole nom=
ber, then is that nōber called *ſuperfluouſe*, oɀ *abundaunt.*
As 12. whoſe partes are 1. 2. 3. 4. ꝑ. 6. whiche make 16.

So. 20. hath foz his partes. 1. 2. 4. 5. 10. whiche
make. 22. Likewaies. 120. hath theſe partes.

1.   2.   3.   4.   5.   6.   8.   10.   ⎫
60. 40. 30. 24. 20. 15. 12.   ⎬ whiche make 240.

Nombers
Diminute.

And if the partes make leſſe then the whole nom=
ber, whoſe partes thei be, then is that number called
Diminute, oɀ *Defectiue.* As. 8. hath theſe partes. 1. 2. 4.
whiche make but. 7.

So. 16. hath theſe partes. 8. 4. 2. 1. and thei make
onely. 15.

Likewaies. 32. whoſe partes are. 1. 2. 4. 8. 16. and
make but. 31.

Scholar. In all theſe nombers I note that you re=
ken one, foz a parte of eche one of theim : whiche be=
foze I thought you had denied.

Maſter. 1. canne neither multiplie noɀ deuide, and
therfoze compoundeth no nomber. But one maie in=
creaſe addition, and therefoze where partes be added
together, there. 1. maie well be called a parte.

And this ſhall ſuffice foz the diuiſion of euen nom=
bers

of *Arithmetike*.

bers Abſtracte.

Now to ſpeake of *odde nombers*, ſome of thē are com= pounde, ꝭ ſome *vncompoūde*. Thei are *compounde*, whiche maie bee diuided into any other partes then vnities. As. 9. whiche is cōpounde of. 3. And. 15. that is made of. 5. and. 3. Alſo. 21. is compounde of. 7. and. 3. And ſo furthe. But. 3. 5. 7. 11. 13. 17. 19. 23. 29. and ſuche like, bee odde nombers *vncompounde*. Foꝛ thei are not made of any other then of vnities. *Odde nōbers Compounde.*

Vncompoūde. Nombers Relatiue.

Here muſt you vnderſtande by *compoſition* the mul= tiplication of the partes of nombers together, as you remembꝛe, befoꝛe was declared.

Scholar. I conſider it ſo. And I remembꝛe all that you haue taught me, foꝛ the diuiſiō of nōbers *abſtracte* and abſolute. What ſaie you now of nōbers *relatiue*? *Nombers Relatiue.*

Maſter. Some tymes their *relation* hath regarde to their partes, namely, whether theſe. 2. that bee ſo compared, haue any common parte, that will diuide theim bothe. Foꝛ if thei haue ſo, then are thei called *nombers commenſurable*. As. 12. and. 21. bee *nombers com= menſurable* : foꝛ. 3. will diuide eche of theim. *Commenſu= rable.*

Likewaies. 20. and. 36. be *commenſurable*, ſeyng 4. is a commō diuiſoꝛ foꝛ theim bothe. But if thei haue no ſuche common diuiſoꝛ, then are thei called *incommenſu= rable*. As 18 and 25. Foꝛ 25 can bee diuided by no nom= ber moꝛe then by. 5. And. 18. can not be diuided by it. *Incommen= ſurable.*

In like maner. 36. and. 49. are *incommenſurable* : Foꝛ 49. hath no diuiſoꝛ but. 7. And 7. can not diuide. 36.

Scholar. Doe you meane then, that *incommenſura= ble nombers*, haue no cōpariſon noꝛ *proportion* together?

Maſter. Naie, nothyng leſſe. Foꝛ any. 2. nombers maie haue compariſon ꝭ *proportion* together, although thei be *incommenſurable*. As. 3. and. 4. are *incommenſu= rable*, and yet are thei in a *proportion* together : as ſhall appeare anon.

But firſt I will declare vnto you, the varieties of
B.i.   *proportion*

The seconde parte

Proportion.

proportion, wherein there maie be double conferēce : J
meane of the leſſer to the greater, oʒ of the greater to
the leſſer.

*Of greater
inequalitie.*

WHē the greater is cōpared to the leſſer, it is called
a *Proportion of the greater inequalitie.* As 6 to 2. oʒ 5 to 3.

*Of leſſer in-
equalitie.*

And when the leſſer is conferred to the greater, it
is called a *proportion of the leſſer inequalitie.* As. 3. to. 5.
oʒ. 2. to. 6.

Scholar. And what if J would cōpare two equalle
nombers together?

Maſter. That is accoumpted alſo a pʒopoʒtion of
many men : and is called the *proportion of equalitie.* And
then ought the firſt diuiſion of pʒopoʒtion to be, thus

Pʒopoʒcion of
- Equalitie.
- Inequalitie.
  - The greater.
  - The leſſer.

So pʒopoʒtiō of the greater inequalitie, is diuided
into. 5. ſeuerall kindes : whereof. 3. be *ſimple,* and. 2. o=
ther *compounde.* The firſte kinde is, when a greater
nomber containeth the leſſer diuerſe times : as twiſe,
oʒ thriſe, oʒ oftener. So. 6. containeth. 3. twiſe : and
it containeth. 2. thriſe. This pʒopoʒtion is called

Multiplex.

generally, *multiplex,* that is to ſaie, many folde : but
ſpecially it is named, accoʒdyng to the tymes that it
conteineth the leſſer. So that if it contein hym twiſe,
then is it named *dupla,* oʒ double. As 2 to 1 and 4 to. 2.

And if it containe it thriſe. As. 3. to. 1. and. 6. to. 2. it
is called *tripla,* oʒ triple.

Jf it containe it. 4. tymes, then is it *quadrupla,* oʒ
quadruple.

Of theſe and of diuerſe other ſoʒtes in this kind al=
ſo, here are the names bʒiefly ſet doune with exāples.

Dupla

of Arithmetike.

Dupla.	4. to. 2: 6. to. 3: 10. to. 5: 18. to. 9.	$\frac{2}{1}$ Double.
Tripla.	6. to. 2: 9. to. 3: 12. to. 4: 18. to. 6.	$\frac{3}{1}$ Triple.
Quadrupla.	4. to. 1: 8. to. 2: 12. to. 3: 16. to. 4.	$\frac{4}{1}$ Fowerfolde
Quintupla.	5. to. 1: 10. to. 2: 15. to. 3: 20. to. 4.	$\frac{5}{1}$ Fiuefolde.
Sextupla.	6. to. 1: 12. to. 2: 18. to. 3: 24. to. 4.	$\frac{6}{1}$ Sixefolde.
Suptupla.	7. to. 1: 14. to. 2: 21. to. 3: 28. to. 4.	$\frac{7}{1}$ Seuenfolde.
Octupla.	8. to. 1: 16. to. 2: 24. to. 3: 32. to. 4.	$\frac{8}{1}$ Eightfolde.
Noncupla.	9. to. 1: 18. to. 2: 27. to. 3: 36. to. 4.	$\frac{9}{1}$ Ninefolde.
Decupla.	10 to. 1: 20. to. 2:30. to. 3:40. to. 4.	$\frac{10}{1}$ Tennefolde.
Vndecupla.	11. to. 1: 22. to. 2: 33. to. 3.	$\frac{11}{1}$ Aleuenfolde.
Duodecupla.	12. to. 1: 24. to. 2: 36. to. 3.	$\frac{12}{1}$ Tweluefolde.

And so infinitely.

Beside this there is an other kinde of proportion, when the greater containeth the lesser, more then ones, and not twise : and that maie be in 2 sortes. For if the greater containe the lesser, and any one parte of hym, that proportion is called *Superparticulare.* For example, take. 5. to. 4. Sith. 5. doeth containe. 4. and his quarter. Likewaies. 6. to. 5. is in the same kinde of proportion : although, not of the same speci= all sorte. For 6. comprehendeth. 5. and his fifte parte.

Superparti- culare.

So that for a more speciall distinction, eche of these and many other, haue their seueral names, accordyng to that parte, whiche thei doe containe. As if it con= taine the halfe more, it is named *Sesquialtera.* In whi= che proportion are these nombers folowyng.

Sesquialtera.

3. to. 2: 6. to. 4: 9. to. 6: 12. to. 8: 15. to. 10. $|1\frac{1}{2}$.

But if the greater comprehende the lesser, and his thirde parte, then is that named *Sesquitertia* proporti= on. As in these.

Sesquitertia.

4. to. 3: 8. to. 6: 12. to. 9: 16. to. 12: 20. to. 15. $|1\frac{1}{3}$. And when the fifte, sixte, seuenth, or eight part doeth make the proporcion, or any other part els, the name is taken of that same parte. As for briefnesse I will here sette examples.

B.ii.   *Sesquiquarta.*

The seconde parte

Sesquiquarta.	5. to. 4 : 10. to. 8 : 15. to. 12.	$1\frac{1}{4}$	A quarter moze.
Sesquiquinta.	6. to. 5 : 12. to. 10 : 18. to. 15.	$1\frac{1}{5}$	a fifte moze.
Sesquisexta.	7. to. 6 : 14. to. 12 : 21 to. 18.	$1\frac{1}{6}$	a sixte moze.
Sesquiseptima.	8. to. 7 : 16. to. 14 : 24. to. 21.	$1\frac{1}{7}$	a seuenth moze.
Sesquioctaua.	9. to. 8 : 18. to. 16 : 27. to. 24.	$1\frac{1}{8}$	an eight moze.
Sesquinona.	10. to. 9 : 20. to. 18 : 30. to. 27.	$1\frac{1}{9}$	a nineth moze.
Sesquidecima.	11. to. 10 : 22. to. 20 : 33. to. 30.	$1\frac{1}{10}$	a tenth moze.
Sesquiundecima.	12. to. 11 : 24. to. 22 : 36. to. 33.	$1\frac{1}{11}$	a leuenth moze.
Sesquiduodecima.	13. to. 12 : 26. to. 24 : 39. to. 36.	$1\frac{1}{12}$	a twelueth moze.

And so as farre as you liste to procede in suche pzo=
poztion : where one parte of the lesser, is the iuste dif=
ference and excesse, betwene it and the greater.

Superpartiēs　　But if the difference be. 2. partes, 3. partes, oz moze
partes : the propoztiō is named *superpartiente.* As. 5. to
3. And. 10. to. 6. Foz as, 5. containeth. 3. and. $\frac{2}{3}$. of it : so
10. holdeth. 6. and. $\frac{2}{3}$ of it.

Scholar. Now I perceiue some vse also, of the di=
stinction betwene a parte and partes in nomber : Of
whiche at the beginnyng you did speake. But how
many kindes are there of this sozte?

Master. There are infinite kindes in this sozte of
propozcion, as well as in the other. But foz exam=
ples sake, I will set furthe some of the moste common
nombers : that therby you maie gather the fozmes of
the reste. And these be thei, with their names.

	Tertias.	5. to. 3 : 10. to. 6 : 15. to. 9 : 20. to. 12.	$\frac{2}{3}$
	Quintas.	7 to 5 : 14. to 10 : 21. to. 15 : 28. to. 20	$\frac{2}{5}$
Superbipartiens.	Septimas.	9 to 7 : 18 to 14 : 27. to. 21 : 36. to. 28.	$\frac{2}{7}$
	Nonas.	11 to 9 : 22 to 18 : 33. to. 27 : 44. to. 36	$\frac{2}{9}$
	Vndecimas.	13. to 11 : 26 to 22 : 39. to. 33 : 52 to. 44	$\frac{2}{11}$

of *Arithmetike*.

	Quartas.	7. to. 4 : 14. to. 8 : 21. to. 12 : 28. to. 16.	$\frac{3}{4}$
	Quintas.	8. to. 5 : 16. to. 10 : 24. to. 15 : 32. to. 20.	$\frac{3}{5}$
	Septimas.	10 to. 7 : 20. to. 14 : 30 to. 21 : 40 to. 28.	$\frac{3}{7}$
ſupertripartiēs	*Octauas.*	11. to. 8 : 22. to. 16 : 33. to. 24.	$\frac{3}{8}$
	Decimas	13. to. 10 : 26. to. 20 : 39. to. 30.	$\frac{3}{10}$
	Vndecimas.	14. to. 11 : 28. to. 22 : 42. to. 33.	$\frac{3}{11}$
	Decimastertias.	16. to. 13 : 32. to. 26 : 48. to. 39.	$\frac{3}{13}$
	Decimasquartas.	17. to. 14 : 34. to. 28 : 51. to. 42.	$\frac{3}{14}$

	Quintas.	9. to. 5 : 18. to. 10 : 27. to. 15 : 36. to. 20.	$\frac{4}{5}$
	Septimas.	11 to 7 : 22 to. 14 : 33. to. 21 : 44. to. 28.	$\frac{4}{7}$
Superquadru	*Nonas.*	13 to 9 : 26. to. 18 : 39. to. 27 : 52. to. 36.	$\frac{4}{9}$
partiens.	*Vndecimas.*	15. to. 11 : 30. to. 22 : 45. to. 33.	$\frac{4}{11}$
	Decimastertias.	17. to. 13 : 34. to. 26 : 51. to. 39.	$\frac{4}{13}$
	Decimasquintas.	19. to. 15 : 38. to. 30 : 57. to. 45.	$\frac{4}{15}$

	Sextas.	11 to 6 : 22 to 12 : 33. to. 18 : 44. to. 24.	$\frac{5}{6}$
	Septimas.	12. to. 7 : 24. to. 14 : 36. to. 21.	$\frac{5}{7}$
	Octauas.	13. to. 8 : 26. to. 16 : 39. to. 24.	$\frac{5}{8}$
	Nonas.	14. to. 9 : 28. to. 18 : 42. to. 27.	$\frac{5}{9}$
Superquintu	*Vndecimas.*	16. to. 11 : 32. to. 22 : 48. to. 33.	$\frac{5}{11}$
partiens.	*Duodecimas.*	17. to. 12 : 34. to. 24 : 51. to. 36.	$\frac{5}{12}$
	Decimastertias.	18. to. 13 : 36. to. 26 : 54. to. 39.	$\frac{5}{13}$
	Decimasquartas.	19. to. 14 : 38. to. 28 : 57. to. 42.	$\frac{5}{14}$
	Decimasſextas.	21. to. 16 : 42. to. 32 : 63. to. 48.	$\frac{5}{16}$

	Septimas.	13. to. 7 : 26. to. 14 : 39. to. 21.	$\frac{6}{7}$
	Vndecimas.	17. to. 11 : 34. to. 22 : 51. to. 33.	$\frac{6}{11}$
Superſextu=	*Decimastertias.*	19. to. 13 : 38. to. 26 : 57. to. 39.	$\frac{6}{13}$
partiens.	*Decimasſeptimas*	23. to. 17 : 46. to. 34 : 69. to. 51.	$\frac{6}{17}$
	Decimasnonas.	25. to. 19 : 50. to. 38 : 75. to. 57.	$\frac{6}{19}$
	Viceſimastertias.	29. to. 23 : 58. to. 46 : 78. to. 69.	$\frac{6}{23}$

Scholar. I vnderſtande by theſe examples, ſome=
what of their reaſons : but I perceiue, you doe not fo=
lowe their naturalle o꜖der, without interruption, in
                          B.iii.   theſe

The seconde parte

these of the laste kinde.

Master. To thintente you maie the better vnder=
stande good ground in that omission, J wil set furthe
here those omitted nombers : That you maie see how
thei woulde expresse some other proportiõ here named
And therfore thei doe seme rather to be omitted, then
in deede so to be.

Marke theim well.

Superbipartiens.

Secundas.	4. to. 2 : 8. to. 4.	$\frac{2}{1}$
Quartas.	6. to. 4 : 12. to. 8.	$1\frac{1}{2}$
Sextas.	8. to. 6 : 16. to. 12.	$1\frac{1}{3}$
Octauas.	10. to. 8 : 20. to. 16.	$1\frac{1}{4}$
Decimas.	12. to. 10 : 24. to. 20.	$1\frac{1}{5}$

Scholar. In deede here J see, the firste is double
proportion. The seconde *sesquialtera,* the thirde *sesqui=
tertia,* the fowerth *sesquiquarta,* & the fifte *sesquinquinta.*

Master. So marke these other.

supertripartiens

Secundas.	5. to. 2 : 10. to. 4.	$2\frac{1}{2}$
Tertias.	6. to. 3 : 12. to. 6.	$\frac{2}{1}$
Sextas.	9. to. 6 : 18. to. 12.	$1\frac{1}{2}$
Nonas.	12. to. 9 : 24. to. 18.	$1\frac{1}{3}$
Duodecimas.	15. to. 12 : 30. to. 24.	$1\frac{1}{4}$

Scholar. The firste of these J knowe not, but all
the other are named before.

Master. The firste is a compounde proportion (as
anon J will declare) and is namen *duplasesquialtera.*

But now will J sette furthe all the other omitted
names.

Secundas.

of *Arithmetike.*

superquadru=partiens.		
Secundas.	6. to. 2 : 12. to. 4.	$\frac{3}{1}$ *Tripla.*
Tertias.	7. to. 3 : 14. to. 6.	$2\frac{1}{3}$ *Duplasesquitertia.*
Quartas.	8. to. 4 : 16. to. 8.	$\frac{2}{1}$ *Dupla.*
Sextas.	10. to 6 : 20. to. 12.	$1\frac{2}{3}$ *subipartiēstertias*
Octauas.	12. to. 8 : 24. to. 16.	$1\frac{1}{2}$ *sesquialtera.*
Decimas.	14. to 10 : 28. to. 20	$1\frac{2}{5}$ *supbipartiēsquintas*
Duodecimas.	16. to. 12 : 32. to. 24	$1\frac{1}{3}$ *Sesquitertia.*
Decimasquartas.	18. to. 14 : 36. to. 28	$1\frac{2}{7}$ *supbipartiēsseptimas*
Decimassextas.	20. to. 16 : 40. to. 32.	$1\frac{1}{4}$ *Sesquiquarta.*
superquin=tupartiens.		
Secundas.	7 to 2 : 14 to 4.	$3\frac{1}{2}$ *Triplasesquialtera.*
Tertias.	8 to 3 : 16 to 6.	$2\frac{2}{3}$ *Duplasuperbipartiens tertias.*
Quartas.	9 to 4 : 18 to 8.	$2\frac{1}{4}$ *Duplasesquiquarta.*
Quintas.	10 to 5 : 20 to 10.	$\frac{2}{1}$ *Dupla.*
Decimas.	15 to 10 : 30 to 20.	$1\frac{1}{2}$ *Sesquialtera.*
Decimasquintas.	20 to 15 : 40 to 30.	$1\frac{1}{3}$ *Sesquitertia.*
supersextu=partiens.		
Secundas.	8 to 2 : 16 to 4.	$\frac{4}{1}$ *Quadrupla.*
Tertias.	9 to 3 : 18 to 6.	$\frac{3}{1}$ *Tripla.*
Quartas.	10 to 4 : 20 to 8.	$2\frac{1}{2}$ *Duplasesquialtera*
Quintas.	11 to 5 : 22 to 10.	$2\frac{1}{5}$ *duplasesquiquinta.*
Sextas.	12 to 6 : 24 to 12.	$\frac{2}{1}$ *Dupla.*
Octauas.	14 to 8 : 28 to 16.	$1\frac{3}{4}$ *supertripartiensquartas.*
Nonas.	15 to 9 : 30 to 18.	$1\frac{2}{3}$ *superbipartienstertias.*
Decimas.	16 to 10 : 32. to 20.	$1\frac{3}{5}$ *supertripartiensquintas.*
Duodecimas.	18 to 12 : 36 to 24.	$1\frac{1}{2}$ *sesquialtera.*
Decimasquartas.	20 to 14 : 40 to 28	$1\frac{3}{7}$ *supertripartiensseptimas.*
Decimasquintas.	21 to 15 : 42 to 30.	$1\frac{2}{5}$ *superbipartiensquintas.*
Decimassextas.	22 to 16 : 44 to 32.	$1\frac{3}{8}$ *supertripartiensoctauas.*
Decimasoctauas.	24 to 18 : 48 to 36.	$1\frac{1}{3}$ *sesquitertia.*
Vicesimas.	26 to 20 : 52 to 40.	$1\frac{3}{10}$ *supertripartiensdecimas.*
Vicesimassecūdas.	28 to 22 : 56 to 44.	$1\frac{3}{11}$ *supertripartiensvndecimas*

Scholar. I see well that these proporcions, bee a=
greable with some other name : and therfore might
seme superfluouse in this place.

　　　　　　　　　　　　　　　　　　Master.

The seconde parte

Master. Not onely superfluousely, but also false=
ly should thei bee placed here : seynge thei doe belong
to other places of right.

Scholar. Why doe you not name theim all by
Englishe names?

Master. Bicause there are no soche names in the
Englishe tongue. And if I should giue theim newe
names, many would make a quarrelle againſt me,
for obscuryng the olde Arte with newe names : as
some in other cases all redy haue doen.

Scholar. Yet I praie you declare those doubtfull
names of compounde proportions.

Master. As there is one kinde of proportion, that
is named *multiplex*, or manyfolde, whiche doeth con=
taine the leſſer diuerſe times exactly. And two other
whiche doe containe the leſſer ones, and some parte
or partes of the same : So those kindes maie be com=
pounded together. As when the greater number con
taineth the leſſer, twiſe, or thriſe, or oftener : and yet
more ouer some parte or partes of the same. So. 8.
containeth 3 twiſe, and his $\frac{2}{3}$. And 10 comprehendeth
3. thriſe and his $\frac{1}{3}$.

The firſte example is generally called *multiplex su=*
perpartiens : bicause the greater containeth the leſſer
certaine tymes, and some partes of it beſides. But
more specially it is called *dupla superbipartiens tertias*,
that is, double with $\frac{2}{3}$ more.

The seconde example is generally referred to *mul=*
tiplex superparticularis : bicause in it the greater com=
prehendeth the leſſer oftentymes, (as here thriſe) and
his $\frac{1}{3}$ more. And therfore specially it is called *tripla ses=*
quitertia.

But as I doe intende briefly to ouer runne this
parte : so will I by tables set forthe the kindes of thē
with their examples.

The

of Arithmetike.

*The table of proportion of the
greater inequalitie.*

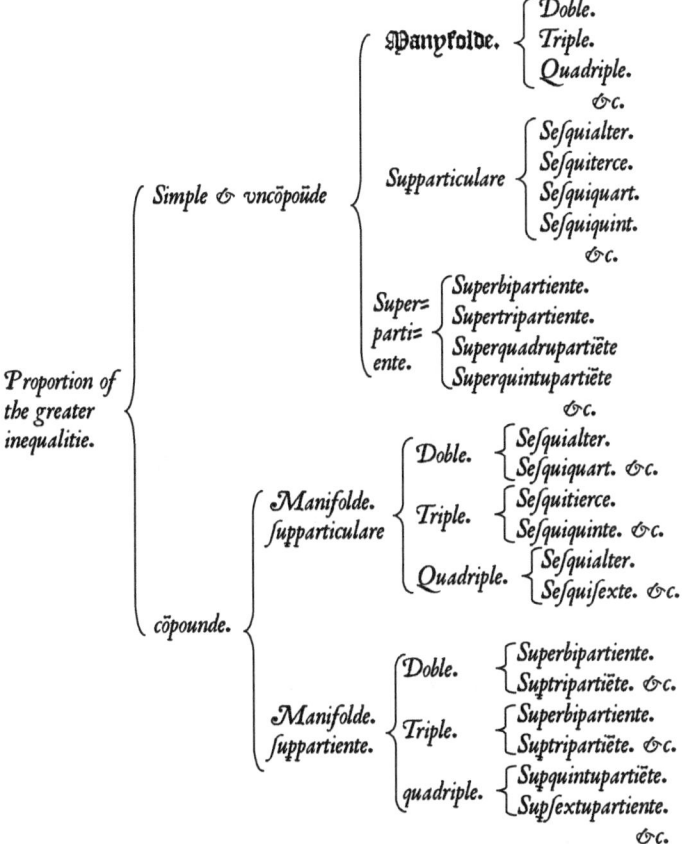

The seconde parte

Examples of eche compounde kinde,
mentioned in the former table.

Manifolde. Superparti= culare.	*Double.*	*Sesquialter.*	5 to 2.
		Sesquiquarte.	9 to 4.
	Triple.	*Sesquitierce.*	10 to 3.
		Sesquiquinte.	16 to 5.
	quadriple	*Sesquialter.*	9 to 2.
		Sesquisexte.	25 to 6.
Manifolde suppartïete	*Double.*	*Superbipartiente tierces.*	8 to 3.
		Superipartiente quartas	11 to 4.
	Triple.	*Superbipartiente tierces.*	11 to 3.
		Suptripartiente quartas.	15 to 4.
	quadriple	*Supquĩtupartïete quartas*	29 to 6
		supsextupartïete septimas.	34 to 7

Scholar. What moꝛe is there to bee learned of thefe pꝛopoꝛtions? Foꝛ by thefe foꝛmes, J maie eafe= ly gather the balure oꝛ rate of any pꝛopoꝛtion.

Mafter. This maie ftande foꝛ their numeration : faue that mofte aptly thei ought to bee fette as fracti= ons, in their leafte tearmes : as you haue here diuerfe eramples.

Scholar. You meane that double *sesquialter* muft be wꝛitten thus $\frac{5}{2}$, and fo of the refte.

Mafter. Oꝛ els thus $2\frac{1}{2}$. and fo *triple sesquiquinte* in this foꝛte : $3\frac{1}{5}$, oꝛ thus $\frac{16}{5}$ and fo of all other.

And foꝛ farther wooꝛke, you fhall bnderftande that pꝛopoꝛtions maie bee added, fubtracted, multi= plied and diuided : and berie ftraunge woꝛkes therby acchiued :

of Arithmetike.

acchiued. For of the Arte of Proprotiōs, dependenth all the subtilties, and fine workes, not onely of *Arithmetike*, but also of *Geometre* : besides farther mater that as now I will not touche. But as for the workes of *Proportions*, I will omitte them til an other tyme : consideryng not onely the troublesome condition, of my vnquiete estate : but also the conuenient order of teachynge, whereby it is required that the extraction of rootes, should go orderly before the arte of Proportions : whiche without those other, cā not be wrought.

Therfore will I now onely declare these kindes of proportion, whiche yet are not spoken of : to the intente that you maie haue here, the generall diuision of numbers, somewhat sufficiently touched.

As you see that betwene any two numbers, there maie be a conference of proportion : so if any one proportion be continued in more then. 2. nombers, there maie be then a conference also of these proportions, in their seuerall termes : and that conference or comparison is named *Analogie* : whiche some delighte to call proportionalitie : As in this example, where 3 nobers beare like proportion in their progression : 4. 6. 9. You see that 6. to 4. is in proportion *sesquialter* : and so is 9. to 6. and therfore is there a like proportion betwene the. 2. laste, as there is betwene the. 2. firste.

Analogie.

Scholar. This I consider well by progression in *Arithmetike*.

Master. Likewaies where fower termes or more be set in order of proportion, as here 2. 6. 18. 54.

Scholar. I perceiue this wel : for here the proportiō is triple. But what saie you to this forme of comparison in Proportion? As 6. is to. 2 : So. 30. is to. 20. Is it not all one kinde of *Analogie*?

Master. It is one kinde of Analogie generalle, whiche maie be called *directe Analogie* : bicause the first is compared to him that doeth folowe nexte : & so eche

Directe analogie.

C.ii.   other

The seconde parte

other is still referred to that, that foloweth nexte. But
this is the difference : that in the firste, there is a con=
tinuaunce of collation : and one terme is compared
with twoo nombers : But in that forme of example,
whiche you put, there is no nomber compared twise :
For the first is referred to the seconde, and the second
to the thirde. And so haue thei seueralle names to di=
stincte theim a sonder.

Continuall
Proportion.

Wherfore whē the first nomber is referred to the
seconde, and that seconde to the thirde : the proporti=
on is called *continualle* : and it maie consiste betwene
3. termes. As 5. 15. 45. doe procede in a continuall tri=
ple proportion. For as 5. is to 15 : so is 15. to 45. as you

Discontinual
Proportion.

doe see. But when I saie thus : as 5. is to 15. so 6. is to
18. Here is a triple proportion, but not continualle.
For the seconde terme beynge 15. is not compared
with the thirde terme, that is 6. And therfore is it cal
led a proportion *discontinualle*.

Scholar. Now I perceiue certainly their distinc=
tion : For in twoo poinctes these examples doe agree,
and differ in a thirde poincte.

Firste thei agree in that (as you saied) that the for=
moste is referred to the other that foloweth it nexte :
And secondly, thei agree in this also, that bothe are
compared in a triple proportion. But in this thei dif=
fer, that the seconde terme, doeth not beare like pro=
portion to the thirde, as the thirde doeth to the fourth
or the firste to the seconde.

Master. Farther more there is to bee noted, that
in discontinualle proportion, there can bee no fewer
then fower terms, or numbers : and so by euen for=
mes still, as. 6. or. 8. and so forthe. Where as in conti=
nuall proportion, your termes maie bee of any num=
ber, euen or odde : aboue. 2.

And although I might saie more of the diuersities
of proportion : as of *Proportion conuersed* or *indirecte Pro=*
portion,

of Arithmetike.

portion interchaunged, compounde Proportion, parted Propor=
tion, reuerſed Proportion, and Proportion by equalitie. Yet
I thinke better to procede for this tyme, to the other
kindes of nomber, and to reſerue the explication of
proportions to their peculiare place.

Scholar. As you knowe the beſt order, ſo it ſhal be
mete that you doe bſe your owne iudgement therein.

Of figuralle nombers.

Maſter.

HE nexte kinde of nom=
bers are called *Figuralle nom=
bers* : bicauſe thei doe, or maie
repreſente ſome figure : And
are euer conſidered in relati=
on to thoſe formes.

Some of them haue a com=
pariſon and relatiō to length
onely, and therefore are na=
med *Linearie nombers* : whiche name, although it maie
bee referred moſte aptly to ſuche numbers, as will
make no other forme duely, yet maie it alſo be applied
to any number abſtract. Sith all ſoche numbers maie
be conſidered as the ſides of other figuralle numbers.

*Nombers li=
nearie.*

Secondly, numbers maie be conſidered, according
to ſoche formes as thei make other in progreſſion, or
in multiplication : And thoſe maie well be named *Su=
perficiall nombers, or Flatte nombers.* Whereof there bee
as many varieties, as there bee diuerſities of figures
in *Geometrie.* As numbers *Triangulare, Quadrate, Cinke=
angeled, Siſeangeled :* and ſo furthe. Alſo numbers *circu=
lare, diametralle, & likeflattes,* all whiche nombers haue
bothe lengthe and breadthe : and thereof bee named
ſuperficiall nombers.

*Superficiall
nombers.*

*Flatte nom=
bers.*

C.iii.   Beſide

The ſeconde parte

Beſide theſe there are other numbers, whiche are made of many multiplications, and thei are called *ſounde nombers* : bicauſe that as by the firſte multiplication, thei take lengthe and bꝛeadthe, like flatte numbers, ſo by the ſecond multiplication, thei take depthe alſo : And thereof be thei named *bodily nombers, oꝛ ſound nombers.*

The leaſte of them all is commonly called a *Cube, oꝛ a Cubike nomber* : And the other in their degrees ſeuerally named, as thei bee made by ſeueralle multiplicatiõs. Foꝛ accoꝛdynge to the number of their multipliplications, thei take their names. And all that haue like number of multiplications, are of one kinde, and bere one name : as well in flatte numbers, as in ſounde.

But conſideryng the infinite multitude of thoſe figuralle numbers, I thinke beſte to ſpeake of theim onely in this place, whiche haue muche pꝛofitable vſe in this arte. And, of thoſe, emong infinite flatte numbers, I will take onely fower. That is to ſaie, *ſquare nombers, longeſquares, diametralle nombers,* and *likeflattes.*

Square nombers are thoſe, whiche maie be diuided by ſome one number, and haue the ſame number foꝛ the quotiente : that is to ſaie, that a ſquare nõber is made by the multiplication of any number into itſelf, as 10 multiplied by. 10. maketh. 100. That. 100. is a ſquare number : whiche. 100. if I doe diuide by. 10. the quotiente will be. 10. alſo.

Scholar. So, 4. multiplied by. 4. maketh. 16 : and that muſt be a ſquare number by like reaſon.

Maſter. So it is.

Scholar. And if I multiplie. 9. by. 4. is not that a ſquare number : Seyng fower ſemeth to make all nõbers ſquare by multiplication.

Maſter. Conſider this well, that a ſquare number doeth make ſuch a ſquare in number, as a *iuſte ſquare* doeth make in *Geometrie* : That is ſuche a one
whoſe

of *Arithmetike*.

whoſe ſides are equalle. For and if the one ſide be lon=
ger then the other, that figure in *Geometrie* is called a
long ſquare, and ſo it is named in number, a *long ſquare*
alſo.

Now if I ſette doune the figure of your number,
as you termed it, and ſette. 4.
for the one ſide, and. 9. for
the other, this will the figure
ſhewe.

Where you ſe a plain longſquare :
yet is the whole number that amoun=
teth of this multiplication : truely na=
med a ſquare number, as here you
maie ſee. But then is the ſide or roote
of it, neither. 4. nor. 9. but. 6.

Scholar. Now I vnderſtande it : and the better by
this figuralle example. And here alſo I haue learned   *A roote.*
what a *Roote* is : for you ſeme to expounde it, to bee the
ſide of a figuralle number.

Maſter. Euery flatte nomber, and euery ſounde
number alſo haue their ſides : But no flatte number,
ſaue onely ſquares haue a roote : bicauſe a roote in
flatte numbers, is a number multiplied by it ſelf.

And in ſounde numbers, thei onely haue rootes,
whiche bee made by many multiplications, of ſome
one nūber by it ſelf : other by that, whiche riſeth of it.

As when I ſaie, twoo tymes, twoo twiſe, maketh
8. that number is a ſounde number : and is named a
Cube. And ſo. 3. tymes. 3. thriſe, doeth make. 27. whiche
is alſo a *Cube*.

And generally, any number that is made by ſuche
2. multiplications, is called a *Cube*, or *Cubike* number.   *A cube.*
And the number of that multiplication , whiche com=
monly is named the multiplier, is in this poincte cal=   *A cubike*
led the Cubike roote of that number.   *roote.*

Wherfore, thus alſo maie you define a Cubike nō=   *A cubike*
                                                    ber,   *nomber.*

The seconde parte

ber : to bee suche a number, as beeyng diuided by his roote, shall haue for the quotiente the square of the same roote.

Scholar. Hereby I perceiue, that one multiplication, of any number by it selfe, doeth make a square number. And twoo multiplications in that sorte, doe make a Cubike number.

What if I doe multiplie any number thrise, or fower tymes, or oftener in that sorte, are there proper names for suche numbers?

Master. Yes indeede : as I will declare anon.

But firste before we attempte the other sounde nombers, it shall bee mete, that we doe declare those twoo sortes of flatte numbers, whiche I named before : that is diametralle numbers, and like flattes.

A diametral nomber.

A *Diametralle nomber,* is suche a number as hath twoo partes of that nature : that if thei bee multiplied together, thei will make the saied *diametralle nomber* : And the squares of those twoo partes, beeyng added together, will make a square nōber also : whose roote

A diameter.

is the *diameter* to that *diametralle nomber.*

As 12 is named a *diametralle nomber,* for that he hath twoo partes, that is. 3. and. 4, whiche beeyng multiplied together, doe make 12. that is the firste number. And if their squares be added together, thei wil make a thirde square : and the roote of that number will bee the *diameter* to that platte forme of 12. As in this example you see.

The one side is. 4. and the other side is 3 whiche bothe multiplied together, doe make 12. Then take the square of fower whiche is 16 and the square of. 3, which is. 9. and put them

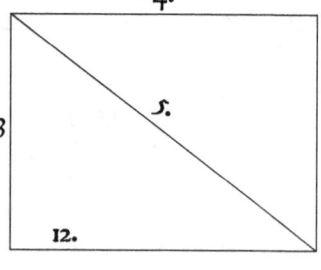

together

of Arithmetike.

together, and thei will make. 25. whose roote, beyng 5. is the *diameter* of that platte forme.

Scholar. That doe I perceiue well, bicause it is confirmed by the. 33. theoreme of the pathewaie.

Master. Yet take an other example. In this platte forme of. 60. you see the one side to bee. 5. and the other side to bee 12. Now take the square

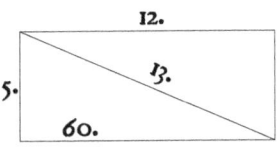

nomber of. 12. whiche is. 144. and the square of. 5. whiche is. 25. and put them together : so will it make 169. whiche is a square nomber : and hath. 13. for his roote.

Likewaies. 120. is to be accoumpted a *diametralle nomber.* For so muche as it hath twoo partes : that is 8. and. 15. whiche beeyng multiplied together, doe make the firste nomber. 120. And the square of those twoo partes (that is. 64. for. 8 : and. 225. for. 15.) beyng bothe added together, doe make. 289. whiche is a square nöber : and hath for his roote. 17. And therfore that. 17. is the *diameter* to that *diametralle nomber.* 120.

Like examples infinite might I giue you. But these for explication of the name, maie suffice.

Scholar. I doe well vnderstande the examples : saue that I knowe not how to finde the roote of the laste square nomber, whiche amounteth by the addi=tion of the former twoo squares together.

Master. That arte will I teache you anon. But we maie not forgette firste to ende all the difinitions of soche names, as I minde to write of.

Whereof yet there resteth *like flattes* : whiche maie *Like flattes.* bee as well taken for triangular figures, as for qua=drate figures.

So that any of them, when the sides of one plat forme, beareth like proportion together, as the sides

D.i.   of

The seconde parte

of any other flatte forme of the same kinde doeth, then
are those for=
mes called *like*
flattes. As in
these. 2. longe
Squares : bi=
cause the sides

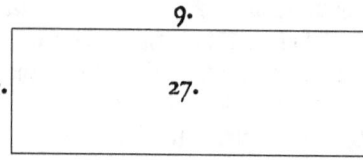

of them bothe, are in one
proportion (for. 6. is tri=
ple to. 2 : as well as. 9. is
triple to 3.) Therfore are
the whole figures called
like flattes.

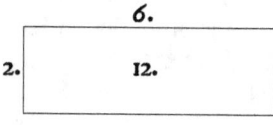

And so of due conueniencie, their nombers (that
expresse their quantities, whiche here are. 27. and 12)
be called by the like names, *like flattes.*

Farthermore in triangles (as here you se) if the si=

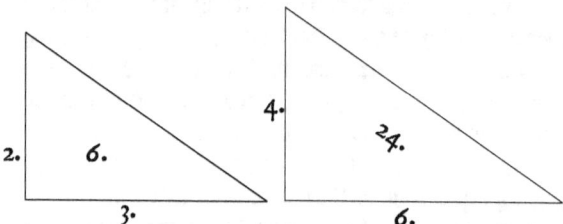

des of the one beare like proportion together, as the
sides of the other doe : then are thei called *like flattes* al=
so. And their nombers, that declare their quantities,
in like sorte are named *like flattes.*

Scholar. I pearceiue here : As 4 is to 2 : so. 6. is to
3. bothe beyng in a double proportion. And therfore 6
and. 24. are to be called *like flattes.*

Master. You vnderstande it well.

And thus haue we briefly ouer runne the diuision
of nomber, into his principalle kindes : And haue set
forthe the definitiōs of eche of them, with examples.
                                                              The

of *Arithmetike*.

The vſe of them you ſhall ſe largely in the pꝛactiſe of this arte.

But to the intent you maie the better obſerue and regarde theſe twoo laſte kindes of nombers : whiche are commonly neglected of artes men, I will ſhewe you ſome vſe of them, with their pꝛoperties.

Firſte, all *diametralle nombers* doe ſette foꝛthe a tri= *Of diame-*
angle, hauyng all three ſides knowen : whiche thyng *tralle nom-*
as it doeth ſerue to many and wonderfull purpoſes : *bers.*
ſo can it be found in no other nombers, then onely in *diametrall nombers.*

Foꝛ although in figures *Geometricalle*, you maie e= uer moꝛe vnfallibly finde one line, that will make a ſquare, equall to the twoo ſquares of any other twoo lines (as in the patthe waie you doe ſee it taught) yet the meaſure certaine of thoſe ſides, are not knowen.

Wherfoꝛe in nomber that is not poſſible alwaies to be doen : neither can it be doen with any other nõ= bers, then onely *diametricall nombers.* Yet maie other nombers go very nigh. As namely in theſe examples

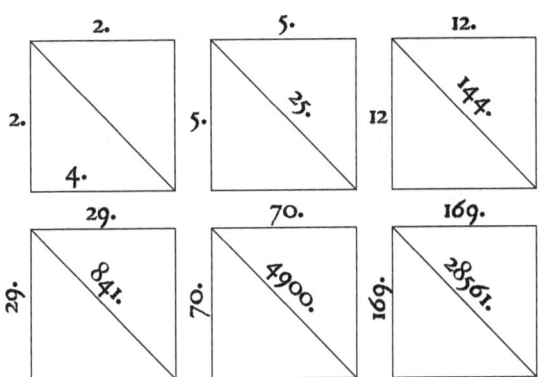

of ſquare nombers : whoſe double, I take foꝛ the ſqua=
D.ii.   res

The seconde parte

res of the sides,
bicause thei are
equall : and thei
make. 8. 50. 288.
1682. 9800.
57122. &. 332928.
All whiche dif=
fer onely by an
vnitie, from a
square nomber.

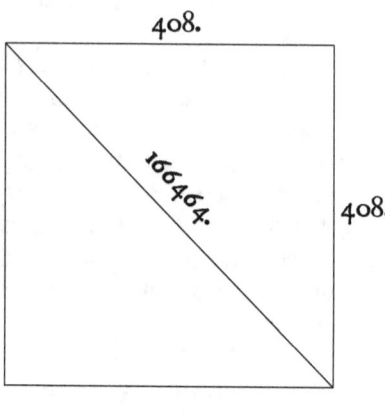

408.

408.

For nine is a
square nomber
and so are these
other folowyng.

　　　49. 289. 1681. 9801. 57121. &. 332929.
whose rootes be. 7. 17. 41. 99. 239. 577.

　　　Whiche examples if you doe consider well hereaf=
ter, thei will helpe you to gesse at the nighefte rootes
of nombers that be not square. And also for doblyng
of squares, in a square forme : within an vnspeake=
able nerenesse.

　　　For as in doblyng of this greater square. 166464.
there riseth. 332928. whiche wanteth one of a iufte
square. You se easely, that as that one is but a smalle
portion to the whole square : So yet, that one wan=
teth not in the roote, but in the whole square : where
by you maie perceiue, that it is a very smalle and vn=
senfible parte of one, that wanteth in the roote.

　　　Scholar. It muft seme by reason of multiplicati=
on : that it is scarse the. 10000. parte of one.

　　　Mafter. You saie truthe.

　　　Scholar. But how shall I finde the diameter of
soche nombers?

　　　Mafter. That is easily doen, if you knowe firfte
certainly that your nomber is a diametrall nomber.

　　　And secondarily, if you knowe the true partes of
　　　　　　　　　　　　　　　　　　　　　　　it :

of Arithmetike.

it : whiche you should vſe in this caſe.

Scholar. Will not any twoo ſoche partes ſerue, whiche by multiplication will make the whole nom= ber?

Maſter. You maie by the former examples, eaſily ſe the contrary. For 12 is a *diametrall nomber* : and hath theſe partes (as it is ſone perceiued). 2. 3. 4. 6. Yet if you take. 2. and. 6. for the ſides of it, thei will not make a *diameter* in knowen nomber.

Scholar. That I vnderſtande : for the ſquare of 2. beyng. 4. added to. 36. whiche is the ſquare of 6. doeth make. 40. whoſe roote muſt bee greater then. 6. and leſſe then. 7. And therfore. 40. can haue no roote in whole nomber.

Maſter. Neither yet in broken nombers : for that is a generalle rule : that if any whole nomber haue a roote, that roote ſhall be a whole nomber. So that if the roote can not bee founde in whole nomber : you ſhall neuer finde it in broken nombers.

And for more certaintie of that I ſaied before, that all partes be not apte for the ſides of a *diametralle nom= ber*, to finde out the *diameter* : marke well the ſeconde example, whiche is. 60. and hath theſe partes.

2. 3. 4. 5. 6. 10. 12. 15. 20. 30.

So that beginnyng with the two extremeſte, that is. 2. and. 30. thei will by multiplication make. 60.

And likewaies any two nombers, equally diſtant from thoſe extremes : As. 3. and 20. Likewaies. 4. and 15 : other. 5. and. 12. And in like maner. 6. and. 10. All thoſe couples by multiplication doe make. 60. Yet none of them are apte ſides to finde the *diameter* by, but onely 5 and. 12. For of the other ſides beyng mul= tiplied ſquarely (that is by thē ſelfes) and thoſe ſqua= res beyng added together, there wil not riſe a ſquare nomber. As you ſhall better vnderſtande, when you

D.iii.   haue

The seconde parte

haue learned to knowe square nombers, by extractiō of their rootes.

Yet in the meane ceason J will set forthe certaine notes, to knowe the *diameter*, and the apte sides, in all *diametralle nombers*.

1. And firste J saie : that as thei are three nombers in all (J meane the twoo sides, and the *diameter*) so all waies if the firste or leaste side bee odde, then shall there be twoo of them odde nombers. And the *diame= ter* shall euer bee the other of the odde nombers : that is to saie, the greateste of them.

2. Secondarily. It is true that all *diametrall nombers* are euen nombers. And no odde nomber can bee a *di= ametralle nomber*.

3. Thirdly, J saie, that all odde nombers aboue one, maie be the lesser side in soche *diametralle nombers*.

But euen nombers doe not serue so generally : for thei onely maie stand in soche place, whiche be grea= ter then. 4 : As. 6. 8. 10. 12. 14. 16. 18. 20. &c. And none other euen nombers then soche as maie be diuided by 4. maie be the greater side in any *diametralle nomber*.

4. Fourthly. If the lesser side bee an odde nomber, then ordinarily, the square of it is iuste equalle with that that amounteth by the addition of the *diameter*, to the greater nomber. As in the firste example, 3. is the lesser nomber, and. 4. is the greater : vnto them bothe the *diameter* is. 5. Now. 3. hath for his square 9. and so moche is made by the addition of. 4. and. 5.

Again in the seconde example, the lesser nomber is 5. and his square is 25. The greater nomber is 12, and the *diameter*. 13. Put. 12. and. 13. together, and thei make. 25. whiche is equalle with the square of the lesser.

Likewaies. 7. and 24. multiplied together maketh 168. whiche is a *diametralle nomber*. And bicause the square of the lesser side (whiche here is. 49) must bee
equalle

of Arithmetike.

equalle to the greater side, and the *diameter* added to=
gether : therfoze seyng. 25. added to. 24. maketh. 49.
that. 25. must nedes bee the *diameter* to the fozesaied
nomber.

By thefe rules (if you doe marke them well) you
maie fone perceiue, how to make any *diametralle nom=
ber* : if the leffer side bee giuen vnto you, and bee an
odde nomber. Yet foz your eafe, I will giue you this
plaine rule.

When any odde nomber is pzopounded : as the
leffer side of a *diametralle nomber*, and you would finde
the other side, and the *diameter* alfo : oz els the *diame=
teralle nomber*, that maie haue foche a side : multiplie
that pzoponed nomber by it felfe, and it will make a
fquare nomber, and will be an odde nomber : fo that
of it you fhall finde no iufte halfe. Therfoze take you
thofe twoo nombers, that are nexte vnto the halfe of
it : The leffer fhall alwaies bee an euen nomber, and
fhall be the feconde side of the *diametralle nomber* : The
other nomber whiche is the greater, fhall alwaies be
an odde nomber : and fhall bee the *diameter* of that nö=
ber whiche you desire. Foz example marke wel thefe
fozmes that doe folowe.

If three bee pzopounded as the one side of a *diame=
tralle nomber* : And you would knowe, what maie bee
the other side : and what is the *diametralle nomber* : And
thirdly, what is the *diameter* to that nomber : Doe, as
I faied befoze : multiply. 3. by it felf, and it will make
9. whiche is a fquare nomber, and an odde nomber :
and therfoze hath no iufte halfe. But the nigheft nö=
mbers to the halfe, are. 4. and. 5.

Therfoze I faie, that. 4. whiche is the leffer of thē
twoo, is the feconde side of the *diametralle nomber* : and
5. beyng the greater of them, is the *diameter* it felf.

Scholar. Now is it light inough to perceiue that
the *diametralle nomber* is. 12 : feeyng. 3. multiplied by
fower

The seconde parte

4. maketh. 12.

Master. So is it.

Again, if. 5. be assigned for one side of a *diametralle nomber*, and you obserue the former worke you maie easily finde the other side, and the *diameter*.

First you see, that the square of 5. is. 25. and it hath no halfe. But. 12. and. 13. are the. 2. nombers nighest his halfe : wherfore. 12. shall bee the seconde side : and 13. must be the *diameter*. And the *diametralle* nōber is. 60.

Likewaies, if. 7. be set for the lesser side, the grea= ter side shall be. 24. and the *diameter*. 25.

Scholar. Touching this I nede no more instruc= tion : the thyng is so manifeste.

Master. Then shewe your knowlege by an exam ple, or twoo.

And first I appoincte 9 for the lesser side of a *diame= tralle nomber*, whereunto I would haue you to assigne the other side, and the *diameter*. &c.

Scholar. I followe your precepte, and multiplie 9. by it self, whereof commeth. 81. whose halfe is be= twene. 40. and. 41. Therfore must. 40. be the other side : and 41 the *diameter*. And here the *diametralle* nom= ber is. 360.

Master. Proue the like : where. 15. is the lesser nomber.

Scholar. 15. multiplied square maketh. 225 : whose nighest halfes are. 112. and. 113. of whiche the first is the seconde side, and the later is the *diameter* : and the *diametralle nomber* is. 1680.

Master. What shall be the other nombers : where 21. is the lesser side?

Scholar. 21. yeldeth in square. 441. whose por= tions nighest his halfe, are. 220. and 221 : And so ap= pereth their offices, and the *diametralle nomber* is 4620

Master. So maie you saie that vnto. 27. being the lesser side : the greater side shall be. 364. and the *dia=*

ter

of Arithmetike.

ter. 365. bicauſe the ſquare of. 27. is. 729. And the *dia=*
metralle nomber is. 9828.

    Scholar. So muſt it be, by your rule.

    Maſter. Not onely the rule doth teache you that
it is ſo, but alſo the nature and figure of ſoche *flatte*
nombers. As here you ſee.

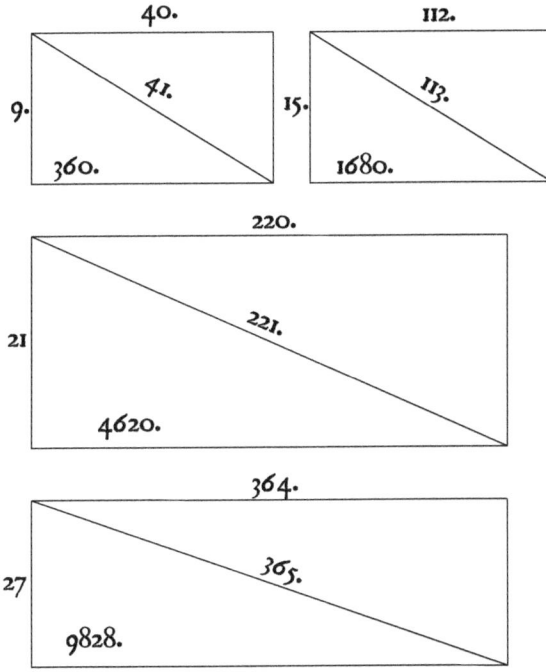

    But to the intente you maie the better vnderſtand
the nature of theſe nombers : J wil ſet foʒthe here the
like ſides with other nombers : whereby you maie
knowe, that one ſide maie ſerue to diuerſe *diametralle*
                              E.i.   *nombers*

The feconde parte

nombers. Wherfoꝛe marke thefe foꝛmes well.

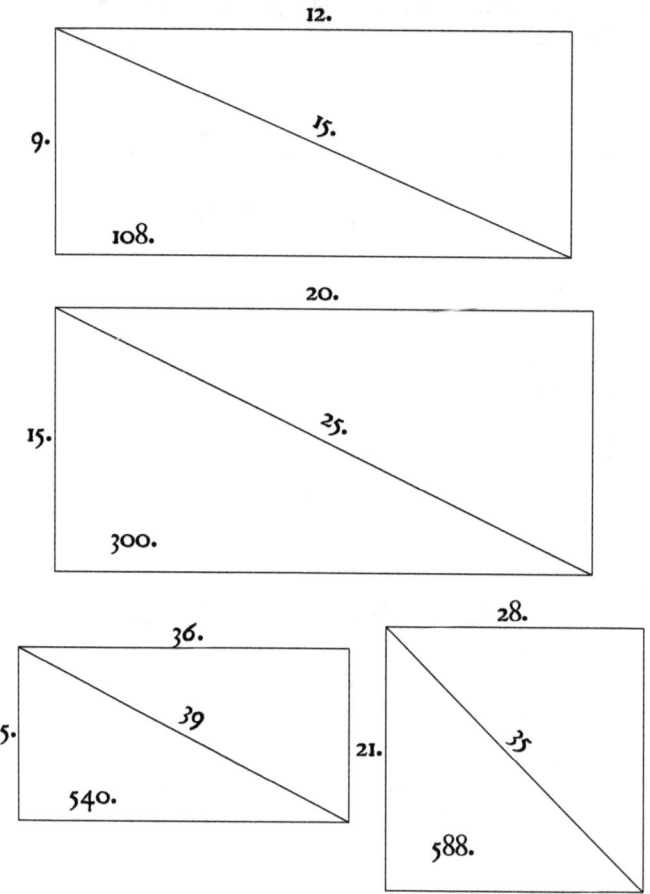

of *Arithmetike.*

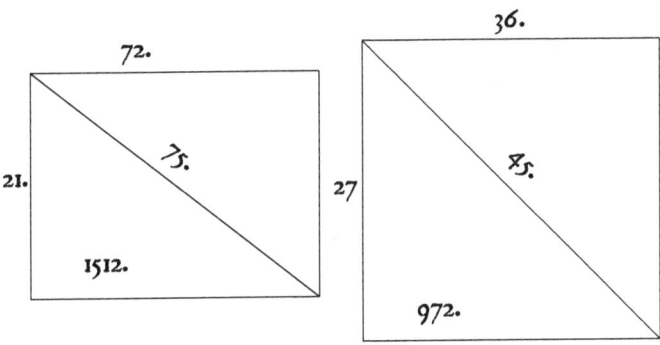

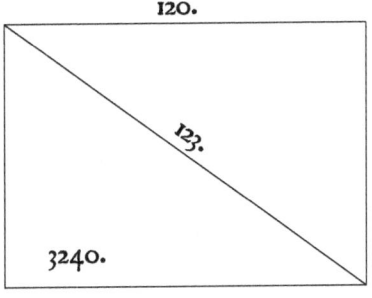

Scholar. Here J see the same. 4 nombers. 9. 15. 21. and. 27. set as the lesser sides : And their greater sides are soche as disagree frō the for= mer rule. And in. 15. 21. and. 27. J see twoo va= rieties, vnlike to the for mer example. But see= yng the sides doe disa= gree, J doe not maruel that the *diametralle nombers* are diuerse from the former.

Master. Examine these nombers, whether thei be true.

Scholar. J must multiplie eche side by it self, and then adde thē together : and if thei make as moche iu= stly, as the *diameter* beyng multiplied square, then are thei true nombers. So J see, that. 9. maketh. 81. and 12 doeth yelde 144 whiche bothe added doe make. 225. And so moche doth 15 make, being multiplied square.

Likewaies, for the second figure 15. bryngeth forth

The ſeconde parte

225. and. 20. giueth. 400. that is by addition. 625.
whiche ſomme doeth amounte alſo, when. 25. is mul=
tiplied ſquare.

The thirde figure hath. 15. alſo for the one ſide,
whoſe ſquare is. 225. and for the other ſide. 36. whiche
maketh in ſquare. 1296. And thei bothe together giue
1521. And ſo many commeth of 39 multiplied by it ſelf
in ſquare.

Again for the fourthe figure. 21. maketh. 441. and
28. doeth yelde. 784. whiche bothe beyng added, doe
amounte vnto. 1225. And ſo moche doeth there ariſe
by. 35. multiplied into it ſelf.

The fifte figure hath. 21. alſo, and his ſquare is
441. and the ſeconde ſide beyng. 72. maketh in ſquare
5184. So that bothe thoſe ſquares doe make. 5625.
And the like nomber is made by. 75. multiplied in
ſquare forme.

Now in the ſixt figure 27 beyng multiplied ſquare
bryngeth forthe. 729. And. 36. likewaies multiplied
doeth make. 1296. and that with the other will make
by addition. 2025. whiche ſomme (as is well ſeen)
doeth come of the multiplication of. 45. by it ſelf.

In the ſeuenth figure. 27. multiplied ſquare, doeth
giue. 729 : and the other ſide (whiche is. 120.) doeth
bryng forthe. 14400. Theſe bothe ioyned together
doe make. 15129. And the like ſomme is gathered by
the multiplication of. 123. ſquarely.

So that all thoſe figures doe appere true.

But how thei maie agree with your former rule,
I can not ſee.

Maſter. That rule did I make for nobers vncom=
pounde. For nombers compounde haue not onely in
their owne name, the vſe of that rule, but alſo thei fo=
lowe the forme of thoſe nombers, of whiche thei bee
compounde.

So. 9. beyng compounde of. 3. foloweth the forme
                                                        of

of Arithmetike.

of. 3. And therfore as. 3. hath. 4. fo; to make the feconde
fide with hym, fo. 9. (beepng thrife. 3.) fhall haue. 12.
(which is thrife. 4.) fo; a matche fide with hym.

Likewaies. 15. beyng compounde of. 5. and. 3. fhall
haue their fo;mes in the makyng of the *diametralle nõ=
bers*. Fo; as. 3. hath. 4. fo. 15. (beepng fiue tymes. 3.)
fhall haue. 20. (whiche is fiue tymes. 4.) fo; the fe=
conde fide.

Again, as. 5. hath. 12. fo fhall. 15. (beepng three ty=
mes. 5.) haue. 36. (that is three tymes. 12.) fo; his fe=
conde fide.

Likewaies. 21. beyng compounde of. 3. and. 7. fhall
haue bothe their fo;mes.

And. 27. whiche is compounde of. 3. and. 9. fhall
haue all the varieties of their fo;mes.

Scholar. I fee it is euen fo, and that in the *diame=
ter*, as well as in the feconde fide. But the *dimetralle
nomber* doeth varie moche in them.

Mafter. Yet doe thofe nombers agree in a mar=
ueiloufe good p;opo;tion. Fo; if you doe confider the
p;opo;tion of bothe the fides in one figure, to bothe
the fides in an other figure : and adde thofe twoo p;o=
po;tions together, the addition of theim doeth make
the nomber that rep;efenteth the p;opo;tiõ betwene
their twoo *diametralle nombers*. Whiche thynge I will
now onely touche, as b;iefly as maie bee, to giue you
occafion to marke it better hereafter : Sith this place
doeth not fully ferue fo; it. As. 3. and. 4. beepng the
twoo fides of a *Diametralle nomber*, doe make. 12. So if
9. and 12 be the fides of a *diametralle nomber*, that nom=
ber muft be. 9. tymes. 12. that is. 108. Fo;. 9. is triple
to. 3 : and. 12. is triple to. 4. And bicaufe the addition
of p;opo;tions, is like the multiplication of fractiõs,
I muft multiplie. 3. by. 3. o; els $\frac{3}{1}$ by $\frac{3}{1}$, whiche is all
one, and that will maie. 9.

Likewaies, if 3. and. 4. be taken fo; the fides of the
E.iii.   leffer

The seconde parte

lesser nomber *diametralle*, and. 15. and. 36. for the sides
of the greater nomber : As the lesser nomber shall bee
12. so the greater must be. 540. that is. 45. tymes. 12.

For. 15. vnto. 3. is in a *quintuple* proportion, and is
written thus. $\frac{5}{1}$: and. 36. vnto 4 is a *noncuple* proportion, and is written thus $\frac{9}{1}$. Now if you multiplie these
nombers together, thei will make 45 : whiche declareth thē proportions of the twoo *diametralle nombers*.
And so of all the reste, as you maie easily consider.

Scholar. J praie you, let me examine one or twoo
of thē, in comparison to that firste *diametrall nomber*. 12.

J see that 15 beyng the lesser side, and 20. the greater side, doe make. 300. as their *diametralle nomber* : and
that. 300. is. 25. tymes so moche as. 12. is. Therfore
by your saiyng the proportion of 15. to. 3. and of. 20. to
4. must make. 25. And so it doeth. For eche of them is
a *quintuple* proportion. And it is quickely gessed, that
5. multiplied by. 5. doeth make. 25.

For farther proofe, J take the *diametralle nomber*
1680. whose sides are. 15. and. 112. First J see, that. 15.
to. 3. beareth a *quintuple* proportion : and. 112. to. 4. is
as. 28. to. 1. Therfore J multiplie. 28. by. 5. and it maketh. 140. Then if J multiplie that nomber by. 12. it
will make. 1680.

This is a sufficiente trialle for these nombers.

Of even sides.   But of soche *diametralle nombers*, as haue euen nom
bers for their lesser side, you haue giuen no rule, nother examples, saue onely of. 8. wherfore J praie you
tell me, how shall J finde out the *diametralle nomber*,
with his other side, and the *diameter* in soche euen
nombers.

Master. You shall make it square, as you did in
the other nombers, that wer odde : And of that square
you shall take twoo quarters, whiche you shall alter
in soche sorte, that you shall abate. 1. frō the one quarter, and put it to the other quarter. And so haue you
twoo

of Arithmetike.

twoo nombers, differyng onely by. 2. and bothe be=
yng odde. The leſſer of them twoo, is the greater ſide
of the *diametralle nomber* : and the other is the *diameter*
to it. As. 8. beyng your leſſer ſide, the ſquare of it is
64. whoſe quarter is. 16. from whiche J abate. 1. and
there reſteth. 15. and that is the ſeconde ſide. Alſo J
adde 1. to 16. and it maketh. 17 : whiche is the *diameter*.

Scholar. This is no thyng harde. As by example
J will proue. Jf. 12. bee the leſſer ſide : his ſquare is
144. and the quarter of it is. 36. Then abatyng. 1.
J ſee there will bee. 35. for the other ſide of the *diame=
tralle nomber*. And addyng. 1. to. 36. it maketh. 37. to be
the *diameter*. And if J multiplie. 35. by. 12. it bryngeth
forthe. 420. whiche is the *diametralle nomber*.

Now for proofe of theſe nombers, J multiplie. 12.
by it ſelf, and it maketh. 144. Then J multiplie the
other ſide, that is. 35. by it ſelf, and it yeldeth. 1225.
Thoſe bothe together doe make. 1369. And ſeyng 37
multiplied by it ſelfe, doeth make the ſame nomber.
Therfore are thei all true nombers.

An other example. 10. beyng ſet for the leſſer ſide,
J doe multiplie it ſquarely : and there riſeth. 100.
whoſe quarter is. 25. For whiche J take (as you
taught me). 24. and. 26. And ſo the whole *diametralle
nomber* is. 240. For proofe of the other nombers, J
take. 100. whiche commeth of. 10. multiplied ſquare,
and to it J adde. 576. whiche is the ſquare to. 24. and
thei bothe doe make. 676. And ſo muche amounteth
by the multiplication of. 26. ſquarely.

Maſter. This maie ſuffice for this preſente : if
you marke that the euē nombers haue not onely one
generalle forme, whiche J did expreſſe in the former
rule, but alſo ſoche as be compounde of any other nō=
bers, euen or odde : Haue the like nombers in propor=
tion, for the greater ſide, and for their *diameter* as the
nomber haue, of which thei bee compounde. And
bicauſe

The seconde parte

bicause J will not staie to long on this matter, J will
here set foarthe diuerse varieties of *diametrall nombers,*
whereby you maie gather not onely the true vnder=
standyng of the foamer rules : But also in theim you
maie see other notable cōclusions : and straunge woa=
kes of the natures of nombers.

    Marke well this table foame, with the titles ouer
it : whiche declare the true meanyng of it.

    And where you see one nomber in the firste co=
lumpne againſt twoo, three, oa fower in the other co=
lumpnes, you ſhall vnderſtande that that nomber is
the ſide to ſo many ſeueralle nombers *diametralle.*

The table of diame=
tralle nombers.

The lesser side.	The greater side.	The diameter.	The nöber diametralle
3.	4.	5.	12.
5.	12.	13.	60.
6.	8.	10	48.
7.	24	25	168
8.	15	17	120
9 {	12	15	108
	40	41	360
10	24	26	240
11	60	61	660
12. {	16	20	192
	35	37	420
13	84	85	1092
14	48	50	672
15 {	20	25	300
	36	39	540
	112	113	1680
16 {	30	34	480
	63	65	1008
17	144	145	2448
18 {	24	30	432
	80	82	1440
19	180	181	3420
20 {	48	52	960
	99	101	1980
21 {	28	35	588
	72	75	1512
	220	221	4620
22	120	122	2640
23	264	265	6072
24 {	32	40	768
	45	51	1080
	70	74	1680
	143	145	3432
25 {	60	65	1500
	312	313	7800

The lesser side.	The greater side.	The diameter.	The nöber diametralle
26	168	170	4368
27. {	36	45	972
	120	123	3240
	364	365	9828
28. {	96	100	2688
	195	197	5460
29	420	421	12180
30 {	40	50	1200
	72	78	2160
31	480	481	14880
32 {	60	68	1920
	126	130	4032
	255	257	8160
33 {	44	55	1452
	180	183	5940
	544	545	17952
34	288	290	9792
35 {	84	91	2940
	120	125	4200
	612	613	21420
36 {	48	60	1728
	105	111	3780
	160	164	5760
	323	325	11628
37	684	685	25308
38	360	362	13680
39 {	52	65	2028
	252	255	9828
	760	761	29640
40 {	75	85	3000
	96	104	3840
	198	202	7920
	399	401	15960
			F.I.

The seconde parte

This table maie you extende infinitely. And these thinges maie you se, as thinges of greate admiratiö.

1.   There is no *diametralle nomber*, but it maie be diui= ded by. 12. Wherfoze thei be all euen nombers euen= ly and oddely.

2.   Again, there is no *diametralle nomber*, but it endeth in. o. in. 2. oz in. 8.

3.   Thirdely, there is no *diametralle nomber*, that can haue any moze *diameters* then one.

4.   Yet maie one nomber bee the *diameter* to diuerse other.

As you se 25. is the *diameter* to. 168. and also to. 300. So. 65. is the *diameter* to. 1008. and also to. 1500. Likewaies. 145. is the *diameter* to. 2448. and to 3432.

5.   Fiftely : No square nomber can bee a *diametralle nomber.*

Scholar. These propertieƒ be notable.

To know a diametralle nomber.
But how shall J knowe, when a nomber is pzo= poned, whether it be a *diametralle nomber*, oz not?

Master. In that thyng J finde a tediouse trauell, by any rules, in those that wzite of it. But J wil ease you of moche paine therein.

Firste remember the properties of those nombers.

And if you haue any other figure in the first place, then. o. 2. oz. 8. it is no *diametralle nomber.*

Secondarily, if it maie not bee diuided by. 12. al= though it ende in one of those. 3. figures, it is no *dia= metrall nomber.*

Wherfoze if it haue bothe those twoo properties (whiche an infinite multitude of nombers doe want) and be no square nomber (as none be that ende in. 2. oz. 8. oz with odde cyphers) then sette out all the par= tes of it, in soche sozte, that the lesser parte doe stande directly ouer those greater partes, which beyng mul= tiplied together, will make the whole nomber.

                                                              And

of Arithmetike.

And then examine thofe partes, whiche feme to haue any likelihod : accordyng to the former doctrine.

As for example : if. 72. be proponed to be examined in that forte, I fette his partes in order thus.

2.   3.   4.   6.   8.

36.   24.   18.   12.   9.

Howbeit I neded not to fet doune. 2. nother. 4. for leffer partes, nother thofe other greater partes, that aunfwere to them : For, as I faid before, thei can not bee the leffer fide in any *diametralle nomber.* Wherfore thei nede no examination.

Farthermore, for them that you fhall nede to exa= mine, if the leffer nomber bee an odde nomber, the fquare of it muft contain double to that greater nom= ber (this is coupled with it) and one more.

And if the leffer be an euen nomber (of them twoo that you would examine) then muft the fquare of it containe the greater nomber (that ftandeth by it). 4. tymes, and. 4. more. And this is not onely a fhorter waie, then I fee to be taughte by other artes menne : but it is alfo more certaine, for all nombers not com= pounded of other *diametralle nombers.*

Scholar. By this doctrine it appeareth quickely, that. 72. is no *diametrall nomber.*

For although it doeth ende in. 2. and maie be diui= ded by. 12. yet no couple of nombers here haue thofe properties that is required.

For vnder. 3. is. 24. whiche is to greate : and vnder 6. there is. 12. whiche is to greate alfo.

But vnder. 8. ftandeth. 9 : whiche is to litle, by a greate deale.

Mafter. Then proue in this other nomber. 132.

Scholar. His partes will ftande thus.

The seconde parte

3.	6.	11.
44.	22.	12.

Where I see quickely that it can not bee a *diame=tralle nomber.* For the nombers vnder. 3. and. 6. be to greate : sith no nomber that should bee sette vnder. 3. maie be aboue. 4.

Nother vnder. 6. maie any nomber bee set greater then. 8. As it doeth sufficiently appeare by that that is taughte before.

And vnder. 11. there can bee no lesse nomber pla=ced then. 60 : and therfore. 12. is to smalle.

And herein I perceiue greate helpe by this table, whiche you haue set forthe.

Master. It is well marked of you. But yet trie this other example. 6072.

Scholar. I set doune his partes in order, thus.

3.	6.	8.	11.	12.	22.	23.	24.
2024.	1012.	759.	552.	506.	276.	264.	253.

33.	44.	46.	66.	69.
184.	138.	132.	92.	88.

And here I see a greate sorte of nombers, whiche can not serue to my purpose, bicause those that bee e=uen, and are lesse then. 44. make to litle a square, to be 4. times so moche as the nomber vnder any of thē.

And. 44. maketh to greate a square : wherfore it can be none of the euen nombers.

Again, those that be odde vnder. 23. doe make to li=tle a square, to bee double to the greater nomber vn=der it. And those that bee odde aboue. 23. doe make to greate a square. So that. 23. doeth remain to bee the true nōber for the lesser side : and 264 the greater side.

Master. Bicause exercise is the beste instrument

in

of Arithmetike.

in learnyng : therfoze will I propounde to you one example moze.

What faie you of. 5460? Is it a *diametralle nomber* oz no?

Scholar. I will trie it, by settyng doune his par=
tes thus.

3.	5.	6.	7.	10.	12.	13.	14.	15.
1820.	1092.	910.	780.	546.	455.	420.	390.	364.

20.	21.	28.	30.	35.	42.	52.	60.	70.
273.	260.	195.	182.	156.	130.	150.	91.	78.

And here I se diuerse and many nombers, whiche
at the firste sighte, appere nothyng mete foz this pur=
pose. Foz. 20. is to smalle a nomber, as I maie sone
iudge : and therfoze all other nombers vnder it, must
nedes be to smalle, of foze.

Againe, I see that. 30. is to greate a nomber, and
therfoze, of necessitie, all other nombers aboue it,
must nedes be to greate. So that. 21. other. 28. must
be the true nomber, oz els none.

Wherfoze I examine first. 21. whose square is 441
whiche should bee one moze then double, to the nom=
ber vnder it, that is to saie, it should bee. 521. And so it
is not : Therfoze I refuse it, and examine. 28. whose
square is. 784. And that should bee fower tymes so
moche as. 195. (whiche is the nomber vnder it) and
4. moze. Therfoze I doe *quadriple*. 195. and it ma=
keth. 780. And then I see that it wanteth, but fower
of the other square : wherfoze I take those twoo nom=
bers, I meane. 28. and. 195. foz the true sides of. 5460.
whiche I finde to be a *diametralle nomber*.

Master. By the waie, remēber that you could ea=
sily perceiue, that all nōbers vnder. 20. were to small
foz your purpose : and contrary waies, all aboue. 30,

F.iii.   to

The seconde parte

to be to greate. So that you neded not to sette doune so many partes of your firste nomber.

Wherfore if your nomber bee soche a one, as hath many partes, you maie chose one by gesse, which you thinke will go nigh to serue your purpose : and if you finde it to smalle, then set theim doune onely that bee greater then it, til you finde one other iuste : and then haue you your purpose. Or if you finde any to great, after that whiche was to smalle, and betwene theim none iuste, then is not your nomber a *diametrall nöber.*

But and if the parte whiche you tooke by gesse, be to great, you shall refuse all partes aboue it, and take onely lesser partes, til you finde a iuste parte for your purpose : or els one that is to litle.

And if in descendynge orderly, you finde no iuste parte, before you come to one that is to litle, then is your nomber no *diametralle nomber.*

Scholar. This is a greate ease in shortenynge of worke : whiche I will proue in this nomber. 9786.

Master. If you remembred well your former ru= les, you would not admitte this to be examined for a *diametralle nomber* : bicause it endeth in none of the thre peculiare terminations : that is. 0. 2. or. 8.

Scholar. I confesse my faulte. And therfore I take this nomber. 9780. whose. 20. parte is. 489. But se= yng. 20. doeth make in square but. 400. therfore is it very moche to litle.

Then I take the. 30. parte of it, whiche is. 326. and finde it also to litle.

Thirdely, I take the. 40. parte of it, whiche is 244 $\frac{1}{2}$: and seyng. 40. maketh in square. 1600. I see that it is almoste. 7. tymes so moche as. 244 $\frac{1}{2}$: and therfore is it to greate.

So must the true nomber be betwene. 30. and. 40 : or els there is none at all.

Therfore firste I take. 35. whiche is the middelle nomber

of Arithmetike.

nomber (as the moste apte for a coniecture) and it yel=
deth. 279 $\frac{3}{7}$. And the square of. 35. is. 1225. whiche is
farre moꝛe then the double of. 279 $\frac{3}{7}$.

Therfoꝛe, again I pꝛoue with. 32. whiche giueth
305 $\frac{5}{8}$. And seyng the square of. 32. is. 1024. it is not
4. tymes so moche as. 305 $\frac{5}{8}$. foꝛ that is. 1222 $\frac{1}{2}$.

Wherfoꝛe I take a greater nomber, betwene it
and. 35. And first I take. 33. whiche bꝛingeth foꝛthe
296 $\frac{1}{11}$. wherby I maie see that. 33. is to greate. And
seyng there is no nomber lefte betwene. 32. and. 33.
therfoꝛe I iudge that firste nomber. 9780. to bee no
diametralle nomber.

Master. Examine this nomber. 43200.

Scholar. Bicause I see it to be a greate nomber,
I will begin with a greate parte of it. And therfoꝛe,
I take. 100. whiche yeldeth. 432. And consideryng
that the square of. 100. is. 1000. whiche is farre to
greate, I must seke a lesser nomber.

Master. I will ease you of your paines in that.
Foꝛ bicause here is moꝛe to bee considered. You re=
member that I tolde you befoꝛe, in makyng of *diame=*
tralle nombers, how that some nombers doe followe the
rules of other, of whiche thei be compounde. And far=
thermoꝛe, that soche compounde *diametralle nombers,*
did beare pꝛopoꝛtion to the lesser, as the pꝛopoꝛtion
was of bothe their sides added together.

Scholar. That is true.

Master. Of like reason all soche *diametralle nom=*
bers, must bee excluded from these rules, whiche bee
made peculiarly foꝛ nombers that haue their owne
pꝛoper foꝛmes, and depende not of other.

And yet some common rule must bee giuen, that
maie extende as well to them, as to any other.

Wherfoꝛe let this be it.

That the twoo sides of all *diametralle nombers,* haue
soche a pꝛopoꝛtion together, as here you see expꝛessed

in

The seconde parte

in some one of these formes: if thei bee continued as
here thei be begon.

❡ The Firste order.

$$\frac{3}{4} \; \frac{5}{12} \; \frac{7}{24} \; \frac{9}{40} \; \frac{11}{60} \; \frac{13}{84} \; \frac{15}{112} \; \frac{17}{144} \; \frac{19}{180} \; \frac{21}{220} \;$$

$$\frac{23}{264} \; \frac{25}{312} \; \frac{27}{360} \; \frac{29}{420} \; \frac{31}{480} \; \frac{33}{544} \; \frac{35}{612} \; \frac{37}{684} \; \frac{39}{760} \; \text{&c.}$$

❡ The seconde order.

$$\frac{8}{15} \; \frac{12}{35} \; \frac{16}{63} \; \frac{20}{99} \; \frac{24}{143} \; \frac{28}{195} \; \frac{32}{255} \; \frac{36}{323} \; \frac{40}{399} \; \frac{44}{483} \;$$

$$\frac{48}{575} \; \frac{52}{2704} \; \text{&c.}$$

Here haue I sette the lesser side as the numerator,
and the greater side as the denominator. Whereby
you maie perceiue the cause of their distinction.

For the first order is, when the lesser side, or nom=
ber, is odde.

The seconde order is, when that lesser side is an
euen nomber.

Stifelius doeth set them so, that the numerator stan=
deth for the seconde, or greater side : and the denomi=
nator for the firste nomber, or lesser side. And for the
more delectable contemplation, to behold their forme
of progression, he setteth doune as many whole nom=
bers, as the fraction will giue.

And this is his forme.

❡ The firste order.

$$1\tfrac{1}{3} \; 2\tfrac{2}{5} \; 3\tfrac{3}{7} \; 4\tfrac{4}{9} \; 5\tfrac{5}{11} \; 6\tfrac{6}{13} \; 7\tfrac{7}{15} \; \text{&c.}$$

❡ The seconde order.

$$1\tfrac{7}{8} \; 2\tfrac{11}{12} \; 3\tfrac{15}{16} \; 4\tfrac{19}{20} \; 5\tfrac{23}{24} \; 6\tfrac{27}{28} \; 7\tfrac{31}{32} \; \text{&c.}$$

Where

of Arithmetike.

Where in the firſt oꝛder, you ſe bothe in the whole nombers, and alſo in the numeratoꝛs of the fraction, the naturalle oꝛder of nombers. And in the denomi= natoꝛs, the naturalle pꝛogreſſion of odde nombers.

But in the ſeconde oꝛder, you ſee that the whole nombers go in their naturalle oꝛder, and the nume= ratoꝛs and denominatoꝛs, kepe an *Arithmeticalle* pꝛo= greſſion, by equalle diſtaunce of. 4. ſaue that in the numeratoꝛs, all the nombers bee odde : and in the de= nominatoꝛs, thei be all euen.

Now by this generalle rule, if you finde any twoo partes of any nomber, in one of theſe foꝛmer pꝛopoꝛ= tions, you maie bee ſure that it is a *diametralle nomber.* But foꝛ the moꝛe apte conference of the partes, you ſhall doe beſte to reduce them to their leaſt nombers : as you haue learned in the firſte parte of *Arithmetike.*

So in your laſt nomber, whiche was 43200. you ſhall finde his. 180. parte, to bee. 240. whiche beyng reduced to their ſmalleſt nombers, will bee. $\frac{3}{4}$: wher= foꝛe I am aſſured, that it is a *diametralle nomber.*

Yet one thyng moꝛe ſhall you marke.

If any nomber ende in Ciphers, abate euen Ci= phers, as often as you can (I meane. 2. 4. oꝛ. 6. ꝛc, and if the reſte be a *diametralle nomber,* ſo was the firſt. And therfoꝛe in this laſte example. 432. is a *diametralle nõ= ber,* as well as. 43200.

Alſo if any nomber beeyng diuided by any ſquare nomber, doe make a *diametralle nomber* in the *quotiente,* then was the firſte nomber a *diametralle nomber* alſo.

And this, foꝛ this tyme, ſhall ſuffice foꝛ *diametralle nombers.*

Now will I ſpeake ſomewhat bꝛiefly of *likeflattes :* and then pꝛocede to other *figuralle nombers.*

Of like flattes.

Scholar. I remember you defined them befoꝛe, to bee ſoche flatte nombers, as had one foꝛme of pꝛopoꝛ= tion betwene their ſides.

G.i.   As

The seconde parte

As here 27. and 12. be *likeflattes* : bicause their sides be in one proportion. For as. 9. is to. 3. so 6 is to. 2. bothe beeyng in triple proportion.

Master. You saie well. And that is the cause why thei be called like : for the likenesse in the proportiõ of their sides. Although some menne delite

Squarelike figures.

more to call them *squarelike figures* : bicause thei haue some properties agreable with square nombers (for as *Euclide* saieth in his. 8. booke, and. 18. proposition :

Euery twoo nombers, beeyng likeflattes, haue one meane nomber betwene theim in proporti= on. And the one flatte nomber beareth vnto the other flatte double that proportion, that their sides doe.

For declaration of whiche proposition, marke the twoo flatte nombers before : I meane. 27. and. 12. whose sides are in proportion *Sesquialter* : And the flat nombers themselfes be as $\frac{9}{4}$. or. 9. to. 4 : that is double *Sesquiquarte.* Now doe you double the proportion *Ses= quialter,* and it will make *double Sesquiquarte.*

Scholar. Thus doe I sette them in order. $\frac{3}{2} : \frac{3}{2}$. And I multiplie the numerators together, and the denominators also. (For I remember, you tolde me before, that proportions are added, as fractions are multiplied) and then will it be. $\frac{9}{4}$: euen as you saied.

Master. Again *Euclide* saith in the twenteth pro= position of the same booke.

If any nomber stande as a middle nomber in proportion,

of Arithmetike.

proportion, betwene other twoo nombers, thoſe
twoo are like flattes.

That is to ſaie : if any twoo nombers, beyng mul=
plied together, doe make a ſquare nomber (foʒ none
but ſoche can haue a middle nomber betwene theim)
then are thei *like flattes.*

As. 3. and. 12. multiplied together doe make. 36.
whiche is a ſquare nomber : and. 6. therby appeareth
to bee the middell nomber betwene theim. And ther=
foʒe are. 3. and. 12. *like flattes.*

Likewaies. 3. and. 27. foʒ thei make. 81. whiche is
a ſquare : and their middle nomber is. 9.

And ſo are. 2. and. 8 : 2. and. 18 : 2. and. 50. 2. ɟ. 72
3. and. 48 : 3. and. 75 : 4. and. 9. 4. and. 16 : 4. and
25. 5. and. 20. 5. and. 45 : 6. and. 24 : 6. and. 54.

And ſo of infinite other.

This expoſition is confirmed by the firſte and ſe=
conde pʒopoſition of the nineth boke of *Euclide,* where
he ſaieth thus.

If twoo nombers beyng like flattes, bee mul=
tiplied together, the nomber that thei make,
ſhall be a ſquare nomber.

And if. 2. nombers beyng multiplied together,
do make a ſquare nöber, then are thei like flattes.

By whiche rules it doeth appere, that you cã haue
no pʒogreſſiö *Geometricalle,* but it muſt be made either
of ſquare nombers, oʒ els of *like flattes,* wherby there
appeareth a greate agreablenes, betwene *like flattes,*
and ſquare nombers. And therfoʒe ſaieth *Euclide* al=
ſo in the. 26. pʒopoſition of the eight booke.

Nõmbers that bee like flattes, haue ſoche
proportion together, as one ſquare nomber bea=

The seconde parte

reth to an other.

This maie you proue by any of the former exam=
ples. For. 12. to. 3. is in like proportion, as. 16. to. 4.
or. 36. to. 9.

Also. 27. to. 3. hath like proportion as. 36. to. 4 : or
144. to. 16. other. 81. to. 9.

And farther, if you deuide the one of theim by the
other, the *quotiente* will be a square nomber.

Scholar. That doeth appeare euidentely at the
firste vewe.

For. 12. diuided by. 3. doeth make. 4. And. 75. diui=
ded by. 3. giueth. 25.

So. 54. by. 6. maketh. 9. And. 72. by. 2. yeldeth. 36.
And so I see in the reste, that all the *quotientes* will be
square nombers.

howe like flat
tes be made.
But I desire moche to knowe, how those nombers
be produced. For that I knowe not yet.

Master. Take any twoo square nombers, what
so euer thei bee, and multiplie them by any one nom=
ber, that you liste : and thei will make. 2. *like flattes.*

So. 4. and. 9. multiplied by. 2. doe make. 8. and. 18 :
whiche bee *like flattes.*

Again, if you multiplie them by. 5. thei make. 20.
and. 45. whiche be also *like flattes.*

Scholar. I am perfect inough in this, if that be al.

Master. An other waie you maie make them al=
so : If you take any twoo square nombers, that will
admitte one diuisor and diuide them bothe by it.

As for example. Seyng 9. and. 36. will be bothe di=
uided by. 3. I doe so diuide theim : and their *quotientes*
are, 3. and. 12. whiche are *diametralle nombers.*

So in like maner, if I diuide 196 and 49 (whiche
bothe are square nombers) by. 7. the *quotientes* will be
28. and. 7.

Again, 16. and. 100. beyng bothe square nombers
and

of Arithmetike.

and diuided by. 4. doe make. 4. and. 25. as their *quoti=* *ente,* and thei be *like flattes.*

Scholar. And in these J see an other straunge worke : that if those twoo *like flattes* bee multiplied to= gether : thei will make the greater square, of whiche thei came.

Fo2. 3. tymes. 12. maketh. 36 : and. 7. tymes. 28. gi= ueth. 196 : And so. 4. tymes. 25. b2yngeth fo2the. 100.

Mafter. Jt doeth so happen often times : but it is not alwaies so.

Fo2 if you diuide. 16. and. 100. by. 2. the *quotientes* will be. 8. and. 50. whiche twoo nombers multiplied together, doe make. 400. farre differyng from. 100. So. 36. and. 196. beyng bothe square nombers, and diuided by. 2. doe make. 18. and. 98. whiche be *like flat= tes :* and those *like flattes* multiplied together, doe yelde 1764. whiche is a square nomber, but it is. 9. tymes so greate as is. 196.

Scholar. Yet one doubte J haue : whether all square nombers be *like flattes,* and so bee not diftincte from them?

Fo2 although in the diuision of figuralle nombers you did diftincte them, yet in the examples of *like flat= tes,* you put certain square nombers amongeft other.

Mafter All square nombers are *like flattes,* beyng compared together : and els not. Fo2 as any. 2. square nombers maie be compared together : so maie thei be referred to their rootes, without comparison toge= ther. O2 els thei maie be compared to other nombers that bee not square.

Therfo2e marke these two rules well. that no one nomber can bee called a *like flatte :* but in comparison to some other. Fo2. 2. by hymself is not called a *like flatte,* excepte he bee compared to. 8. o2 to. 18. other to 32. o2. 50. o2 some other foche.

So likewaies. 4. whiche by nature is a square nõ=
C.iii.   ber,

The seconde parte

ber, and alwaies shall bee so : yet is it not accepted as
a *like flatte*, onles it bee referred to some other square
nomber.

Scholar. What if it be compared with. 12. which
you named before to be a *like flatte?*

Master. You remember : one of *Euclide* his rules
(whiche I repeated before) is, that *like flattes* beeyng
multiplied together, will make a square nöber. And
so doeth not. 12. beyng multiplied by. 4.

Scholar. Now I doe vnderstande your woordes
better. So. 3. and. 8. compared together, bee not *like
flattes* : yet eche of them compared to other nombers,
maie be *like flattes*. As. 3. compared to. 12. oz to. 27 : and
8. compared to. 18. oz to. 50.

*Of rooted
nombers.*

Master. Now will we lette these *like flattes* alone
for a tyme : And intreate moze of rooted nöbers. And
first I will tell you somewhat of the names and na=
tures of soche nombers as haue rootes : Then secon=
darily I will teache you the ozder to extract their roo=
tes : And afterwarde will I shewe some parte of the
vse of theim.

A roote.

Wherfoze to begin, where we lefte a little befoze,
the explicatiö of rootes : I saie, that the roote of nom=
ber, is a nomber also : and is of soche sozte, that by sön=
drie multiplications of it, by it self, oz by the nomber
resultyng thereof, it doeth pzoduce that nöber, whose
rooe it is. And accozdyng to the nomber of times that
it is multiplied, the nomber that resulteth thereof, ta=
keth his name.

So that one multiplication maketh a *square nomber*
And twoo multiplications doe make a *Cubike nomber*.

Likewaies. 3. multiplications, doe giue a *square of
squares*. And. 4. multiplications doe yelde a *sursolide*.

And so infinitely.

Foz as multiplication hath no ende, so the nom=
bers amountyng of them be innumerable, and their
                                                    rootes

of Arithmetike.

rootes as infinite. But their names thei take certain
ly, of the nombers that thei doe make.

So the roote of a square nomber, is called a *Square* *A square*
roote : and the roote of Cubike nomber, is named a *Cu=* *roote.*
bike roote : In like sorte that roote is called a *Squared* *A cubike*
square roote, whiche maketh a square of squares in nō= *roote.*
ber. And that roote is a *Surfolide roote,* that yeldeth a *A squared*
Surfolide nomber : in whiche sorte of multiplication, you *square roote.*
maie procede infinitely, as I saied. *A surfolide*
roote.

Notwithstandyng for your ease, I haue set foorthe
here in a table, certain of the moste notable kindes of
rooted nombers.

And to the intente you maie partly conceiue the
reason of their names, I will after the table, set foorth
a brief explication of their names, with the protrac=
ture of the figures, that thei doe resemble in multi=
plications *Geometricalle* : where poinctes, lines, platte
formes, or soundformes bee multiplied : and brynge
foorthe other formes agreable to soche multiplica=
tions.

But first marke the table well : And it will giue
you greate lighte, and aptnesse to vnder=
stande all that foloweth, moche the
better.

For examples are the
lighte of tea=
chyng.

The

The table of rooted numbers.

	The vulgare names.	Rootes.									The authors names.
1.	Rootes.	2	3	4	5	6	7	8	9	10	Rootes.
2.	Squares.	4	9	16	25	36	49	64	81	100	Squares.
3.	Cubikes.	8	27	64	125	216	343	512	729	1000	Cubes.
4.	Squares of Squares.	16	81	256	625	1296	2401	4096	6561	10000	Longe Cubes.
5.	Surfolides.	32	243	1024	3125	7776	16807	32768	59049	100000	Squares of cubes.
6.	Squares of cubes.	64	729	4096	15625	46656	117649	262144	531441	1000000	Cubike Cubes.
7.	Seconde Surfolides.	128	2187	16384	78125	279936	823543	2097152	4782969	10000000	Longe Cubike Cubes.
8.	Squares of squa-red squares.	256	6561	65536	390625	1679616	5764801	16777216	43046721	100000000	Squares of Cu=bike Cubes.
9.	Cubes of Cubes.	512	19683	262144	1953125	10077696	40353607	134217728	387420489	1000000000	Cubes of Cubike Cubes.
10.	Squares of Surfolides.	1024	59049	1048576	9765625	60466176	282475249	1073741824	3486784401	10000000000	Long Cubes of Cubike Cubes.

of Arithmetike.

Here you see diuerse rewes of nombers, and a=
gainst euery rowe twoo names written : one on the
right hande, and the other on the lefte hande, whiche
serue for all the nombers in that rewe.

The names on the lefte hande bee those names,
whiche bee commonly vsed, and attributed to those
nombers.

The names on the righte hande, are names of my
addition, whiche doe aptly expresse the very natures
of the nombers, vnto whiche thei bee assigned : as a=
none I will declare.

And now concernyng the nombers, you see firste
in the hedde of the table, a rewe of nombers set in or=
der, as thei followe in common numbryng, from one
forward. And thei bee called rootes, for that the mul=
tiplication of eche of them, by theimselfes, or by that,
that thereof amounteth, bryngeth forthe all thother,
that bee set vnder them. Of the whiche, the seconde
rewe is called *Square nombers* : bicause that their length   *Square*
and their bredth (whiche I vnderstand by the. 2. nom=   *nombers.*
bers of their multiplication) is equalle.

As. 2. tymes. 2. doeth make. 4. whiche is a
square nomber, and maie bee figured thus.

Likewaies. 3. tymes. 3. maketh. 9. whiche
is a square nomber, and is represented thus.

And here you se, that if you diuide the *Square nomber*
by his roote, the *quotiente* will be the same nöber also.

Scholar. That must nedes be so.

Master. Then in the thirde rewe are placed *Cu=*   *Cubike*
bike nombers : whiche are produced by triple multipli=   *nombers.*
cation. As. 2. tymes. 2. twise, maketh. 8. And. 3. ty=
mes. 3. thrise, yeldeth. 27. So. 4. tymes. 4. fower ty=
mes, giueth. 64. These nombers can not be expres=
sed aptly in flatte, but prospectiuely, as Dice maie be
made in protracture.

H.i.   And

The feconde parte

And thefe are their fozmes.

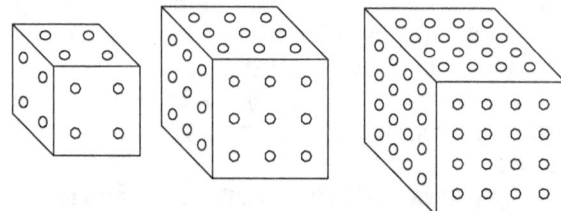

In the firfte figure you fee. 2. erpzeffed in lengthe
bzedthe, and depthe. And in the fecond fozme. 3. is re=
pzefented in all thofe. 3. dimenfions. In the. 4. figure
4. is the roote, and is drawen agreably to that fozme.

Scholar. This is manifefte inough to fighte.

Mafter. Yet reafon ought to waigh it moze er=
actly, then fight can compzehende it. Foz as their tri=
ple multiplication doeth refeble the nature of founde
bodies, fo it might appeare moze iufte erpzeffyng of
their figures, agreably as founde bodies ought : in
whiche every parte can not appeare to fighte, fith di=
uerfe of them loke inwardly. As by thefe. 3. lafte figu=

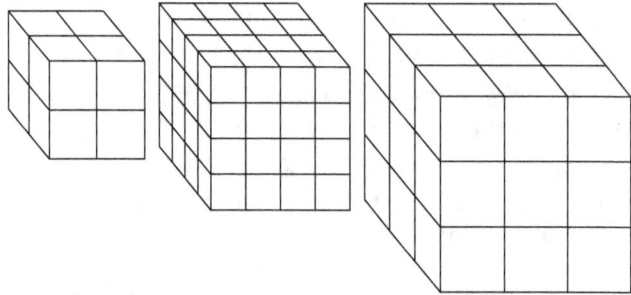

res you maie partely coniecture. Of whiche at this
tyme and in this place, fome men will thinke it an o=
uerfighte to fpeake, and moche moze ouerfighte to
wzite of them any thyng largely. Saue that we maie
vfe theim foz the apter erplication of that triple mul=
plication,

of Arithmetike.

tiplication, wherby thei be made.

So that as it is multiplied thrife, fo the nomber that doeth amounte thereof, hath gotten. 3. dimenſi= ones, whiche properly belongeth to a bodie, oʒ ſound fourme. And therfoʒe is it called a *Cube*, oʒ *Cubike nomber.* Whiche nomber if you diuide by the roote, the *quoti= ent* wil be the ſquare of the fame roote. As I ſaid afoʒe.

But to pʒocede, if you doe multiplie that *Cubike nomber* by his roote, the nomber that riſeth of it, is called a *Square of ſquares* commonly : bicauſe that not onely it is a *Square nomber*, but the roote of it alſo is a *Square nomber.* As you maie perceiue by examina= tion, of all thoſe nombers that be in the fourth rewe, whiche nombers I doe call *longe Cubes* : bicauſe thei make a line of Cubes. And hath in lengthe ſo many Cubes, as the firſte roote doeth containe vnities.

Squares of Squares.

Long Cubes.

This line of Cubes, although it haue foʒ his bʒedthe, and depthe alſo, the thickeneſſe of one Cube, yet bicauſe it hath no nomber of Cubes, in bʒedthe, noʒ in depthe (oʒ generally no nomber of that thyng, whereof it is called a line) therfoʒe maie it tollerably beare the ſimilitude and name of a line. And ſo doe we commonly call lines, thoſe ſmalle coʒdes, whiche are onely long, and haue litle bʒedthe to their length. But yet are thei not without all bʒedthe.

Scholar. And thereof (I thinke men call a line of Bʒickes, and a line of Aſſhelers ſtones, when many bee laied in a rowe, in lengthe : and but one (oʒ fewe) in bʒedthe.

Maſter. You ſaie truthe. And that name doeth continue ſtill, emongeſt all our countrie menne : ſaue that moſte menne doe not call it ſharply a line, but moʒe bʒoder (after tholde Engliſhe language) a laine And ſo men vſe to ſaie, a laine of wine buttes, and a laine of bʒode clothes : and ſoche other like.

And vſe hath ſo largely applied this name, that it

H.ii.   maie

The seconde parte

maie seme no greate absurditie, to name any thynge a line oz laine, that hath moche moze lengthe then bzedthe : and is made by often addition, oz multipli= cation of any one quantitie. But yet foz auoidyng of erroure, it ought to bee limited, whereof that line is named. As in our mater to saie, *a line of vnities : a line of Cubes : a line of Cubike Cubes :* and *a line of Cubike Cubes Cu= bikely :* and so forthe.

In likewaies must we iudge of platte formes, that thei haue no depthe oz thickenesse. When one nom= ber is multiplied by an other, onely twise : that is to saie, in bzedthe and lengthe onely : and is not multi= plied the thirde time by any nomber, to make it beare depthe.

And this must be considered generally, though the nomber so multiplied bee a *Cube*, oz any other sounde nöber. Foz in soche case, that *Cube*, oz sounde nomber, what so euer it be, standeth but as an vnitie.

Scholar. Sir, I doe very well vnderstande the meanyng, and reasonablenesse of those names, line, and square, in any thing. But I knowe not those ter= ms, *Cubike Cubes*, and *Cubike Cubes Cubikely.* Although I se them set in the table, whiche you haue giuen me.

Master. No moze then doe you vnderstande di= uerse other names there, whiche I will therfoze de= clare vnto you.

If you agree to the vse of the name, of a line and a square, in that sorte that you haue consented vnto : then if I multiplie a Cubike nomber by his roote. As to saie. 8. by. 2. oz. 27. by. 3. other. 64. by. 4. then

Squares of squares.

shall I haue a line of *Cubes.* whiche I doe therfoze call *longe Cubes :* but commonly thei bee called *Squared Squares*, oz *Squares of Squares :* and of some men thei are named *Zenzizenzikes*, as square nombers are called *Zenzikes.* Whiche name although in sounde bodies, it hath no vse, yet in pzactice of sounde nombers, it

maie

of *Arithmetike.*

maie and doeth expꝛeſſe ſome pꝛoperties aptly. As
namely that all thoſe nombers, whiche riſe of 4 mul=
tiplications, maie be as well made by twoo multipli=
catiõs. But then the roote of that multiplication ſhal
be a ſquare nomber alſo.

Scholar. So J vnderſtande that. 16. is a nomber
of that ſoꝛte, which here is called *Square of ſquares.* And
yet maie it bee called a ſquare nomber : and is ſo in
deede, in compariſon to. 4. And therfoꝛe, J perceiue,
it is ſet twiſe in the table : ones emongeſt ſquare nom
bers, vnder 4 whiche then is his ſquare roote : And a=
gain it is ſet emongeſt *ſquares of ſquares,* vnder 2 which
in that place ſtandeth as his ſquared ſquare roote.

Likewaies. 64. is twiſe ſet in the ſame table, ones
emongeſt *ſquares,* vnder 8. whiche is his ſquare roote :
And again emongeſt *(Cubike nombers,* vnder. 4. whiche
is his Cubike roote.

Maſter. ẏou ſaie truthe. Although the laſte exa=
ple be not to your purpoſe, concerning *Squared ſquares*
oꝛ *Zenzizenzikes.* And if you did note it onely, foꝛbi=
cauſe it is twiſe ſet in the table : then maie you ſee it
thriſe ſette in the ſame table : foꝛ it is in the ſixte rewe
vnder. 2.

Scholar. So J ſee, wherfoꝛe J might rather haue
takẽ. 81. whiche is a *Zenzizenzike nomber,* and ſo hath
foꝛ his roote. 3 : And alſo it is a ſquare nomber, and
hath. 9. foꝛ his roote.

Maſter. Farther to pꝛocede, if J multiplie thoſe
Squares of ſquares by their roote, thei will make *Surſo=*   Surſolides.
lide nombers.

Scholar. J perceiue by the nombers in the table,
that you meane the leaſte roote of the twoo : bicauſe
vnder. 16. J ſee. 32. in the rewe of *Surſolides.*

Maſter. Reaſon maie driue you to thinke ſo. Foꝛ
the nomber and his roote, muſte beare alwaies one
name. So that if J name. 16. as a ſquare nomber, J

                        P.iii.   muſt

The seconde parte

muſt referre it to his ſquare roote. And if I name it
as a *Zenziȝenȝike nomber* : it muſte bee referred to his
Zeȝiȝenȝike roote. And in like ſoȝt of al other names.

As when I call. 64. a ſquare nomber, & demaunde
what is his roote : you muſte nedes aunſwere by his
Square roote, whiche is. 8. But if I name. 64. as a
Cube, and doe then ſeke foȝ his roote : you muſt vnder=
ſtande his *Cubike roote*, and that is. 4. But if I name
it to bee a *Square of Cubes*, oȝ *zenzicube* : then is. 2. his
roote. As you maie by the table perceiue. And alſo by
the oȝderly multiplication of euery rewe, oȝ oȝder of
nombers by their roote. Foȝ therby amounteth the
nexte rewe.

And ſo maie you increaſe the nombers of thoſe re=
wes, oȝ oȝders, accoȝdyng to the tymes of your mul=
tiplicatiō, as moche as you liſt. And euery oȝder ſhall
beare ſoche names, as agreeth to the nature of their
rootes.

Wherfoȝe thei appeare to bee ouerſene, that call
thoſe foȝmers nombers *Surdeſolides*, ſeing thei are not
any waies *Surde nombers*, but haue their rootes. And
yet, to confeſſe the truthe, I cannot well tell you the
true *etymologie* of their name : except thei be ſo named,
as it were *ſolide* vpon *ſolide*. And that interpȝetation
were to ſtreightly racked. But the name beyng re=
ceiued and well knowen, wee maie moȝe eaſily with
libertie vſe it, then with ſcrupuloſitie, curiouſly ſcā it.

Theſe nombers are ſimple nombers in their kind.
Foȝ thei riſe of. 5. multiplications. And if their roote
bee a digite nomber, then is it the ſame nomber, that
ſtandeth in their firſte place. And if their roote be an
article, then hath that *Surſolide*. 5. tymes ſo many Cy=
phers together in the firſte places, as his roote hath :
and the nexte figure after thoſe Cyphers, is the firſte
figure ſignificatiue of his roote.

Scholar. I ſee it ſo in all theſe nombers that bee

in

of Arithmetike.

in the table.

Maſter. And ſo ſhall you finde it in all others.

And farther if the roote bee a nomber mixte, then the firſte nomber of the *ſurſolide,* is the firſt nomber of the roote. And this J doe tell you foz ſome helpe, in geſſyng at their rootes.

This name therfoze of theim, J meane *Surſolides,* in *Arithmetike,* maie ſerue to admoniſhe you of their roote. But in *Geometrie,* oz in compoſition of ſounde bodies, it ſerueth to no vſe : and therfoze J doe call the agreable to their figure, *Squares of Cubes* : bicauſe thei *Squares of* make a ſquare fozme : but ſo that euery vnitie of that *Cubes.* ſquare, is in it ſelf a *Cube* : As by the figures that fol= owe, you maie well coniecture.

And alſo thei are made by multiplication of a *Cu= bike nomber,* and a *Square nomber* together, bothe ha= uyng one roote : and the *Surſolide* hauyng the ſame roote. Wherfoze reaſon with the nature of their ſounde figure, infozceth me to call the *ſquares of cubes.*

Yet other menne attendyng moze to the nature of their rootes, then to their owne fozmes and nature, doe giue that name to the nexte rewe of nombers, bi= cauſe thei maie be made of multiplication, of any *Cu= bike nomber* by itſelf, that is to ſaie ſquarely.

Scholar. Jt is ſo. Foz. 8. whiche is a *Cubike nomber* multiplied ſquarely maketh. 64. And that. 64. is ſet emongeſte the *Squares of Cubes.*

Maſter. And this commoditie commeth by that name : that it putteth menne in remembzaunce of the ſpedie and eaſie extraction of their roote : As you ſhall learne hereafter.

But J conſiderpng their owne nature and ma= kynge, as ſounde nombers oz bodies : doe call theim *Cubes of Cubes,* oz *Cubike Cubes.*

After theſe nombers in the ſeuenth rewe, there do followe thoſe nombers, whiche commonly are called *bſurſolides,*

The seconde parte

bsursolides, oz bissursolides, that is, seconde sursolides, oz dou=
ble sursolides. But J maie call them seconde squares of cu=
bes, alludyng at the same name. Howbeit if J looke to
their forme and nature, J shall be inforced to call thē,
longe cubes of cubes, oz longe cubike cubes.

And so by like reason, doe J cal the nexte nombers
square cubes of cubes, oz square cubike cubes : whiche other
men doe cal zenzizenzizenzikes, that is *squares of squa=*
red squares.

The nineth rewe of nombers, is commonly called
Cubike Cubes, oz Cubes of Cubes : bicause the *Cubike* rootes
of those nombers are *Cubike* nombers also. But J af=
ter their true nature, doe call them *Cubes of Cubes Cubi=*
kely : oz *Cubes of Cubike Cubes.*

The tenth rewe of nombers is named vulgarely,
Squares of sursolides, bicause thei haue a Square roote,
whiche is of it self a *sursolide* nomber. And foz their fi=
gure *Grometricalle,* J name thē *long cubes of cubike cubes.*

So that J consideryng their nature, that thei be fi=
guralle nombers, am constrained to name theim, ac=
cozdyng to their figure, J meane in this place, where
J doe make explication of their natures and names.

But other men foz aide of woozke, in extraction of
rootes, haue giuen theim soche names, as maie beste
put menne in remembzaunce of redy woozke therein.
Whiche names J will vse also hereafter, in my wzi=
tynges, bicause J will not bee an aucthoz of vnnede=
full singularitie. And yet bicause truthe in nature is
as well to be regarded, as ease in woozkyng, and ra=
ther moze, J could not omitte in this place, the decla=
ration of their true nature and very fozmes.

And so bothe of vs hauyng good reasons, foz those
names, neither maie contempne other, neither con=
tende together.

And although the names that J doe giue, maie
seme to some menne (whiche are scarse apte iudges)
moze

of Arithmetike.

moze odioufe, foz the newe inuention (as thei maie *mes of thefe*
thinke) then nedefull to the pzactife of tharte : yet ſhal *numbers.*
you fee in theim a naturall fequele, and ozderly pzo=
pagation.

Foz all thofe nombers are confidered, in one of. 2.
foznes firfte. That is to faie, other thei bee taken as
nombers abfolute, without any côfideration of mul=
tiplication : And fo thei maie be named nombers one=
ly, without name of relation. Or els thei bee confide=
red as nombers multiplied, and that can be but in. 3.
varieties.

If thei be multiplied but ones, then doe thei make
a line of nombers, oz a liniarie nomber. And that no=
ber hath onely lengthe, without bzedthe, oz depthe :
And therfoze maie be the roote to a *Square*, oz a *Cube*.
But is of it felf, in that confideration, nother *Square*
noz *Cube*.

Secondarily, it maie bee multiplied twife, the one
nomber ſtãdyng foz the lengthe, and the other foz the
bzedthe : and fo is it a *Square nomber*, and therfoze a *flat
nomber*.

Thirdly, it maie bee multiplied thrife, and therby
gette lengthe, bzedthe, and depthe : wherby it is made
a *founde nomber*. And bicaufe the fides bee equalle, it is
fpecially a *Cube* oz *Cubike nomber*.

Now can there be no fowerth waie, that any mul=
tiplication maie increafe : foz there are no moze dimē=
tions in nature.

But if any manne doe multiple the fourthe tyme,
then muſt he accoumpte that he maketh *a line of Cubes* :
and the fifth multiplication maketh a *Square*, in whi=
che euery vnitie is a *Cube* : So the firte multiplication
maketh a *Cube of Cubes*, accoumptyng euery leſſer *Cube*
foz an vnitie. And there is a ſtaie again.

Wherfoze if any man multiplie the feuenth time,
he retourneth againe to the firſte nature of nomber s
                 J.i.   multiplied,

The seconde parte

multiplied, whiche are *liniarie nombers* : And the 8. mul=
tiplication, woorketh as the seconde did, and maketh
flatte nombers. The nineth multiplication agreably
with the thirde, doeth make *Cubes.*

And so infinitely these. 3. woorkes maie bee reite=
rate, but a fourthe forme can neuer be deuised.

And therefore doe J, as reason doeth compell me,
reduce all nombers to those. 3. formes, as their verie
originalle sprynges and fountaines.

But to the intente that you maie the more aptly
iudge of theim, and their natures, J haue here sette
foorthe the formes, whiche thei make in figures *Geo=*
metricalle, or sounde quantities. Admonishyng you to
remember this well. That after any nomber is be=
come a sounde nomber, it is againſt reason, to reduce
him to an abſolute flatte nomber again, and moſte of
all by multiplication. But now marke these figures.

Rootes, or Lines.

4.   3.   2.
|—|—|—|—| |—|—|—| |—|—|—|

6.   5.
|—|—|—|—|—|—| |—|—|—|—|

of Arithmetike.

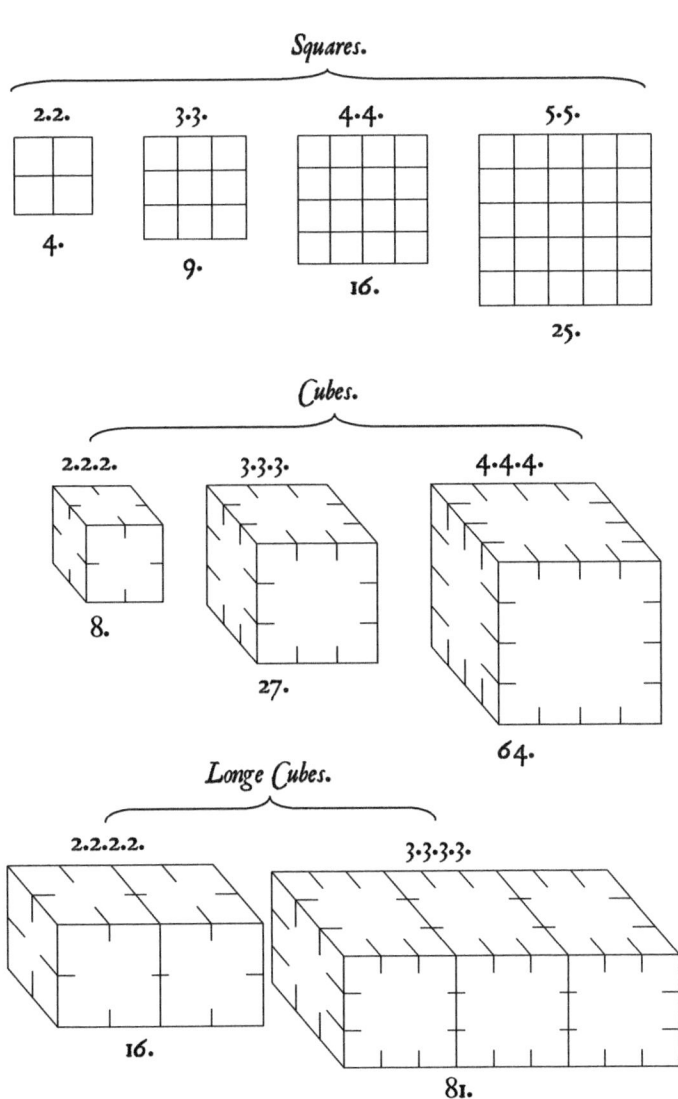

Squares.

2.2.   3.3.   4.4.   5.5.

4.

9.

16.

25.

Cubes.

2.2.2.   3.3.3.   4.4.4.

8.

27.

64.

Longe Cubes.

2.2.2.2.   3.3.3.3.

16.

81.

J.ii   *Squares*

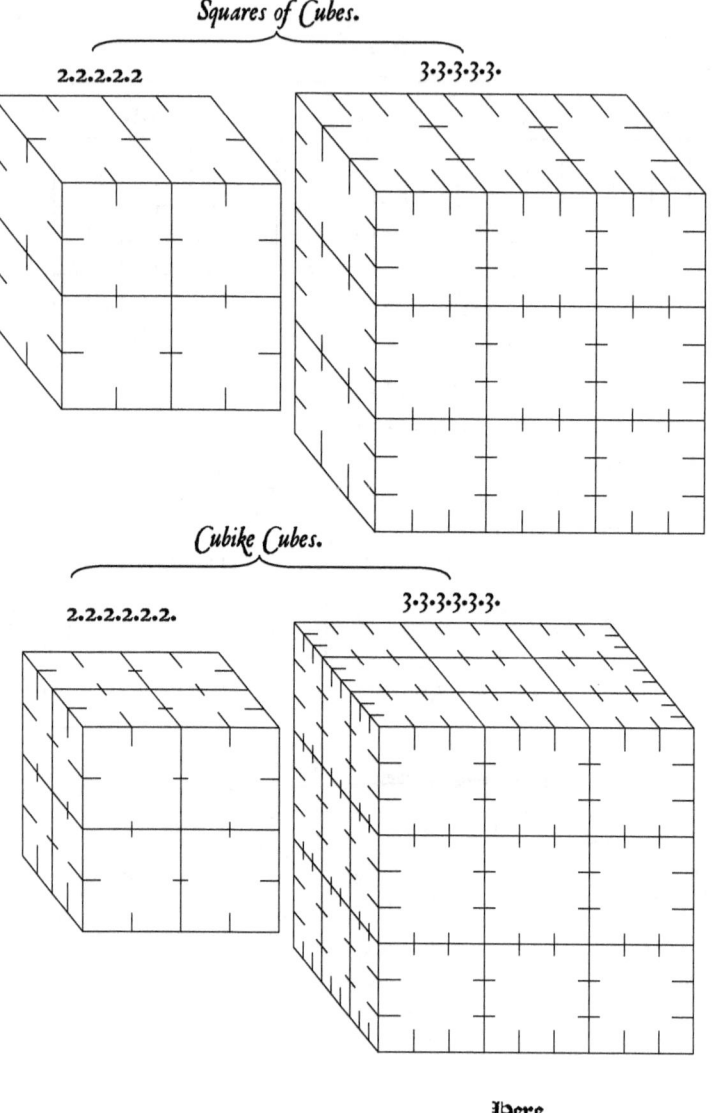

Squares of Cubes.

2.2.2.2.2   3.3.3.3.3.

Cubike Cubes.

2.2.2.2.2.2.   3.3.3.3.3.3.

Here

of Arithmetike.

Here, as you see, I haue set first certaine lines, con=
tainyng soche partes as thei bee made of by multipli=
cation : that is to saie, 2. 3. 4. oʒ 5. And these bee produ=
ced by the firste multiplication, where an vnitie of a=
ny thyng is multiplied by a nomber.

And so an inche multiplied by. 3. maketh. 3. inches :
And a foote multiplied by. 6. maketh. 6. foote : and so
of other measures and quantities, in like soʒte. All
whiche multiplications, doe make onely longe lines,
oʒ measures in lengthe onely, without bʒedth oʒ thic=
kenesse.

And in this multiplication, nother the nomber,
nother yet the vnitie, is accoumpted oʒ called a roote.
But the line that is made therby : maie bee a roote to
any of all the other kinde of nombers befoʒe rehersed,
and sette foʒthe in the table. Foʒ if you multiplie the
same line, by the nomber that his lengthe doeth in=
clude, then there will be made therof, by this seconde
multiplication, a square figure, containyng a square
nomber in it : As you see emongest those figures, the
firste fower to be, whiche are marked with these nom=
bers. 4. 9. 16. and. 25.

Scholar. I perceiue well in eche of thē, that their
lengthe is agreable with their bʒedthe, and so thei
make square figures, but I knowe not what those
nombers doe meane, that be set ouer their heddes.

Master. The quantitie of the nomber, doeth be=
token the valewe of their roote. And the multitude of
the same nomber repeted, doeth declare the nomber of
multiplications, foʒ eche figure.

And therefoʒe the lines, whiche are made by one
multiplication, haue eche of theim their nomber sim=
ply set, ones onely.

The squares haue their nombers double : in token
that thei haue. 2. multiplications. That is, one in
lengthe, and an other in bʒedthe.

J.iii.   The

The seconde parte

The third formes, whiche be *Cubes*, and are made of. 3. multiplications, haue their roote repeted thrise.

And the like nombers did I sette, in the side of the former table, againſt the like quätities. Whiche ſhall helpe you ſomewhat in the extraction of rootes.

Scholar. Now doe I perceiue not onely their na= mes, and multiplications, moche better then I did be= fore : but alſo I vnderſtande better the difference of your names, and their reaſons. For by thoſe figures, whiche you haue ſet in the fowerth place, and doe call theim *longe Cubes*, I ſee their forme doeth agree to that name. For thei are longer, then thei are other brode or depe. And ſaue for their depthe, I might li= ken theim to *longe Squares* in *Geometrie*. Howbeit, o= ther men neglectyng their forme, and lookyng onely to their rootes, doe call them, *Squared ſquares*.

But if you will permitte me, to ſpeake in the de= fence of theim, as a ſimple ſcholar maie ſpeake for af= fection, in the defence of his maſter, it appereth to me, that thei maie well bee called ſquared ſquares : and might be figured thus.

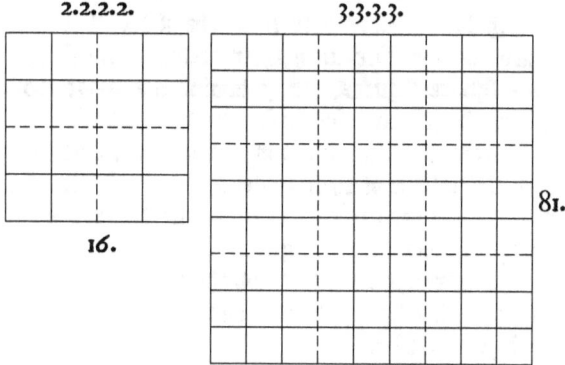

2.2.2.2.   3.3.3.3.

16.   81.

Where the ſmalleſt ſquares, whiche be contained within the pricked lines, beyng taken as rootes, and multiplied

of Arithmetike.

multiplied by thefame nomber again, whiche thei do containe (other els twife by their rootes) will make the whole greater fquares.

And by this figuryng of theim, there doeth appere no inconuenience noz abfurditie, in their vulgare na= mes : but rather a iufte expzeffyng of their naturalle fozmes.

Foz in the firft figure. 2. ftandyng as the fide of the leffer fquare, and multiplied by it felf, doeth make. 4. whiche is the quantitie of the leffer fquare. Then if J multiply that leffer fquare. 4 : by his owne nomber it maketh 16. whiche is the greate and whole fquare : and is a *Square of fquares.*

So in the feconde figure. 3. ftandeth foz the roote of the leffer fquare, contained within the pzicked lines, and if it bee multiplied by it felf, it maketh. 9. whiche is the quantitie of thefame leffer fquare. Then if J multiplie that. 9. by itfelf, it will make. 81. whiche is the quantitie of the greate Square, and is a *Square of fquares.*

Mafter. J commende you well : not onely foz fo diligente excufyng of theim, whiche foz their honefte trauell, deferue moche thankes, but alfo foz that you feke to bzyng manifeft reafon, and fome fhewe, at the leaft, of linearie demonftration foz your purpofe. So that you will not feme to fpeake, without fome good grounde.

But as in deede, your figure doeth truely expzeffe a fquare of fquares, fo it doeth fuppofe the other nom= ber, whiche by ozder of multiplication, doeth go next befoze it, to be a flatte nomber alfo. Foz it is not pof= fible that a founde nomber (as a *Cube* is alwaies) be= yng multiplied by any other nomber, maie lefe the nature of a founde nomber : But fhall continue a founde nomber ftill. And therfoze feeyng the nexte nomber, befoze a *Square of fquares* was a *Cube*, it is not
possible

The seconde parte

possible that a *Square of squares* can be a mere *flatte nom=
ber*, as you haue drawen it.

Wherfore if thei had intended, that a *flatte number*
should occupie the. 4. place, then should thei haue set
some *plat forme* in the third place also. Whiche might
haue been made in this sorte.

And then will it be a *longe Square*, and not a *Cube*.

3.3.3.

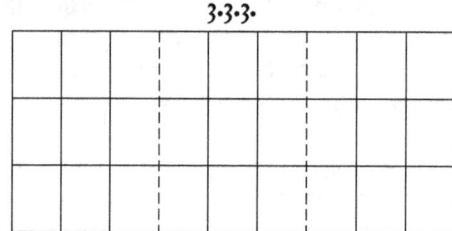

But in as moche as thei doe not admitte this *longe
Square* (whiche by that name hath no roote) therfore
maie not the nomber that foloweth it, bee any other
then a *founde nomber*. For euery *Cubike* forme, beeyng
multiplied by his roote, doeth make a *Square piller*.
Whose length beareth vnto his bredth thesame pro=
portion, that his roote doeth vnto an vnitie.

Scholar. I am very well satisfied now : cōcerning
the names and formes of those nombers. And by this
that you haue saied, I doe farther perceiue, that. 5.
multiplications doeth make the *square of Cubes*, whiche
be set in the fifte place, emongeste the former figures.
And also I vnderstande by the former table, that thei
be called *Surfolides*.

Likewaies I see in the sixte place of the forsaied fi=
gures, *Cubike Cubes*, made by. 6. multiplications. But
commonly the nombers of those quantities, be named
Squares of Cubes. So that for their names, thus farre I
am perfecte inough.

                                                       The

The extraction
of Rootes.

Maſter.

Owe will J ſhewe you,
how you ſhal extract the roote
out of any ſoche nomber.

*The extrac-
tion of rootes*

And firſt J muſt admoniſhe
you, that you ſhal alwaies vn
derſtande, ſoche a roote as the
nomber doeth admit. So that
in a ſquare nomber, you ſhall
ſeke a *Square roote* onely, and
no *Cubike roote*, nother any other kinde.

Likewaies a *Cubike nomber* hath no other roote,
but a *Cubike roote*. Excepte the namebee compounde,
as *zenzicubike*, oz *Squared Cube*. Foz in ſoche there are
2. ſoztes of rootes, accozdyng to the 2. names that thei
beare. That is bothe *Square* and *Cubike roote* : as J
will anon ſhewe you. But firſte J will beginne with
Square nombers, and their rootes.
And this generalle ozder muſte you
obſerue, befoze all other : That you
ſhall haue by harte, in readie memo=
rie all ſoche nombers, whoſe rootes
are digites. Foz as it is ſuperfluous
to ſeke rules foz theim, ſo muſt thei
helpe in all greater nombers, whoſe
rootes are aboue 9. And foz your eaſe
in remembzaunce, J haue here ſette
foozthe a table foz ſquare nombers.
Where in the firſte columpne, you ſe
the rootes ſet, and in the ſeconde pil=
ler, right againſt eche roote, there is
ſet his ſquare. Touchyng which J
nede to ſaie no moze, but that you be
not in any vncertaintie of them, whē

*The table of
ſquarerootes vn=
compounde.*

Rootes.	Squares.
1.	1.
2.	4.
3.	9.
4.	16.
5.	25.
6.	36.
7.	49.
8.	64.
9.	81.

B.i.   you

The extraction

you ſhall nede their aied, whiche ſhall be continually
in vſe of ſearchyng for other greater rootes.

Now for greater nombers, this is the order.

1. Firſt ſet doune the nomber as it is. Then ſette a
pricke vnder euery odde place, J meane the firſte, the
thirde, the fifte, the ſeuenth, and ſo forthe : and ſo ſhall
euery pricke haue. 2. nombers, excepte the laſte, whi=
che ſome tymes hath but one.

2. Secondarily, marke the nombers that belong vn=
to the laſte pricke, toward the lefte hande : And whe=
ther he haue belongyng to it one nomber, or twoo,
looke what the roote maie be of that nomber, if it bee
ſquare. And that roote ſette by a croked line, as you
place the *quotiente* in diuiſion : & cancell all that ſquare
nomber, belongyng to that pricke.

3. But and if the nomber belongyng to that pricke,
bee not a Square nomber, then take the roote of the
greateſte ſquare, whiche is contained in it, and place
the roote as J ſaied before. And the ſquare of it ſhall
you abate from the nomber, that belongeth to that
laſte pricke, and let the reſt be ſet ouer thoſe nombers
cancelled, as you doe in diuiſion. And ſo haue you en=
ded your worke for that pricke.

Scholar. This moche is eaſie inough, if J vnder=
ſtande you rightly.

Maſter. Then proue it in a nomber, or twoo.
And firſt worke with this nomber. 5152900.

Scholar. J muſte marke euery odde place with a
pricke, thus.

And here J perceiue that vnto the firſt
5152900.   pricke, there belongeth 2 Cyphers one=
ly, and to eche of the other. 2. prickes
folowyng, there are appoincted. 2. figures. But the
fourthe pricke hath but one nomber, and that is. 5.

Now accordyng to the ſecond rule, J ſeke the roote
of 5. (for bicauſe there belongeth no more nombers to
                                                              that

The extraction

that pꝛicke) and I ſee, it is no ſquare nomber. Wher=
foꝛe accoꝛdyng to the thirde rule, I take the greateſte
ſquare in it, whiche is. 4. and the roote of. 4. is. 2.
Therfoꝛe I doe ſubſtracte. 4. out of. 5.
and cancell that. 5. and the. 1. that re=
maineth, I ſet ouer. 5. as here you ſee.

    1
    5̸152900 (2.

And the roote. 2. I ſette behinde the *quotiente* line, as
you taught me, and then the nôbers ſtand, as you ſe.

Maſter. You haue doen wel. Proue again in this
nomber. 18766224.

Scholar. Firſt I ſet theim doune
and pꝛicke theim, as here doeth ap=
peare. And now I ſee, that the laſte pꝛicke hath twoo

    18766224.
       ·   ·   ·

nombers belongyng to it, that is. 18. with whiche I
muſt begin. And ſeyng it is no ſquare nomber, I find
16. to be the greateſt ſquare in it : wherfoꝛe I ſubtract
16. out of. 18. and ſet. 2. ouer the. 8.
And the roote of. 16. whiche is. 4.
I ſette behinde the *quotiente* line, as
here is ſeen.

      2
    ₁8766224 (4.
       ·   ·   ·

Maſter. This maie ſuffice foꝛ the firſt wooꝛke.

Now to pꝛocede, you ſhall double your roote, and   4.
put that double vnder the nexte ſpace, towarde the
right hand, that is behinde the nexte pꝛicke. Alwaies
foꝛſeyng, that if the double doe contain moꝛe figures
then one, that the firſt ſhall be ſette vnder that place,
and the ſeconde vnder the nexte figure, towarde the
lefte hande.

Then ſeke a *quotiente*, as you doe in diuiſion, whi=   5.
che ſhall ſhewe how often that double nomber maie
be found in that, that is ouer it, appertainyng to that
place : whiche *quotiente*, you ſhall ſet befoꝛe the firſte
roote, within the *quotiente* line.

But this regarde muſte you haue here ſpecially,
that you maie leaue ouer the nexte pꝛicke, toward the
right hande, as moche as the ſquare of that *quotiente*,

                        K.ii.   with

The extraction

with which you worke, for out of that rest, the square
of that *quotiente* muste bee abated. And then make
bothe subtractions, and note the remainer, if any be,
and place your *quotient*, and then haue you doen with
that pricke alſo.

For the more plaines, I will giue you an example
in your firſte nomber, whiche ſtoode thus, after your
worke was ended.

Here I ſee ouer the laſte pricke
ſaue one. 115. vnder the middell fi=
gure of whiche I muſt ſet the dou=
$$5152900.(2.$$
ble of the former roote. 2. that is. 4. And then I ſeke
how often. 4. is to bee founde in. 11. And I finde that
I maie haue it twoo tymes, and. 3. remainyng. Whi=
che. 3. with. 5. ouer the nexte pricke, doe make. 35. and
that is more then the ſquare of my *quotiente.* 2. Ther=
fore am I bolde to ſette doune that
quotiente : And accordyng to it, to a=
bate twiſe. 4. (whiche is. 8.) out of
$$\begin{array}{l} x3 \\ 5152900.(22. \\ 4 \end{array}$$
11. and there reſteth. 3. Therfore I
cancell. 11. and ſette. 3. ouer it. Then doe I multiplie
the laſte *quotiente* ſquarely : and it maketh. 4. whiche
4. I ſubtracte out of the nomber ouer the pricke, that
is. 35. where. 5. maie ſuffice for this nomber. Ther=
fore I abate. 4. out of. 5. and cancell that. 5. and ſet. 1.
whiche remaineth, ouer the. 5 :
And then will the whole nomber
ſtande thus.
$$\begin{array}{l} x31 \\ 5152900.(22. \\ 4 \end{array}$$
This worke, whiche I haue
wrought now, muſt be repeted as often as there bee
any prickes, or pricked nombers remainyng. Wher=
by you maie eaſily geſſe, that it muſt bee twiſe more
repeated in this example, bicauſe there reſteth yet. 2.
prickes vntouched.

Scholar. Although I thinke, I could doe, as I
haue marked you to doe, yet for more certaintie I
praie

of Rootes.

pꜩaie you woꜩke out this example.

Ǽaſter. Then marke it well.

I ſhall begin againe with doublyng of all, that is within the *quotiente* line. And that double is 44. whi= che I muſt ſet vnder. 312. that remaineth of the laſte woozke. And then will the nom= bers ſtande, as here you ſee.

Then I loke how often tymes maie I finde. 44. in. 312. And I

$$\overset{\overset{\text{\it{\textasciigrave}31}}{}}{8182900}\ (22.$$
$$44$$

ſee it will be abated 7 times, and 4 remain : whiche 4 with the. 9. ouer the next pꜩicke doeth make. 49. And that will ſuffice to extracte the ſquare of my *quotiente.* 7. Foꜩ. 7. tymes. 7. maketh iuſte. 49. Thus ſeyng I maie take. 7. foꜩ my *quotiente*, I woozke with it, as the rule teacheth : abating firſt. 7. times. 44. (that is. 308) out of. 312. and there reſteth. 4. ouer the ſpace befoꜩe the nexte pꜩicke. Whiche. 4. with. 9. ouer the pꜩicke

doe make. 49. out of whiche I a= bate the ſquare of my *quotiente.* 7. (that is. 49.) and ſo reſteth no= thyng, but. 2. Cyphers. And the nomber ſtandeth thus.

$$\overset{\overset{\text{\it{131\textasciigrave4}}}{}}{8182900}\ (227.$$
$$44$$

And ſeyng there remaineth one pꜩicke vntouched, I ſhould repeate theſame oꜩder of woozke againe, by doublyng all the *quotiente*, whiche would bee. 454. and ſettyng it ſo that. 4. whiche is in the firſte place, ſhould be ſette vnder the Cypher, that is without the pꜩicke, and the other figures in oꜩder, towaꜩd the left hand. But all this woꜩke were in vaine, ſeyng there is nothyng lefte, to ſerue foꜩ the ſubtraction.

Yet bicauſe there is lefte one pꜩicked place vntou= ched, I muſt ſet foꜩ it a Cypher in the *quotiente.*

Foꜩ this rule is generalle : that how many pꜩickes ſo euer your ſquare nomber doeth containe, your *quotiente*, oꜩ roote ſhall haue ſo many nombers. Wherfoꜩe this roote muſt be made vp thus. 2270.

                           R.iii.   And

The extraction

And ſo it appeareth that your nomber. 5152900. is a iuſte ſquare nomber. Whiche you maie proue by the orderly proofe of extraction of rootes. That is to multiplie that *quotiente*, or roote (whiche you haue founde) by it ſelf. And if it doe make the firſt nomber exactly, then haue you wrought well.

Scholar. That proofe is as cer=
taine, as can be. And therfore I will
proue, whether it will agree with
this worke. Wherfore multiplying
2270. by it ſelf, I ſee that it yeldeth
the firſte ſomme. As here it doeth ap=
peare. So is this worke approued
good.

2270.
2270.
158900.
454
454
5152900

And now will I attempte the like worke in the ſeconde example. Whiche was. 18766224.

But after the firſte worke was ended, and the greateſt ſquare ſubtracted out of 18. it did remain in this forme.

Now to continue the worke as you did, and as the rule doeth teache, I muſt dou=ble. 4. whiche is the roote, and ſtandeth by the *quoti=ente line* : and muſt ſet it vnder. 7. that ſtandeth in the ſpace, betwene the laſte pricke (whoſe worke is en=ded) and the nexte pricke towarde the right hande. And then will it ſtande thus as you ſee.

```
   2
  18766224 (4.
  ..    .
```

```
   2
  18766224 (4.
  .8    .
```

That doen, I muſt ſeke a *quoti=ente*, that maie declare how often 8. maie bee ſubtracted out of. 27. and that *quotiente* I finde to be. 3 : bicauſe that after I haue taken. 3. tymes 8. (that is. 24. out of. 27. there will remain. 3. whiche 3. with. 6. that ſtandeth ouer the pricke, doe make. 36. And I ſee that nomber to bee greate inough, for the abatemente of the ſquare of my *quotiente* : whiche is but. 3. tymes. 3. that is. 9.

Wherfore

of Rootes.

Wherfore I sette doune. 3.
for my *quotiente*, before. 4. in
the *quotiente* line. And multi=
pliyng 8. by that 3. there riseth

$$2$$
$$237$$
$$18766224(43.$$
$$\cdot 8 \cdot \cdot \cdot$$

24. whiche I doe subtract out
of. 27. that is ouer. 8. and there will remain. 3. That
3. with. 6. ouer the pricke, maketh. 36. out of whiche
I must abate. 9 : whiche is the square of my *quotient*. 3.
and so will there reste. 27. ouer that pricke.

And thus haue I ended. 2. prickes, and yet. 2. more
doe remain : in whiche bothe I must repeate the same
forme of woorke.

Therfore I double the whole *quotiente*, and it ma=
keth. 86 : whiche I set vnder. 276.

And then I seke the *quotiente*, declaryng how many
tymes. 86. maie be abated out of. 276. whiche maie
be. 3. tymes. And for that cause I set. 3. in the *quotiente*
before the. 43.

Then doe I firste multi=
plie. 86. by that. 3. saiyng. 3.
tymes. 8. maketh. 24. which
I abate out of. 27. and there
resteth. 3. And again I saie.
3. tymes. 6. is. 18. whiche I

$$1$$
$$237$$
$$23783$$
$$18766224(433.$$
$$\cdot 86 \cdot$$

abate out of. 36. and there doeth remain. 18.

That doen, I take the square of my *quotiente*, that
is. 9. whiche I doe subtract out of. 12 (for the. 2. ouer
the pricke must borowe. 1. of. 8.) and then will there
remain ouer that pricke. 173.

And thus is that pricke ended.

Now, for the laste pricke in woorke, though he be
firste in place. The double of my *quotiente* is. 866.
whiche I muste sette vnder
1732. As here is doen, where
I leaue out many cancelled
figures, as superfluouse in

$$173$$
$$18766224(433.$$
$$\cdot 866 \cdot$$

this

The extraction

this place.

And then ſekyng foʒ a newe *quotiente*, I ſinde it to be. 2. whiche I ſet with the other nombers in the *quotiente*. And by it I multiplie and ſubtract the 866. ſaiyng : 2. tymes. 8. is. 16. whiche I abate out of. 17. and there reſteth. 1. Again. 2. times 6 is. 12 that I ſubtract out of. 13. and there remaineth. 1. Thirdly, I ſaie. 2. times. 6. giueth. 12. whiche I abate from. 12. and there

$$\begin{array}{l} \mathllap{x\,x} \\ \mathllap{x\,x\,x} \\ 18766224\,(4332. \\ 866 \end{array}$$

is left nothyng. Saue that ouer the pʒicke there ſtandeth 4 whiche is equall with the ſquare of my *quotiët.*

Wherfoʒe abatyng the ſquare of my *quotiente* out of it, there reſteth nothyng at all.

And therby I ſee that. 18766224. is a iuſte *ſquare nomber.* And his roote is. 4332.

The proofe. Maſter. Although I knowe it to bee ſo, yet foʒ your better exerciſe, and full perſwaſion : I would haue you trie it, by ſquare multiplication.

Scholar. That maie I ſone doe.

And ſo I ſinde it to be true.

Foʒ. 4332. multiplied by it ſelf, doeth make. 18766224. As this wooʒke here ſet, doeth ſhewe.

Maſter. Yet bicauſe ſome other ſmall doubtes, maie happen in woʒking, that maie trouble a yong pʒactiſer, I will pʒopounde to you one oʒ

$$\begin{array}{r} 4332. \\ 4332. \\ \hline 8664. \\ 12996. \\ 12996. \\ 17328. \\ \hline 18766224. \end{array}$$

twoo examples moʒe. Wherein you ſhall ſinde ſome varietie, as well in the nomber pʒopounded, as alſo in the *quotiente.*

And firſte to begin, I will you to extract the roote of this nomber. 22071204.

Scholar. I muſt ſet doune the nomber, and note it with pʒickes in euery odde place : Foʒ that rule I perceiue

The extraction

perceiue neuer faileth.

Maſter. No moꝛe doeth any of $22071204.$
the other, although the wooꝛke
maie varie in ſome ſmalle poinctes : whiche yet maie
be greate inough to trouble a young learner.

Scholar. Then accoꝛdyng to the firſte rule, J ſeke
out the greateſt ſquare in. 22. (foꝛ J ſee it is no ſquare
nomber itſelf) and it appereth to be 16. And his roote
4. wherfoꝛe J doe ſette doune. 4. in the *quotiente*, and
then J doe abate. 16. out of. 22.

and the remainer is. 6. whiche J
     6
ſette ouer the pꝛicke, and cancell
     $\dot{2}\dot{2}071204\,(4.$
the. 22. as here is ſeen.

Now goyng on with the nexte pꝛicke, J ſhall dou=
ble the foꝛmer roote in the *quotiente*, and ſette it vnder
the Cypher, betwene the. 2. pꝛickes.

Then do J ſeke how oftē that 8 (whiche is the dou=
ble of the *quotiente*) maie be found in 60 and J finde it
to be 7 times, and 4 remainyng to be ſet ouer the Cy=
pher. So that foꝛ the pꝛicke there remaineth. 47. out
of whiche J ſhould abate the ſquare of my *quotiët*. But
ſeing that. 49 (whiche is the ſquare of 7) can not be ta=
ken out of. 47. there is a newe *quotiente* to be ſought.

Therfoꝛe J take 6. And ſee that it will ſerue. So J
ſet. 6. in the *quotiente* : and by it J
multiplie 8 whereof commeth 48
     $\overset{1}{\cancel{9}}$
That. 48. abated out of. 60. lea=
     $\overset{6}{\cancel{2}}1$
ueth. 12. Therefoꝛe J cancell the
     $\dot{2}\dot{2}\cancel{\emptyset}\cancel{7}1204\,(46.$
60. and ſet. 12. ouer it.
     $\underset{8}{\cdot}\ \ \cdot\ \ \cdot$

Then doe J multiplie the *quotiente*. 6. by it ſelfe :
whereof riſeth. 36. And that abated out of. 127. lea=
ueth. 91. And ſo haue J ended the ſeconde wooꝛke.

Now foꝛ the thirde wooꝛke, J double. 46. and it
doeth yelde. 92. to bee ſette vnder. 911. as J haue put
it here.

And then ſeking foꝛ a *quotient* : J ſe that J maie take

                        L.i.   9.

The extraction

9. Wherfore I set that 9 in the *quotiente* with. 46. and by it I multiply 92 and subtract that, that riseth, in this forme.

$$\begin{array}{r} 7 \\ \cancel{1}8 \\ \cancel{1}9\cancel{0}5 \\ \cancel{6}\cancel{2}\cancel{1}\cancel{3}1 \\ \cancel{2}\cancel{2}\cancel{0}\cancel{7}\cancel{1}\cancel{2}04\,(469. \\ \cancel{9}2 \end{array}$$

Nine tymes. 9. maketh. 81. whiche I abate out of. 91. and there resteth 10. Then 9 tymes 2 giueth 18. whiche I must abate out of. 10. and there will remain. 83.

And now muste I multipie that laste *quotiente*. 9. squarely, wherby will amounte. 81. that shall I sub= tract out of. 832. and there will remain. 751. and so that pricke with his woorke is ended.

Therefore procedyng to the fourthe pricke, I dou= ble all the *quotiente*, whiche will be 938. And set it vnder 7510.

$$\begin{array}{r} 751 \\ \cancel{2}\cancel{2}\cancel{0}\cancel{7}\cancel{1}2 0 4\,(469. \\ 938 \end{array}$$

Then doe I seke a newe *quoti= ente*, whiche I finde to bee. 8. For

8. times. 9. giueth. 72. whiche I abate out of. 75. and there remaineth. 3. Again. 8. tymes. 3. is. 24. and that I deducte out of. 31. and so resteth. 7. When saie I. 8. ti= mes. 8. is. 64. whiche beeyng subtracted from. 70. doeth leaue. 6. And that. 6. with the 4. ouer the pricke maketh. 64. out of whiche I muste withdrawe the square of. 8. that is my *quotient*, and it beyng 64. there resteth nothing. And the whole woorke standeth thus.

$$\begin{array}{r} \cancel{3}\cancel{7} \\ \cancel{7}\cancel{8}\cancel{1}\cancel{6} \\ \cancel{2}\cancel{2}\cancel{0}\cancel{7}\cancel{1}\cancel{2}\cancel{0}4\,(4698. \\ \cancel{9}\cancel{3}8 \end{array}$$

The proofe.

Wherfore I saie that the first nöber 22071204. is a square nöber : and hath for his roote. 4698. As I maie prooue also, by square multiplicati= on. For, as in this example you see : 4698. multiplied by it self, doeth bryng forthe. 22071204.

$$\begin{array}{r} 4698 \\ 4698 \\ \hline 37584 \\ 42282 \\ 28188 \\ 18792 \\ \hline 22071204. \end{array}$$

Master.

of Rootes.

An other example.

Maſter. Yet one example moꝛe ſhall you pꝛoue : and that is this. 901740841.

Scholar. J ſet it doune, and pꝛicke it accoꝛdyng to the rule : And then J ſee ouer the laſte pꝛicke, one onely nomber, that is. 9. whiche hath. 3. foꝛ his ſquare roote. That. 3. J ſet within the *quotiente* line, and therfoꝛe J cancell. 9.

$$9\cancel{0}1740841(3$$

After this J would pꝛoceade with doublynge the roote. 3. and that double ſhould J ſet in the next ſpace, ouer whiche remaineth no nomber, foꝛ. 9. beyng can= celled, the Cypher is nothyng. And ſo am J at a ſtaie.

Maſter. Seyng that you can not ſet the double of your *quotiente* doune there, where no nomber is (oꝛ if it ſo chaunce, as ſome times it doeth, that the nomber ouer it, is leſſer then the double) then ſet a Cypher in the *quotiente*, and ſo haue you doen with that pꝛicke. Foꝛ in ſoche caſe there nedeth no multiplication, noꝛ ſubtraction.

Scholar. Then am J inſtruc= ted fully foꝛ that poincte : The woꝛke is ſo eaſie. J muſt therfoꝛe ſet my nombers thus.

$$9\cancel{0}1740841(30$$
$$60$$

Maſter. And doe you not ſee, that the double of the *quotiente*, is greater then the nomber ouer it?

Scholar. J was ſo mindfull of the one halfe of the rule, that J foꝛgate the other halfe.

But now J ſee, J muſt ſet an other Cypher yet in the *quotient*. And then ſhall J ſet the double of all that, in the thirde ſpace, after this ſoꝛte.

And nowe pꝛoceadynge to ſearche foꝛ anewe *quotiente*, J ſee that. 2. ſhall ſerue me.

$$9\cancel{0}1740841(300.$$
$$600$$

Wherfoꝛe J ſette. 2. in the *quotiente* line, with. 300. And by it ſhall J multiplie the double afoꝛeſaied : ſaiyng. 2. tymes. 6. maketh. 12.

L. ii.   to

The extraction

to bee abated out of. 17. and the remainder will bee. 5.

Then shall I ouerpasse the twoo Cyphers, bicause thei make nothing by mul=tiplication : and so cōmyng to the pricke, I bate the

$$5 \quad 4$$
$$9\,\cancel{0}\,1\,7\,4\,0\,\cancel{8}\,4\,1\,(3002.$$
$$6\,0\,0$$

square of my *quotiente* : whiche is 4 out of. 8. and there resteth. 4. Therfore I cancell. 8. and set doune. 4. and so haue I ended that pricke. And haue but one worke more behinde.

Wherfore I set doune the nombers, with the dou=ble of al the *quoteiente*, thus. And then I loke for a new *quotiente*, whiche I finde to be. 9. by it therfore I mul=tiplie, first 6 and it maketh

$$5 \quad 4$$
$$9\,\cancel{0}\,1\,7\,4\,0\,\cancel{8}\,4\,1\,(3002.$$
$$6\,0\,0\,4$$

54. that doeth abate the 54. ouer it. Then omit I the 2 Cyphers, and multiplie 4. by 9 whereof there com=meth. 36. whiche I abate out of. 44. beyng ouer it, and there remaineth. 8. That. 8. with. 1. ouer the pricke maketh. 81. out of whiche I muste abate the square of. 9. beyng also. 81. And so is nothyng lefte, whereby it appeareth, that. 90174 0841. is a square nomber, and his roote is. 30029. The proofe of it doeth confirme the same. For 30029 multiplied by it self, doeth brynge forthe. 901740841.

30029.
30029.
270261.
60058.
90087
901740841.

Master. This shall suffice for soche nombers as bee fully square. Other nombers there bee infinite, whiche be not square, and therfore haue thei no square rootes. Yet of=ten tymes it happeneth, that we shall bee occasioned to searche for the nigheste nomber, that maie resem=ble their rootes.

Wherfore in soche case, this shall you doe. Firste extracte

of Rootes.

extract the roote, as if it wer a ſquare nõber. And that roote wil ſerue foʒ the greateſt ſquare, that is in your foʒmer nomber : and there will be a remainer beſide. Of whiche remainer with the *quotient*, you ſhal make a fraction, in this ſoʒte.

Set the remainer ouer the line, foʒ the numeratoʒ, and the double of the roote (that you haue founde) ſet vnder the line, foʒ the denominatoʒ. And this ſhall be a ſufficiente pʒeciſeneſſe in greate nombers, foʒ any common wooʒke.

Scholar. I will by an example, taken by chaunce, pʒoue this rule. Foʒ it ſemeth to haue no difficultie. Wherfoʒe I take. 296882.

And this, I am aſſured, can be no ſquare nomber. Foʒ, I remẽber you told me befoʒe, that no ſoche nomber might be a ſquare, which had 2 foʒ his firſt figure.

Then to ſearche his nigheſte roote, I place it, and pʒicke it thus.

And vnder. 29. I finde the greateſte roote to bee. 5. whiche I ſet in the
quotiente line, and cancell 29 ſettyng 4

$$\begin{array}{l} 4 \\ 2\cancel{9}6882(5. \end{array}$$

ouer it. After that I double it, and there cometh 10. & that double I ſet in the nexte ſpace vnder 46. Then finde I a newe *quotiente*, whiche is 4 and by it I mul-tiplie. 10. whereof amounteth 40. to be abated out of 46. And ſo remaineth. 6. Again I
multiplie. 4. by it ſelf ſquarely, and
there riſeth. 16. whiche I abate frõ
18. (ſeyng. 8. is to ſmall) and the remainer will be. 2. So ſtandeth the

$$\begin{array}{l} 4\,5\,2 \\ 2\cancel{9}\cancel{6}882(54. \\ 1\cancel{0} \end{array}$$

whole nomber, as you ſe. Wherfoʒe I double the *quo-tiente*, whiche is. 54. And it yeldeth. 108. that muſt be ſet vnder 528 as I haue here doen.

Then I looke foʒ a *quotiente*,
how often I maie abate. 108. out
of. 528. And I ſee it will be but. 4.

$$\begin{array}{l} 4\,5\,2 \\ 2\cancel{9}\cancel{6}882(54. \\ 108 \end{array}$$

L.iii.   tymes

The extraction

tymes. Wherfoze J set. 4. in the *quotiente,* with the o=
ther nombers, and then doe J woozke with it : Firste
multiplipng. 4. and. 1. together, whereof cōmeth but
4. whiche J abate out of. 5. And there remaineth. 1.

Again J multiplie. 8. bp. 4. whereof commeth. 32.
that doe J subtract out of. 128. and there will remain
96. Then shall J take the square of mp *quotiente.* 4.
whiche is 16. And that must J abate out of 962. And
so remaineth. 946. of whiche
nomber set as the numeratoz,
with the double of the roote, set
foz the denominatoz, J shall
make a fraction in this sozte.
$\frac{946}{1088}$. whiche is almoste. $\frac{9}{10}$.

$$\begin{array}{r} 194 \\ 5266 \\ 296882(544. \\ 108 \end{array}$$

Master. Pou haue doen wel. And so pou perceiue
that the nigheste roote of pour fozmer nomber is
$544\frac{473}{544}$. Foz those fractions are all one.

And hereby also pou maie vnderstande, that if the
remainer ouer pour nomber bee euen, pou maie take
halfe of it foz the numeratoz, and the whole *quotiente*
foz the denominatoz.

So maie pou take the quarter of the remainer (if it
will so bee parted) foz the numeratoz, and the halfe of
the roote foz the denominatoz.

And in like maner generallp, if the remainer and
the roote in the *quotiente,* bee *nombers communicante,* di=
uide them so, that the diuisoz of the remainer, be euer
double to the diuisoz of the *quotiente roote.* And so maie
pou easilp reduce that fraction, to his least termes.

The firste
proofe.

But now foz pzoofe of this woozke, there be twoo
waies : the one is certain, and the other but in a nere=
nesse. Foz as the roote of soche nombers, is not a pze=
cise roote : So if pou mutiplie that roote bp it self, it
will make a nomber, verp nighe to that fozmer nom=
ber, but not exactlp the same.

Whiche faulte some men thinke to redresse, bp ad=
dpng

of Rootes.

dyng of. 1. to the denominatoꝛ : and yet that amende=
mente fometymes increafeth the erroure.

But bicaufe you fhall not wante a fure pꝛoofe, doe
thus. Multiplie the *quotiente,* oꝛ *Roote* of whole nom=   *The feconde*
bers by itfelf, and vnto the nomber that amounteth   *proofe.*
thereof, adde the whole remainer. And if then it make
your firfte nomber, your wooꝛke was well doen : els
haue you miffed.

Scholar. That maie I pꝛoue here quickly. The
quotiente in whole nombers was. 544. whiche beyng
multiplied fquarely, doeth yelde. 295936. vnto whi=
che nomber, if I doe adde. 946. that did
remain, it will amounte to. 296882.
and that was the nomber pꝛoponed to
me : wherfoꝛe it appereth that the woꝛke
was well doen.

Mafter. You fhall neade no moꝛe
examples, foꝛ this foꝛme of wooꝛke.

544.
544.
2176.
2176.
2720.
295936.

But one other waie wil I fhewe you,
how you fhall geffe verie nighe vnto the roote. And   *An other*
you fhall go as nighe as you will defire, in any pꝛac=   *waie to finde*
tike woꝛke. If you defire to geffe within leffe then $\frac{1}{10}$.   *the nigheſte*
of one, then fet befoꝛe your nomber. 2. Cyphers. And   *roote.*
if you would not erre $\frac{1}{100}$. then fet doune 4. Cyphers :
But and if you lifte to fette doune. 6. Cypers befoꝛe
your nomber, you fhall not miffe $\frac{1}{1000}$ of an vnitie frō
the true roote. And if you lift to go any higher in pꝛe=
cifeneffe of partes, adde ftill euen Cyphers.

Scholar. I would faine pꝛoue this foꝛme, in the
fame example, whiche I wꝛoughte lafte : Bicaufe I
would fe the agremente betwene the bothe woꝛkes.

Mafter. Go to. Your confideration is reafonable.
And bicaufe the partes maie the better agree, fette
doune. 6. Cyphers. And then fhall your roote expꝛeffe
thoufande partes of the whole nomber.

Scholar. I fette doune the nomber, and pꝛicke it
thus.

The extraction

thus. Whereby J perceiue
that J ſhall haue theſame oꝛ=   29688̇2000000(
der of wooꝛke, and the ſelfe
ſame nombers that J had befoꝛe, till that J come to
the Cyphers and their pꝛickes.

Maſter. Truthe it is. And therfoꝛe maie you in
ſoche a caſe ſette doune onely the remainer, with the
Cyphers. Oꝛ els cancell all the nombers, ſaue the re=
mainer, and the Cyphers : and ſet the foꝛmer whole
roote, without the fraction, in the *quotiente.*

Scholar. Then will
it ſtande thus.   946
                                    29688̇2000000(544.
Now accoꝛdyng to the
rule J will pꝛoceade : as if
this whole nōber wer the firſt nomber pꝛoponed vnto
me. And therfoꝛe J doe double al the *quotiente,* whiche
maketh. 1088. and that doe J ſet vnder. 9460. And
then ſhall J ſeke a *quoti-*   4
ente, that maie declare   78
how often tymes, that   1829
double is cōtained in the   94666
nomber ouer it. And J ſe   29688̇2000000(5448.
it will bee. 8. wherfoꝛe J   1088
ſet doune. 8. in the *quoti-*
ente,* and by it J multiplie the double and ſubtracte it,
in this ſoꝛte : ſaipng 8. tymes. 1. out of 9. leaueth. 1. re=
mainyng. Again. 8. times. 8. (that is. 64) out of 146
will leaue. 82. Then farther J abate. 8. tymes. 8. out
of. 820. and there reſteth. 756. And laſt of all, J take
the ſquare of the *quotiente,* whiche is alſo. 64. out of
7560. and there will remain. 7496. And ſo haue J
doen with the firſte pꝛicke of the Cyphers.

A notable   Maſter. Conſider now that by thoſe. 2. Cyphers
conſideratiō   you haue gotten 8 into the *quotient* moꝛe then you had
befoꝛe. And all your foꝛmer nomber of the roote, re=
moued by it into one place higher, then it was befoꝛe
                                                        So

of Rootes.

So that, where by the firſt woꝛke, your roote was 544. and almoſte $\frac{9}{10}$: by this woꝛke you haue founde it to bee. $\frac{5488}{10}$, and. $\frac{3748}{5488}$ of $\frac{1}{10}$: whiche is verie nighe the same nomber, that you had befoꝛe.

Scholar. In deede, if J reduce the fractions, it wil bee. 544. $\frac{8}{10}$ and. $\frac{937}{1362}$ of $\frac{1}{10}$: whiche is in one fraction, $\frac{11833}{13620}$ aboue. 544.

Maſter. Marke this triall. And vſe the like after euery twoo Cyphers are ended : And you ſhall ſee a goodly agremente of the wooꝛkes together.

Scholar. In the meane tyme, to pꝛocede with the foꝛmer woꝛke, J ſet doune the nomber with the remainer, and the doble of the quotiente, as here appeareth.

$$7496$$
$$296882000000(5448.$$
$$10896$$

And ſearchyng foꝛ a newe quotiente, J finde that it will be. 6.

Therfoꝛe J ſette doune. 6. in the quotiente with the other nombers. And by that. 6. J doe multiplie the double of the whole quotiente, and ſubtract it oꝛderly, ſaiyng : 6. times. 1. be=yng abated out of. 7. leueth. 1.

Likewaies. 6. ty=mes. 8. maketh. 48, whiche J ſhall abate out of. 49. and ſo re=

$$5$$
$$968$$
$$10120$$
$$749644$$
$$296882000000(54486.$$
$$10896$$

ſteth. 1. Then 6. times. 9. (whiche is. 54.) muſt be ſub=tracted out of. 1016. and there will remaine. 962. Againe J ſhall abate. 6. tymes. 6. (that is. 36.) out of 9620. and there is lefte. 9584. Then take J the ſquare of my quotiente, whiche is alſo 6 times 6, oꝛ 36. and that J muſt abate out of. 40. and there reſteth. 4. And thus is the ſeconde pꝛicke of the Cyphers ended.

And now J finde in the quotiente not. $\frac{8}{10}$ as J did in
M.i.   the

The extraction

the laſte woozke befoze this. But J finde. $\frac{86}{100}$ whiche goeth moze nighe to $\frac{9}{10}$. Foz $\frac{90}{100}$ would be $\frac{9}{10}$: and $\frac{80}{100}$ is equalle with $\frac{8}{10}$. And J maie eaſily ſe, that $\frac{86}{100}$ is moze nigher to $\frac{90}{100}$ then to $\frac{80}{100}$: beſide the remainer, whiche will make $\frac{47962}{54486}$ of $\frac{1}{100}$, oz els $\frac{47962}{5448600}$ of one.

Maſter. J ſee, a well willyng mynde can marke diligentlp, and learne ſpedilp : wherfoze go fozwarde with pour woozke.

Scholar. J muſte ſette doune the double of all my *quotiente*, whiche will be. 108972. And it will ſtande thus.

Wherfoze J doe ſeke foz a newe *quo=* *teiente*, and J finde it to be. 8. whiche. 8. J

$$95802$$
$$296882000000(54486.$$
$$108972$$

ſet in the *quotiente*, with the other nombers, and by it J woozke after my rule, ſaiyng : 8. tyme. 1. is. 8. whiche J abate from. 9. and there reſteth. 1. Then take J. 8. tymes. 8. (that is. 64.) out of. 158. and the remainer will bee. 94.

Again J ſubtract. 8. tymes. 9. (beeyng. 72.) from .940. and there is lefte. 868. Farthermoze J take. 8. times. 7. (whiche is. 56) out of. 82. and there re= ſteth. 26. Then doe J withdrawe. 8. tymes. 2. oz. 16. out of. 60. And there remaineth. 44.

$$8624$$
$$19486$$
$$958024$$
$$296882000000(544868$$
$$108972$$

Laſt of al J take 8 times 8 oz 64 (whiche is the ſquare of my laſt *quotiente*) out of 862440 and the remainer will be. 862376. And ſo haue J ended all my woozke.

And now J haue foz the roote $\frac{544868}{1000}$ that is. 544. and $\frac{868}{1000}$ beſide $\frac{431188}{544868}$ of $\frac{1}{1000}$ oz in leſſer termes $\frac{107797}{136217}$ of $\frac{1}{1000}$ that is $\frac{107797}{136217000}$ of one : Whiche beyng reduced into one fraction with the $\frac{868}{1000}$ will make $\frac{118344153}{136217000}$.

Maſter. Pou haue doen well.

And

of Rootes.

And here you see, that you drawe nigher & nigher still, to the very roote, if it might haue any. For $\frac{868}{1000}$ is a nigher nomber to $\frac{9}{10}$, then is $\frac{86}{100}$ as that was nigher then. $\frac{8}{10}$.

And if you would worke with moze Cyphers, you should perceiue still, that it would drawe nigher and nigher. But this maie suffice for examples sake.

Scholar. Then I praie you tell me, what is the chief vse of this rule : and for what maters it serueth.

Master. One yere will not suffice, to expresse the commodities of it. It serueth so many waies, in buil= ding : in proiection of plattes, for measuring of groud Timber, oz stone : And also in warre, for kramyng of battailes, oz makyng of diuerse engines, and gene= rally for all woozkes of *Geometrie* and *Astronomie*. But for to satiskie you partly, I will sette forthe twoo oz three questions, that depende of this woozke of extrac= tion of square rootes.

And firste of a battaile : bicause it semeth to serue leaste for that purpose.

A question of an armie.

A captaine generall hauyng three greate armies, would caste theim into three square battailes, but he knoweth not how many men, he shall set in the krote of eche battaile.

The nombers of the three armies, are for the firste 5625: For the second 9216: And for the third 15129.

Scholar. I dooe perceiue easily, that for eche of these nombers, I muste searche out the square roote, and then haue I the fronte, oz flanke. Sith bothe are equalle in a square battaile.

Wherfore I set doune the first nomber thus, with his prickes. And then vnder the first pricke towarde the lefte hande, I finde the grea= teste roote to bee. 7. seeyng the greateste square is. 49. That roote doe I set within the *quoti= ente* line : and his square doe I abate from. 56. and so

5.6.25(

M.ii.   remaineth

The extraction

remaineth. 7.

Then doe I double that roote, and sette the double vnder. 72. and see that the newe *quotient* will bee. 5. And there will remaine. 25. whiche is the iuste square of the last *quo=tiente.*

$$\begin{array}{r} 7 \\ 5625(75 \\ 14 \end{array}$$

Wherby it is euidente, that his first armie contai=ned a square nomber, and the roote, or side of it is 75. And so many menne shall be in the fronte of the firste battaile, and as many in the flanke.

Now for the seconde battaile, I seke the square of 9216. and finde it to bee. 96. As in the example I haue wrought it.

For the firste nomber is. 9. seyng it is the greateste square roote, that can bee founde in. 92. And so is the double

$$\begin{array}{r} 8 \\ 113 \\ 9216(96 \\ 18 \end{array}$$

of it. 18. and the *quoteiente* for it. 6. as it appeareth ma=nifestly inough.

Wherfore I saie that the second battaile shal haue in euery ranke. 96. men.

And now for the thirde battaile, I sette doune the nomber, accordyng to this rule : and I finde the firste roote to be. 1. bicause. 1. tymes. 1. ma=keth. 1. And his double is. 2. whiche I abate twise from the nomber ouer it : and after double those bothe nom=bers, whiche make. 24. And finde that to be abated. 3. tymes.

$$\begin{array}{r} 1 \\ 17 \\ 15129(123. \\ 24 \\ 2 \end{array}$$

And so haue I gathered that the nomber is square and the roote 123. Accordyng to whiche nomber, that thirde battaile must be marshalled.

Master. Seyng you are so redy in this poincte so sone. Tell me how many menne, shall be sette in the fronte, if all these. 3. armies be ioined into one square battaile.

Scholar. Firste I must adde all. 3. nombers toge=ther.

of Rootes.

ther. And so will thei make. 29960. as here by example doeth appere.

5625
9216
15129
29960

But this nomber can bee no square nomber, bicause it hath one odde Cypher in the firste place : for I remember your saiyng, that square nombers can not be=gin with odde Cyphers. Wherfore this nomber will not make a square battaile.

Yet wil I proue, what maie be the frot of the grea=teste square battaile, that maie be made of that nöber.

And for that purpose I pricke the nombers, and finde the greateste roote in. 2. to be. 1 and the same nöber to bee the square also. Then double I that roote, and place his double vnder. 9. that is vn=pricked : and serchyng for a *quotiente*,

```
1 1 3
1 5 0 4 1
2 9 9 6 0 (173
2 3 4
```

I finde it to be. 7. with whiche I woorke by the rule, and so doeth remain for the nexte pricke. 10.

Then doe I double that. 17. whereby commeth 34 whiche I set vnder. 106. And for it I finde. 3. to be the meteste *quotiente* : with whiche if I woorke accordyn=gly, there will remaine. 31. as the excesse aboue the greateste square.

Wherby it appeareth that. 29929. is a square nö=ber : and hath. 173. for his roote. And that should bee the fronte of this greate battaile.

Master. Now will I proue you with an other question of like sorte.

A Prince hath an armie verie greate. With whi=che he passeth in a Uallie, so that in marchynge the fronte can be but. 18. menne. And by that meanes the flancke containeth. 449352.

The seconde question of an armie.

After that the armie is passed that valie, the kyng mindyng to occupie all the beste grounde, willeth the battaile to be set square. How would you doe it?

Scholar. First I multiple the flancke, by the front.

M.iii. And

The extraction

And ſo J finde the whole nomber to be. 8088336.

That nomber doe J pꝛicke
as my rule teacheth me, and
J finde the firſt roote to be. 2.
and his ſquare. 4. whiche firſt
J ſubtracte out of. 8. and ſo re=
ſteth. 4. Then doe J double
that *quotiente*, and finde that double. 8. tymes in the
ſomme ouer it.

$$\begin{array}{l} 2223 \\ 484471 \\ 8088336(2844. \\ 45668 \\ 8 \end{array}$$

And ſo doe J pꝛocede till J haue founde out all the
4. figures, accoꝛdyng to the. 4. pꝛickes vnder that nō=
ber. And then the roote appeareth to be. 2844.

The thirde question of an armie.

Maſter. Yet one queſtion moꝛe, foꝛ to exercife
your penne, will J pꝛopounde of a like mater.

A generalle hath three armies, to the nomber of
28289. men : and none of thoſe three armies is apte
to make a ſquare battaile, yet he is appoincted by his
ſoueraigne, to ſette theim in three ſquare battailes.

Theſe be the. 3. nombers of the. 3. armies. Jn the
firſte there are. 10296. men : Jn the ſeconde. 9493 :
and in the third. 8500. Now let me ſee how you can
caſt them into three ſquare battailes.

Scholar. J thinke it reaſonable, to take the grea=
teſte ſquares of the firſt and ſecond nombers, and the
exceſſe of them bothe, to put to the thirde nomber.

Maſter. So are you not ſure that the third nom=
ber, will be a true ſquare.

Scholar. Then knowe J not how to doe it.

Maſter. Take the greateſte ſquare in the thirde
nomber alſo. And note thoſe three exceſſes, and their
rootes alſo.

Then put one to euery roote, and marke the ſqua=
res that will riſe of them.

Thirdly, ſubtract the firſte 3. nombers, out of thoſe
3. newe ſquares, and note the difference of eche of the
firſte nombers, from thoſe ſquares : and ſo haue you. 3.
nombers

of Rootes.

nombers of ercesse, and. 3. other of wante.

Now compare those ercesses and wantes well to=
gether : and you shall easily see from whiche you shall
take any nomber, and to whiche you shall adde any.

Scholar. In the firste nöber the greatest square is
10201. and therby the ercesse is. 95. and the roote 101.

In the second nomber the greatest square is. 9409
and his roote 97. So is the ercesse. 84.

And in the thirde nomber, the greateste square is
8464 : and the roote of it. 92. Wherfore the ercesse
appeareth to be. 36.

And thus haue I founde the. 3. ercesses.

Now for to finde the 3 defaultes or wantes, I adde
one to eche roote, and multiplie theim square : and so
of. 102. I finde the square to bee. 10404. and if I
subtracte the firste nomber, whiche is. 10296. out of
it, there will remain. 108. for the firste wante.

Then for the seconde roote. 97. I take. 98. whose
square will bee. 9604. out of whiche I abate the se=
conde nomber, whiche is. 9493. and there is left 111
as the wante of the seconde nomber.

Thirdly, I take 93 for the newe roote, nert aboue
92. and I finde his square to bee. 8649. from whi=
che when the thirde nomber. 8500. is abated, the de=
faulte appeareth to bee. 149. And thus haue I the. 3.
defaultes or wantes, and also the. 3. ercesses. Whiche
for ease of comparyng, I set in order thus.

	A.	*B.*	*C.*	*A. B,* and *C.* beto=
Excesses.	95.	84.	36.	ken the order of
Wantes.	108.	111.	149.	the 3 first nöbers.

And here I compare the ercesses with the wantes,
to see if any. 2. ercesses will make vp the others want
And I see by a lighte proofe, it will not serue.

As for the wantes, I doe not compare theim to the
ercesses,

The extraction

exceſſes, for J ſe that euery one want, is greater then any one exceſſe. And therefore. 2. wantes are farre to greate aboue any one exceſſe. And ſo am J at a ſtaie.

Maſter. Therfore although that rule bee gene= ralle, yet where it faileth, this ſhall you doe.

Take the. 2. wantes, of any. 2. nombers, and adde theim firſte together, and then abate theim from the thirde nomber : and if the remainer be a ſquare nom= ber, then haue you gotten your purpoſe.

Scholar. That will J proue here. And firſt J take the wantes, of the. 2. firſte nombers, whiche make 219. And that doe J abate from the thirde nomber 8500. and there remaineth. 8281. whiche as J ſee, maie be a ſquare nomber. And therfore J proue it, in my tables, and J finde it ſo to bee. And. 91. to bee the roote of it.

Wherfore J ſaie to the queſtion, that theſe ſhall be the nombers of the 3 battailes, as here J haue ſet thē.

The firſte battaile. 10404. and his fronte. 102.
The ſecond battaile. 9604. and his fronte. 98.
The third battaile.   8281. and his fronte. 91.
The ſomme of all ⎫
the. 3. battailes. ⎭ 28289.

And bicauſe theſe nōbers are not onely ſquare, but alſo their whole ſomme doeth agree, with the ſomme of the 3 ſeuerall armies, you maie be ſure that thei are well parted, accordyng to the intente of the queſtion.

But bicauſe ſoche queſtions, haue more difficultie then commoditie, to them that are not mete, to be tra= uelled in ſoche marſhall affaires, J wil leaue that ma= ter to marſhall men, and will come to lower maters in warre.

A question
of ſcalyng.

A citie ſhould bee ſcaled, beyng double diched. And the inner diche. 32. foote broade. And the walle. 21. foote high. The capitain commaundeth ladders to be
made

of Rootes.

made of that iuſte lengthe, that maie reche from the vtter brow of the inner diche, to the toppe of the wal=as in this figure is partly expreſ=ſed.

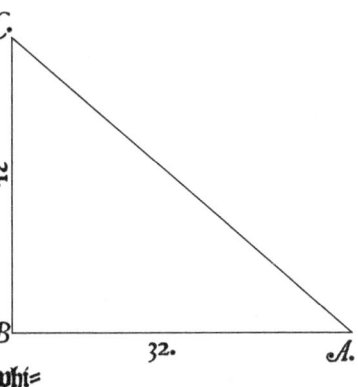

Where the line A. B. ſtandeth for the bredth of the diche. And the line, B. C. for the heighte of the walle. Nowe I demaunde, what ſhall be the lēgth of the line. A. C : whi=che here doeth repreſente the ladder?

Scholar. This figure doth occaſiõ me to remēber the 33. theoreme of the pathewaie, whiche ſaith thus.

In all righte anguled triangles, the ſquare of that ſide, whiche lieth againſt the righte angle, is equalle to the twoo ſquares of bothe the o=ther ſides.

Wherby I vnderſtand, that I muſt multiply thoſe twoo ſides ſquarely, that is, eche of theim by it ſelfe. And then addyng thoſe. 2. ſquares together, I muſte extract the roote of that whole nomber : whiche roote ſhall be the true lengthe of the ſlope line.

Wherefore, firſte I multiplie. 32. by it ſelf, and there riſeth of it 1024.

Againe, I multiplie. 21. by it ſelf, and it yeldeth. 441. Theſe bothe ſommes, beyng added to=gether, doe make. 1465. whiche nomber maie bee ſquare, bicauſe it begin=

```
      21
      21
      21
  ─────
      42
     441
```

```
      32
      32
  ─────
      64
      96
    1024
```

The extraction

neth with. 5.

Master. It is no square nomber, as it appeareth at the firste sighte. For although the firste nomber be 5. yet in soche nombers it is requisite, that the seconde figure should be. 2. els can it not be square: and here, you see, that the seconde figure is. 6. so that it can not be a square nomber.

Wherefore you shall seke the nigheste roote, that you can finde in it, and take that for your purpose.

Scholar. Here is my woorke set forthe.

And so it appeareth well that the nigheste roote is. 38. $\frac{21}{76}$, whiche is lesse then a quarter of a foote, aboue

$$\begin{array}{c} 2 \\ 5\,8\,1 \\ 1\,4\,6\,5\,(38\frac{21}{76} \\ 6 \end{array}$$

38. foote and that must be the lengthe of the ladder.

Master. Yet one question more will I propound agreable to the firste forme.

A questiõ of encampyng.

A capitaine generalle hauynge three armies, in three seueralle battailes, in the firste. 4900. menne, in the seconde. 2401. And in the thirde. 2500. (so that the greateste armie, is as moche as bothe the other, excepte one manne) is inforced to ioine all three battailes in one. But is in doubte, whether he maie haue good and conueniente grounde to encampe thẽ, in battaile forme. Wherefore consideryng, that all. 3. battailes together, are but double to the greateste of the. 3. alone. The capitaine desiryng a mete grounde for his armie, so ioined in one square battaile, is in doubte, what square of grounde will serue his purpose. But sure he is, that it muste bee double to the grounde, that the greateste armie of the 3. did occupie and that was square euery waies. 210. foote. Wherfore his demaunde is, how many foote square, shall the side of that grounde bee, that is double to the former square platte, whose side was. 210. foote euery waie?

Scholar.

of Rootes.

Scholar. Firſte I muſt multiplie. 210. by it ſelf, and ſo haue I the iuſt platte of grounde, of. 44100. foote, that muſt I double, and it will be. 88200. And out of this nomber, ſhall I ſeke the nigheſte ſquare roote. For a iuſte ſquare, I ſe, it is not : by reaſon that after the euen Cyphers, there foloweth. 2, whiche is one of thoſe figures, that can not beginne any ſquare nomber.

Wherefore, ſekyng for the nigheſte roote, I finde it to bee 296. $\frac{121}{148}$ that is almoſte. 297. foote euery waies ſquare. And ſo moche muſte the ſquare ſide of that grounde bee, whiche ſhould ſerue for that whole ar= mie.

$$14$$
$$1458$$
$$42124$$
$$88200(296\frac{242}{296}$$
$$4$$
$$58$$

And hereby I doe perceiue, the ouerſighte of many men : whiche being required to double a ſquare platte do double the ſide of it, thinking the mater eaſily doen.

But if thei marke it well, thei maie perceiue, that thei doe make, by that meanes, a ſquare fower ti= mes ſo bigge as their firſte ſquare was. As by this figure, any man maie ſee.

For if 2. be the ſide of the ſquare then is the ſquare 4. But if I dou= ble the ſide, and make it. 4. the ſquare thereof will be 16. whiche is. 4. tymes. 4. and not onely double.

So that the roote of the double platte, ſhould bee the roote of. 8. whiche is ſomewhat leſſe then. 3. and therfore moche leſſe then. 4.

Maſter. You maie perceiue the ſame, with the reaſon of it, by the 18. propoſition of the. 8. booke of *Euclide*, as it is before alleged.

But now for to ſhewe the larger vſe of this rule,

P.ii.   I

The extraction

I demaunde this queſtion.

*A queſtion
geographical.*

There be. 2. tounes, as *Chicheſter* and *Yorke* whiche lye Southe and Noꝛthe, and betwene them. 220. miles. A thirde toune as *Exceſter*, lieth plaine VVeſte frõ *Chicheſter*. 120. miles. I deſire to knowe the iuſte diſtaunce of *Yorke* from *Exceſter*.

Scholar. I muſt ſet thoſe. 3. tounes, in foꝛme of a Triangle, with their diſtaunces : As here is repꝛeſented. VVhere *A.* ſtãdeth foꝛ *Exceſter*, *B.* foꝛ *Chicheſter*, & *C.* foꝛ *Yorke*.

And then accoꝛdyng to the rule, I multiplie. 120.

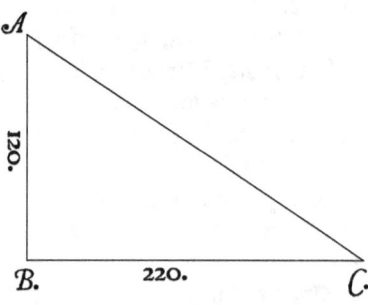

ſquarely : and it maketh. 14400. Likewaies I doo multiplie. 220. and it yeldeth. 48400.

Theſe bothe nombers I ſhall ioyne in one, and ſo haue I. 62800. whoſe roote is very nigh. 250. miles and $\frac{3}{5}$ of a mile.

And that is the true diſtaunce of *Yorke* and *Exceſter*.

By this example I gather, that this rule doeth helpe to *Geographie*, foꝛ to drawe the true platte of any countrie.

$$\begin{array}{l} \not{z} \ \ 3 \\ \not{6}\not{2}8\dot{0}\dot{0}(250\tfrac{3}{5} \\ \ \ 4 \end{array}$$

Maſter. If I ſhould ſtande in pꝛopoundyng exãples of this rule vnto you, vſyng but one foꝛ euery arte and ſcience, and foꝛ euery different kinde of commodiouſe pꝛactiſe : it would make a greate booke.

And therefoꝛe omittyng that, till occaſion ſerue otherwaies, I will pꝛoceade to the extraction of *Cubike rootes*.

Df

of Rootes.

Of Cubike rootes.

Hen any *Cubike nomber* is propoun=
ded, whose roote you should extract
After the nomber is written doune
orderly : you shall set a pricke vnder
the firste figure : and vnder the. 4.
and so vnder euery third figure, o=
mittyng still. 2. figures vnpricked.

And looke how many prickes, your nomber hath,
so many figures shall the roote of your nöber contain

Then to begin the searche, for the firste figure of
the roote (in this order) you shall looke what maie be
the roote of the nomber, belongyng to the laſt pricke
toward the lefte hande. And that roote shall you sette
by a *quotiente* line, as you did in ſquare rootes.

And if the whole nomber ouer that pricke, be a *Cu
bike nomber*, you shall cancell it all. But if it bee no *Cu
bike nomber*, then ſubtracte out of it, the greateſte *Cube*
in it, and cancell the whole nomber, and set the reſte
ouer it : as you did in ſquare rootes.

But conſideryng, that you ought to haue in ready
remembraunce, all thoſe *Cubike rootes*, whiche be digi=
tes, with the *Cubes* that thei make : for without theim
you can not procede in this woorke. I thinke it good
to set forthe here in a table, all thoſe rootes with their
Cubes, that therby you maie be the more
adſured in tyme of your worke. For els
a litle miſtakyng, might be the occaſion
of a greate erroure.

And now for this firſt rule I ſaie, as
I ſaied of *Square rootes*, this shall be euer
more the firſte woorke, and shall not be
repeted in any one *Cubike* nöber. Where=
as all the other rules folowyng, ſhal be
ſo often repeated, as there are prickes in

1.	1.
2.	8.
3.	27.
4.	64.
5.	125.
6.	216.
7.	343.
8.	512.
9.	729.

I.

The extraction

your nomber.

2.   And of theim this is the firſte : that you ſhall triple the firſte roote. And that triple ſhall you ſet vnder the nexte nomber, toward the righte hande, befoze that pzicke, whiche you did laſte ende.

3.   Then multiplie that triple, by the ſame *quotiente*. And ſet it doune vnder the firſt triple : and that nom= ber ſhall be called your diuiſoz.

4.   Thirdly, loke out a *quotient*, that maie declare how often the diuiſoz is in the nomber ouer it.

In whiche doyng, you muſt haue this regard, that betwene that pzicke that is ended, and the nexte that ſtandeth toward the right hande, you muſt ſubtracte 2. other nombers. That is to ſaie, the ſquare of the laſte *quotiente*, multiplied by the formet triple. 10. ty= mes : and the *Cube* of the ſame *quotiente*.

Scholar. This rule is very obſcure in woozdes.

Maſter. Then will I terme it thus.

2. & 3.   Take the ſquare of your whole *quotiente*. 300. ty=
4.   mes : and that ſhal be your diuiſoz. Then ſeke a newe *quotiente*, declaryng how often that diuiſoz, maie bee founde in the nomber, that doeth belong to the nexte pzicke. But ſo that the ſquare of that newe *quotiente*, multiplied by the laſt *quotiente*. 30. tymes : and alſo the *Cube* of that newe *quotiente*, ioyned all in one ſomme, maie be taken out of the ſame nomber. And if you vn= derſtande this, there reſteth no moze difficultie.

Scholar. I truſt by exaple, to vnderſtand it better.

Maſter. Then take you this exaple. 26463592 whiche I ſhall ſet doune and pzicke, as I taught you befoze : and as you maie here ſee. Where the. 3. pzic= kes declare vnto me, that the roote will haue. 3. figures,

                                        26463592(

And then vnder the pzicke that is nexte the lefte hande, whoſe nomber is. 26. I finde the greateſte *Cubike* *nomber* to bee. 8. and his roote. 2.

                                                       Foz

of Rootes.

For. 27. whiche is the nexte *Cube*, is to greate.

Therfore I set. 2. in the *quotiente*, and his *Cube*, be=yng. 8. I doe abate out of 26. and so remaineth. 18.

That. 18. I doe sette ouer. 26. whiche I muste cancell : and then standeth the nomber, as here you doe see.

$$
\begin{array}{r}
1\,8 \\
\cancel{2}6\,463592(2
\end{array}
$$

This is that firste woorke, whiche is not repeted.

Then to procede forward, I doe triple the *quotiente* 2, and so haue I. 6. whiche I shall set vnder. 4. beyng the nexte nomber, on the righte hande of the pricke that is ended.

And that triple must I multiplie, by the first *quoti=ente*, wherby amouteth that nomber, that must be the diuisor : and it is in this woorke 12. whiche must be set vnder the same triple : as here I haue placed it.

$$
\begin{array}{r}
1\,8 \\
\cancel{2}6\,463592(2 \\
6 \\
12
\end{array}
$$

Then shall I seke for a newe *quotiente*, declaryng how often ty=mes. 12. maie be founde in the nomber ouer it, that is 184. And I see it maie be in appearaunce. 15. tymes, but more then. 9. you shall neuer take for a *quotiente* : wherefore it appeareth, that I maie boldly take. 9. whiche I shall sette in the *quotiente* with the firste. 2. And then shall I multiplie. 12. whiche is the diuisor, by. 9. and thereof commeth. 108. to bee sette vnder 184. benethe the line, whiche shall euermore be drawen vnder the diuisor.

Now muste I take the square of my laste *quotiente*. 9. (whiche is 81.) and multiplie it by the triple of the former *quotiente* (that is by 6.) and so haue I. 486. to be sette one place more toward the right hande.

$$
\begin{array}{r}
2 \\
1\,8074 \\
\cancel{2}6\,463592(29 \\
6 \\
\hline
12 \\
\hline
108 \\
486 \\
729 \\
\hline
16389
\end{array}
$$

Last

The extraction

Laſte of all, J ſhall multiplie the laſte *quotiente* (*u= bikely* : and that maketh. 729. whiche muſt be ſet, yet one place moꝛe toward the right hand, that is to ſaie, vnder the nexte pꝛicke. And then ſhall J adde thoſe 3 ſommes into one : wherby will riſe. 16389. to be ſub= tracted out of. 18463. and ſo will remaine ouer that pꝛicke. 2074.

And the wooꝛke of that pꝛicke is doen.

This oꝛder of woꝛke, if you marke well, you haue learned the whole arte of extraction of (*ubike rootes*.

Foꝛ how greate ſo euer your nomber be : you ſhall not haue any newe kinde of wooꝛke.

But yet bicauſe J did teache you befoꝛe, the ſame wooꝛke in other wooꝛdes, J will wooꝛke the ſame e= xample again, accoꝛdyng to theſe wooꝛdes.

And firſte, after that the nomber is ſet doune, and the firſt (*ubike roote* taken, and the (*ube* abated. Then take the ſquare of that roote. 300. tymes, that is in this example. 4. tymes. 300, whiche maketh. 1200. and that ſhall be your diuiſoꝛ. This nomber, and all other in this wooꝛke, ſhall you ſet doune ſo, that the firſte nomber, ſhall be vnder the nexte pꝛicke, toward the righte hande.

Then ſeke your *quotiente*, with the foꝛmer cautele, and it will bee. 9. Wherefoꝛe multiplryng. 1200. by. 9. there will amounte 10800 : to be ſet vnder the line.

After this, J ſhall take the ſquare of. 9. (whiche is the new *quotiente*), and multiplie it by. 2. (whiche was the laſte *quotiente* befoꝛe) 30. tymes. So muſte J

	1 8
	26463592(29
	1 2 0 0
	10800
	4 8 6 0
	7 2 9
	16 3 8 9.

multiplie 81. by. 60. and it will make. 4860. whiche J place oꝛderly.

Then ſet J doune the (*ube* of the *quotiente*, whiche maketh

of Rootes.

maketh. 729. And so are the. 3. nombers placed, and agree with the former woozke, in all thinges, saue in 2. poinctes. For here the triple of the *quotiente*, is not set doune, but kepte in memozie. And again, here are diuerse cyphers, whiche are not in the former wozke.

Scholar. Sir, I perceiue, that the Cyphers dooe nothyng els, but set the nombers in their due places. And the triple of the *quotiente*, is supplied in wozke by 2. multiplications. First by. 300. and then by. 30. So that it is all one in effecte.

And by the one wozke I vnderstande the other the better : when I compare theim bothe together. But yet I pzaie you, ende the woozke that you began.

Master. To continue that woozke, firste I must set doune the nombers, as thei should remaine, after 16389. is abated out of. 18463. and then will thei stande thus.

$$2074$$
$$2\cancel{6}4\cancel{6}\cancel{3}592(29$$

Then shall I repeate the foz= mer wozke, by settyng doune the triple of all the *quotiente*, whiche will be. 87. and that must be placed vnder. 45.

Nexte that I shall multiplie that. 87. by. 29. and there will come. 2523. whiche must be the diuisoz.

Wherfoze I seke foz a new *quotiente*, that maie shewe me how often. 2523. is contained in. 20745. And it will bee. 8. That 8 doe I set in the *quotient* and by it I multiplie. 2523. and it giueth. 20184 whiche I sette doune, as here you see.

$$2074$$
$$2\cancel{6}4\cancel{6}\cancel{3}592(298$$
$$87$$
$$\overline{2523}$$
$$20184$$
$$5568$$
$$512$$
$$\overline{2074592.}$$

Then doe I multiplie that *quotient* squarely, and that wil be 64. Whiche I shall multiplie by the triple, that is 87, and there will amounte. 5568. to be set one place moze towarD the righte hande.

D.i.   Last

The extraction

Laſt of all, I muſt take the *Cube* of. 8. that is. 512, and it ſhall bee ſette yet one place moze towarde the righte hande.

And then by additiõ, I ſhall bzyng thẽ all into one nomber : and it will bee. 2074592. whiche is equall with the whole nomber aboue, that is vncancelled. And therfoze if I abate the one out of the other, there will remain nothyng.

Wherefoze I ſee, that the firſte nomber, is a iuſte *Cubike* nomber. And his roote is. 298.

Scholar. I haue marked you ſo well, that I truſt to doe the like, without erroure.

But I pzaie you woozke this laſte parte alſo, by your ſeconde rule, as you did woozke the other : that I maie ſee the due agremente of theim bothe : and alſo perceiue the righte vſe of this woozke, the better by that other fozme.

The ſeconde woorke.

Maſter. I muſt in that caſe ſette doune the nom= bers, as thei were ſet in the o= ther woozke. And then I ſhall multiplie al the *quotiët*, whiche is. 29. by it ſelf ſquarely, and it

```
      2074
2̷6̷4̷6̷3̷592(29.
```

will make. 841. whiche muſt be multiplied by. 300. And ſo there amounteth. 252300. to be ſette doune, as here you ſee.

Then I ſhall ſeke out a *quo= tiente*, declarynge how often 252300. maie bee founde in 2074592. And that *quotiente* will bee. 8 : whiche I ſet in the *quotient* roome, with the other nombers.

```
      2074
2̷6̷4̷6̷3̷592(298
      252300
     2018400
      55680
        512
     ─────────
     2074592
```

And then I dooe multiplie the diuiſoz by the *quotiente*, and thereof riſeth 2018400 whiche I ſet vnder a line, as you maie ſee.

Nexte that, I doe multiplie the newe *quotient*, by it ſelf,

of Rootes.

ſelf ſquarely, whereof commeth. 64. and
that ſquare of the laſt *quotient*, I ſhall mul=
tiplie by. 870. whiche is. 10. times the tri=
ple of the former *quotiente*. 29 : and thereof
commeth. 55680. whiche I ſet doune al=
ſo orderly.

8
8
64
870
4480
512
55680

Laſte of all, I multiplie. 8. (that is the
laſte *quotiente Cubikely*, and it maketh. 512.
whiche alſo I ſet doune in coueniēt order.

And then ſhall I adde theim all together. And ſo
haue I the ſame ſomme, that I had before in the other
former woorke, and it is. 2074592.

Scholar. I neade no more inſtruction for this : I
thinke my ſelf ſo cunnyng, by occaſion of your exam=
ples, whiche you haue wroughte ſo in double forme.

Maſter. That maie you proue, by this nomber
47832147.

Scholar. Firſte I ſhall pricke it, as you taughte *An other*
me, omittyng ſtill. 2. nombers.   *example.*

And then out of the nomber ouer the laſte pricke,
I ſhal ſeke out the *Cubike roote*, and abate the *Cube* ther=
of, out of the ſame nomber, and ſet the remainer ouer
it, cancellyng the reſte.

And ſo in this nomber, I finde
in. 47. the greateſte *Cube* to bee. 27.
and the roote of it 3. Wherfore I a=
bate. 27. out of. 47. and finde the reſte to be. 20. ther=
fore I cancell. 47. and ſet. 20. ouer it. And the. 3. whi=
che is the roote, I ſet in the *quotient*. And ſo is the firſt
woorke eanded.

$$20$$
$$47832147(3$$

Then doe I triple that *quotiente*, and it maketh. 9.
whiche I ſet doune vnder. 8.

Again I multiplie that. 9. by. 3. and it yeldeth. 27.
whiche I ſet vnder the triple, and take it for my diui=
ſor.

Wherefore I ſhall now ſeke a *quotiente*, that maie
            D.ii.   declare

The extraction

declare how often. 27. is in. 208
and J see, it will bee. 7. tymes.
Therfore J sette doune. 7. in the
quotiente : and by it J multiplie 27
and it maketh. 189. whiche J set
vnder the line : and then J dooe
multiplie. 7. by it self, whiche
maketh. 49. ↄ that square doe J
multiplie by the triple of the for=

```
        20
47832147(37.
9
27
189
441
343
23653
```

mer *quotiente*, that is, by. 9. and it yeldeth. 441. whi=
che J set one place more toward the righte hande.

Last of all, J take the *Cube* of. 7. whiche is. 343. and
that doe J sette doune, yet one place more toward the
righte hande.

These. 3. sommes beyng added together, doe make
23653.

Master. That will be hardely abated out of a les=
ser somme.

Scholar. J see now my errour. J must take a lesse
quotient : whiche thyng J might haue perceiued by the
seconde nomber. For thei twoo wer to greate, before
the thirde was added.

So that J should haue taken but. 6. for the *quotiente*
And then would the firste nomber haue been but 162
and the seconde. 324. and the
thirde. 216. but that their pla=
cyng would make theim to be of
other values, saue the last of the.

Therfore, J set euery one in
his due roome : and adde theim
together, and there amounteth
19656. to bee subtracted out of
20832. and the remainer will
be 1176. And thus is that pricke
with his woorke eanded.

```
       I
     20176
47832147(36.
9
27
162
324
216
19656
```

Then for the nexte pricke, J repeate the same very
forme

of Rootes.

fo2me of wo2ke again. Firſt ſettyng doune the triple of the whole *quotiente,* whiche is. 108. ſo that it ſhall ſtande vnder. 11761. o2 vnder. 761. accoumptyng fi= gure fo2 figure.

That triple muſt I multiplie againe by the whole *quotiente.* 36. and it will make. 3888. whiche nomber I muſte take fo2 my diuiſo2.

Wherfo2e I ſeke how ma=
ny times, I maie finde that di=
uiſo2 in. 11761. and I ſee, it
will bee. 3. tymes. Therfo2e I
ſet. 3. as my *quotiente,* in his due
place : and by that *quotient* I do
multiplie. 3888. and ſo haue I
fo2 my firſte nomber. 11664.

$$\begin{array}{l} 1176 \\ 4.7832147(363 \\ 108 \\ 3888 \\ \hline 11664 \\ 972 \\ 27 \\ \hline 1176147 \end{array}$$

Againe I doe multiplie the
laſte *quotiente.* 3. ſquarely, and ſo haue I. 9. whiche I ſhall multiplie by the triple of the fo2mer *quoteient,* and it yeldeth. 972. that ſhall be ſet mo2e nigher the right hande, by one place.

Thirdly, I take the *Cube* of. 3. whiche is. 27. and that doe I ſet yet one place mo2e towarde the righte hande.

Then doe I adde thoſe 3 ſommes into one, and thei make. 1176147. whiche is equalle ſomme, with all the nombers ouer it, that be vncancelled.

Wherfo2e I ſaie that. 47832147. is a *Cubike* nom= ber, and the *Cubike roote* of it is. 363.

Maſter. Now doeth the o2der of teachynge re= quire, that I ſhould inſtructe you, how to extracte the nigheſte *Cube roote,* out of any nomber, that is not a true *Cube.* As this nomber fo2 example maie ſerue. 694582951.

Where firſte I muſte extracte the nigheſte roote, as I taughte you, fo2 the nigheſte *Square rootes,* in nõ= bers that are not ſquare : and then ſhall I note the re=

The nigheſte roote in a nõ= ber not Cu= bike.

D.iii.　　mainer :

The extraction

maíner : whíche J shall set foz the numeratoz. And hís
denomínatoz shall be founde as J wíll tell you anon.
But fírste doe you wozke the example, to hís nígheste
roote ín whole nombers.

     Scholar. J set ít doune, and
pzícke ít, and fínde the greateste
Cube ouer the laste pzícke to bee
512. and the roote of ít ís. 8.

$$1\,8\,2$$
$$694582951(8.$$

     Wherfoze J set doune. 8. ín the *quotiente*. And J a=
bate. 512. out of. 694. and so resteth 182. and the foz=
mer 694. cancelled.

     Then to pzocede, J must tríple that roote. 8. and ít
maketh. 24. whíche. 24. J set vnder. 1825. And then
J doe multíplíe that agaín, by the *quotiente* oz roote. 8
and ít maketh 192. to be set vnder the saíed tríple. 24 :
as the diuísoz. Foz whíche J seke a newe *quotient*, and
ít wíll be 8. That. 8. J set ín the *quotiente* place, and by
ít J multíplíe the diuísoz. 192. and there ríseth. 1536.
to be set vnder the líne, ín conueníente ozder.

     Nexte J multíplíe the *quotiente* squarely : whíche
yeldeth. 46. and that square J multíplíe agaín by the
tríple, and so haue J. 1536. also. But thís must stand
moze fozwardly by one place.

     Last of all J take the *Cube* of the *quotient*. 8. and that
ís. 512. whíche J set vnder the
other twoo sommes, and that
by one place moze fozwardly.

     Now gatheryng all these. 3.
somes ínto one, theí wíll make
169472 whíche J shall abate
out of. 182582. and so remaí=
neth there. 13110. And that
pzícke wíth hís wozke eanded.

1 3	
1 8 2̸ 110	
694582951(88	
2 4	
1 9 2	
1 5 3 6	
1 5 3 6	
5 1 2	
1 6 9 4 7 2	

     Therfoze hauyng one other
space to wozke, J must repeate.
the same ozder of wozke agaín, by tríplyng the whole
quotiente

of Rootes.

quotiente. 88. and that will bee. 264. And againe I muſt multiplie that triplede nomber, by the ſaied *quo-*
tiente, and it will make. 23232. whiche ſhall bee the diuiſoʑ.

Wherfoʑe I ſeke a newe *quotiente,* whiche is eaſily per ceiued to be. 5. That. 5. doe I ſet in the *quotiente,* and by it I dooe multiplie the deuiſoʑ 23232. and there amounteth 116160, as the firſte nom=
ber, to bee ſet vnder the line.

Againe I ſhall multiplie the *quotient* ſquarely, whiche giueth. 25. and that ſquare

$$1428$$
$$1\cancel{3}\cancel{1}\cancel{1}\cancel{0}826$$
$$6\cancel{9}4\cancel{5}8\cancel{2}9\cancel{5}\cancel{1}(885.$$
$$264$$
$$23232$$
$$\overline{116160}$$
$$6600$$
$$\underline{125}$$
$$11682125$$

ſhall I multiplie by the triple. 264. and ſo will there riſe. 6600. to bee ſette, as the ſeconde nomber vn=
der the line : and one place moʑe foʑwardly, towarde the righte hande.

Laſt of all, I ſhall ſette vnder them bothe, and one place moʑe towarde the righte hande, the *Cube* of. 5. whiche is. 125.

And then ſhall I adde all thoſe. 3. ſommes together of whiche commeth. 11682125. to bee abated out of 13110951. and ſo the remainer will bee. 1428826. Whereby I ſee, that the firſte nomber that was pʑo=
poned, I meane 694582951 is no *Cubike nomber* but the greateſte *Cube* in it is. 693154125. and his roote is. 885.

And ſo, I ſee, all other nombers of like kinde muſt bee wʑoughte.

But now foʑ the remainder, how ſhall I dooe to bʑynge it vnto a fraction, that maie aptly expʑeſſe the nigheſte roote in that ſoʑte?

Maſter. There bee as many waies, as there bee wʑiters almoſte, foʑ euery manne deuiſeth, how to
bʑyng

The extraction

bꝛynge it moſte nigheſte to a true roote, if any ſoche were : whereof *Cardane* his rule is this.

Multiplie the roote ſquarely, and againe by 3. and that nomber ſhall be the diuiſor vnto the remainer,

Where he might haue vſed moꝛe plaineſſe in woꝛ= des, if he had ſaied : and that nomber ſhal be the deno= minatoꝛ, to the remainer. Wherefoꝛe as here your roote is. 885 ſo is the ſquare of it 783225 and the tri= ple of that is. 2349675. So would that fraction bee

$$\frac{1428826}{2349675}.$$

But how nigh this doeth go to the truthe, J leaue it till an other tyme.

Scheubelius doeth allege an other reaſon, and infer= eth an other oꝛder, diuerſe frõ this, and ſoche as im= pugneth this, ſaiyng :

Triple the roote, and the ſquare of it alſo, and adde bothe thoſe nombers together, and .1. more : And ſo haue you a denominator for your numeratour.

The numeratoꝛ euermoꝛe is vnderſtãd to be the re= mainer. By whiche meanes the fractiõ in this woꝛke would bee $\frac{1428826}{2352231}$: whiche is a leſſer fraction by a good deale, then is the foꝛmer fractiõ, after *Cardanes* foꝛme.

But bicauſe at this pꝛeſente, J maie not ſpende ſo moche time, to ſcan their ſeueralle opinions, where= in eche of theim, pleaſeth hymſelf well : the one alle= ging demonſtration (whiche ſcarſely ſerueth) and the other namynge it a ſecrete, as it is woꝛthie to bee : J will pꝛocede to a thirde waie, moꝛe certain then ether of theſe bothe. And that is by addition of certain Cy= phers, to the remainer, in ſoche ſoꝛte, that thei muſte all waies bee ternaries, as. 3. 6. 9. oꝛ. 12. ꝛc. And then ſearche

of Rootes.

searche fozward with the like ozder of wozke, as you vsed befoze.

In this maner of pzactise, looke how many pzickes your ciphers hath (oz els how many ternaries of Ci= phers, there bee set to your nomber) so many figures shall the numeratoz of your fraction contain. And the denominatoz shall euermoze, contain 1. moze. Wher= of the laste onely shall bee an vnitie, and all the other shall bee Cyphers.

That is to saie, that if J adde but 3. Ciphers to the nomber, the fraction shall contain certain. 10. partes And if J adde. 6. Cyphers, it shall expzesse. 100. par= tes. So. 9. Cyphers maketh the denominatoz to bee 1000. partes : And 12. Cyphers geueth 10000 partes.

Foz example. J will adde to our laste nomber that remained. 12. Cyphers. And then will the nomber be 1428826000000000000. vnto whiche J set no moze pzickes, then serueth foz the ciphers, bicause J haue passed all the other pzickes, in my fozmer woozke.

And now to continue my woozke, J shall triple all the fozmer _quotiente_, and it will be 2655. whiche nom= ber J shall place, as here you see it set. And then shall J multiplie that triple, by the foz=

$$1428826000000000000($$
$$2655$$
$$2349675$$

mer _quotiente_. 885. whiche will yelde. 2349675. to be set vnder thesaied triple : as J haue sette it here also. And this nomber shall be the diuisoz.

Then shall J seeke foz a _quotiente_, whiche can bee none other then. 6 : wherefoze J sette. 6. in a _quotiente_ line, and by that. 6. J dooe multiplie thesaied diuisoz 2349675. and it giueth. 14098050. to be the firste nomber vnder the line.

After that, J take the square of thesaied _quotiente_, whiche is. 36. and by it J multiplie the triple. 2655.

The extraction

wherby is made
95580, to be the
seconde nomber
vnder the line : ɛ
set, as it ought,
one place moꝛe
towarð ẙ righte
hande.

```
        18064984
14288260000009000900000(8856
        2655
      2349675
     14098050
        95580
          216
     1410761016
```

Laſt of all foꝛ
the thirðe nomber I take the *Cube* of theſaieð *quotiente*
whiche is, 216, anð place it as you ſee, with his firſte
figure vnder the pꝛicke.

Then ðoe I aððe thoſe, 3, nombers into one, whi=
che maketh, 1410761016, Anð that beyng ſubtra=
cteð out of 1428826000, ðoeth leaue 18064984,
Anð ſo is the wooꝛke of the firſte pꝛicke eanðeð.

Whereby it appeareth, that the fraction is ſome=
what moꝛe then $\frac{6}{10}$ oꝛ $\frac{3}{5}$: as it ſhall bee trieð better, by
the wooꝛkes that ſhall enſue.

Therefoꝛe I pꝛoceðe to the nexte pꝛicke. Anð firſte
I triple that whole *quotiente*, whiche yelðeth, 26568,
to bee ſet, as it is often befoꝛe repeateð, anð therefoꝛe
neðeth not hereafter to bee teðiouſly rehearſeð.

That triple ſhall I multiplie again, by the whole
quotiente (as here
I haue ſette it in
wooꝛke, bicauſe
the nomber is

```
18064984000900000 (88560
      26568
    235286208
```

```
        26568
         8856
       159408
       132840
       212544
       212544
     235286208
```

greate, anð not eaſily wꝛoughte by
memoꝛie) anð it ðoe I ſet in his ðue
place, as you ſee.

But then ſeeyng that ðiuiſoꝛ is
greater then all the nomber ouer it,
I ſhall ſet a Cypher in the *quotiente* :
in token that the ðiuiſoꝛ, can not be
abateð ones out of the nomber ouer
it.

of Rootes.

it. And so is the woozke of that pzicke eanded, with=
out any moze trauell.

Wherfoze to go fozward, I triple all that *quotiente*
and set it doune, as the rule would, & as here is seen.

```
          1594819256457
          18064984000000000000(885607
               265680
             23528620800
           164700345600
               13018320
                     343
           16470164743543
```

Then dooe I multiplie that triple, by the whole
quotiente, whereof cometh. 23528620800. and that
shall bee the diuisoz. And the *quotiente* foz it will be. 7.

So then if I multiplie that diuisoz by. 7. there will
amounte. 164700345600. foz the first nomber to
be set vnder the line.

And foz the nexte woozke, I shall multiplie. 49.
(whiche is the square of the newe *quotiente*) with the
triple of the fozmer *quotiente,* and it
will bzyng fozthe. 13018320. whi=
che shall bee the seconde nomber, to
bee set vnder the line.

The thirde nomber shall bee the
Cube of. 7. whiche is. 343.

```
   265680
       49
  2391120
  106272
 13018320
```

And those. 3. sommes added together, will make
16470164743543. whiche is to bee abated out of
18064984000000. and then shall there remain
1594819256457. And so haue I eanded. 3. pzickes
of the Cyphers. And thereby maie saie, that the frac=
tion is $\frac{687}{1000}$ and somewhat moze : That is somewhat
moze then $\frac{3}{5}$.

Scholar. I see by the fraction, that it is $\frac{3}{5}$ and $\frac{7}{1000}$

The extraction

beſide the quantitie of the remainer. But J pɹaie you eande the woozke of that other pɹicke, which dooeth remaine.

Maſter. J muſte triple all the *quotiente* : whereby will riſe. 2656821. whiche muſte bee multiplied by

```
        2656821
         885607
       18597747
      159409260
      13284105
      21254568
      21254568
     2352899275347
```

theſaied *quotiente* : and thereof will pɹocede the diuiſoɹ, being 2352899275347. And his *quotiente* will bee. 6.

Wherefoɹe firſte J ſet. 6. in in *quotiente* line, with the other nombers : and then doe J mul= tiplie the diuiſoɹ by that *quoti= ente*, and it bɹyngeth fooɹthe 14117395652082. Foɹ the

firſte nomber to be ſette vnder the line.

```
     183078734793024
     1894829286487000(8856076
          2656821
        2352899275347
      14117395652082
           95645556
                216
     14117740521663976
```

And again the ſquare of 6. being multiplied by the triple, will yelde. 95645556 : whi= che ſhall bee the ſeconde nomber vn= der the line.

The thirde nomber ſhall be. 216. bicauſe it is the *Cube* of. 6. And thoſe 3. nombers beeyng added together,

```
      2656821
           36
     15940926
      7970463
     95645556.
```

doe make. 14117740521663976. to be abated out of. 1594819256457000. And ſo doeth there re= maine. 183078734793024.

Wherefoɹe

of Rootes.

Wherefoȝe it doeth appeare, that beſide the firſt. 3 nombers of the roote, that is. 885. the reſte (that is 6076.) ſtandeth foȝ the numeratoȝ of a fraction, and the denominatoȝ vnto it is. 10000.

So that the nigheſte roote is. 885 $\frac{6076}{10000}$. beſide the fraction that doeth remaine : whiche would make but $\frac{1}{8}$ of $\frac{1}{10000}$.

Scholar. This is a ſufficiente pȝeciſenes. And ſo J iudge it ſufficiently taughte.

Wherefoȝe J pȝaie you pȝopounde ſome queſtiõs, that doe require this arte, foȝ their ſolution.

Maſter. J am contente. And let this be the firſte.

The Grecians giuen to idle banketting, and ſoche like wantonneſſe, did pȝocure thereby ſoche moȝtalle ſickeneſſes : that the quicke were ſcarſe hable to burie the dedde. Wherefoȝe conſultynge with their God= des, foȝ redȝeſſe thereof, thei receiued aunſwere, that when thei would double the Altare, whiche was of Cubike foȝme, thei ſhould bee deliuered from that pla= gue. Meanynge that learnyng is a due meane, to de= liuer realmes from plagues and enoȝmities. But to the queſtion, what ſaie you? Jf the ſide of a Cube be. 3. foote (as that altare might bee) how many foote ſhall the ſide be of that Cube, whiche muſt be double vnto it.

A queſtion of doublyng a Cube.

Scholar. This J cõſider. That firſte J muſt finde the quantitie of the Cube, that is pȝoponed. And then ſhall J double that quantitie. Thirdly, J muſte ex= tracte the Cubike roote, of that double nomber.

So in this queſtion, the ſide of the knowen Cube is 3. and therfoȝe the whole Cube is. 27. whoſe double is 54. And the Cubike roote is. 3. and $\frac{27}{27}$ by Cardanes rule : That is. 4. whiche is plainly falſe, foȝ. 4. is the roote of. 64. and not of. 54. But by Scheubelius rule, it wil be. 3 $\frac{27}{37}$ that is. 3 $\frac{3}{4}$ almoſte : whiche is moche nigher the truthe. Foȝ. 3 $\frac{3}{4}$ multiplied Cubikely, doeth make. 52. $\frac{47}{64}$ whiche is to litle by a good deale, that is by. 1 $\frac{17}{64}$.

The extraction

whereas, $3\frac{17}{37}$ doeth make a leſſer ſomme : that is to ſay
but $51\frac{25959}{50653}$, and ſo wanteth, $2\frac{24694}{50653}$, And although bothe
theſe ſommes goe nigher to the truthe, then *Cardanes*
rule, whiche miſſeth, 10, wholy : Yet maie it be eaſily
ſeen, that *Scheubelius* rule is not ſo good, as he would
it were, And the worſe here, foz the addynge of that
one moze,

Maſter, You are lepte verie ſodenly from a ſcho=
lar, to a comptroller, And yet I can not but pzaiſe your
diligente obſeruyng of ſoche thynges,

Proue now by the Cyphers, how it will frame,

Scholar, I ſette doune the nomber with, 6, Cy=
phers, and pzicke them thus,

Then dooe I take the greateſte
Cubike nomber in, 54, whiche is, 27
and that I doe abate from 54, and
ſo reſteth, 27, the roote of the *Cube* is, 3, whiche I ſette
in the *quotiente* line,

$$27$$
$$54000000(3$$

And then I triple, 3, whiche maketh, 9, that muſte
be multiplied by the *quotiente* againe, and ſo commeth
27, to be the diuiſoz, And his *quotiente* ſemeth to be, 9,

Wherfoze woozkyng with it,
the firſte nomber is, 243, and the
ſeconde is, 729, that is, 81, mul=
tiplied by 9, whiche is the triple,

Againe, the *Cube* of, 9, is, 729,
And all thei together, dooe make
32319 whiche ſome is to greate,
and therfoze I muſt take a leſſer
quotiente, As I mighte haue per=

$$27$$
$$54000000(39$$
$$9$$
$$27$$
$$\overline{243}$$
$$729$$
$$729$$
$$\overline{32319}$$

$$27$$
$$54000000(38$$
$$9$$
$$27$$
$$\overline{216}$$
$$576$$

ceiued well inough by the ſecond
nomber, if I had marked it in time,

But now amendyng my ouer
ſight, I take, 8, foz the *quotiente*,
And woozkyng with it I ſee, the
firſte nomber vnder the line will
be

of Rootes.

bee. 216 and the seconde. 576. And here all readp J
espie mp ouersighte again.

Wherfore J take. 7. to be the *quotiente*. And by it J
multiplie the diuisor, and so haue
J. 189. for the firste nomber.

And for the seconde nomber, J
doe worke with. 49. whiche is the
square of the *quotiente*, multiplied
bp. 9. that is the triple : and it pel=
deth. 441.

Thirdlp, J take the *Cube* of. 7.
whiche is. 343. And then addpnge
al. 3. nombers together, J finde the
somme to bee. 23653. whiche is to bee abated out of
27000. and so resteth 3347. Wherbp J see, that. $3\frac{7}{10}$
with somewhat more is the roote that J should finde.

But for farther triall, J triple all the *quotiente*, and
finde therebp. 111. whiche J mul=
tiplie bp thesame *quotiente* again,
and so commeth 4107. to bee the
diuisor. And his *quotiente* will bee
8. as it semeth : and so the first nõ=
ber will bee. 32856. And the se=
conde shall bee. 7104. but those. 2. are to greate, as it
is manifeste all readie.

Wherfore J take 7 for the *quotiente*. And by it mul=
tiplipng the diuisor, there riseth
28749.

And for the seconde somme,
there is founde. 5439.

And for the thirde some. 343.

All whiche. 3. sommes ioined
in one, doe make. 2929633.
And that beepng abated out of
the higher somme. 3347000,
doeth leaue. 417367.

$$\begin{array}{l}
3 \\
27\,347 \\
54000000(37 \\
9 \\
\underline{27} \\
189 \\
441 \\
\underline{343} \\
23653
\end{array}$$

$$\begin{array}{l}
3347000(378 \\
111 \\
\underline{4107} \\
32856 \\
7104
\end{array}$$

$$\begin{array}{l}
417367 \\
3347000(377. \\
111 \\
\underline{4107} \\
28749 \\
5439 \\
\underline{343} \\
2929633
\end{array}$$

Wherefore

The extraction

Wherefoze I maie boldly ſaie, that the fraction is $\frac{77}{100}$ and moze, by the poztion of the remainer, whiche is nighe $\frac{1}{700}$.

And it is ſone ſeen that $\frac{75}{100}$ are equalle to $\frac{3}{4}$: where=foze $\frac{77}{100}$ ſhall be moze then $\frac{3}{4}$.

And ſo dooeth *Scheubelius* rule erre moze, then I thought befoze.

So is your queſtion aunſwered, that the ſide of the double *Cube*, ſhall be, 3. foote and $\frac{77}{100}$ and $\frac{1}{7}$ of $\frac{1}{100}$.

Of the rootes of fractions. Maſter. Foz the rootes of fractions, I ſhall nede to ſaie no moze but this : that if the numeratoz and de=nominatoz bothe be *Squares*, oz *Cubes*. &c. then maie you finde in that fractiõ the like roote. But if any of bothe doe ſwarue from that name, then hath that fraction no ſoche roote.

As $\frac{10}{27}$ is nother *Cubike* noz *Square*, bicauſe his partes dooe not agree in *Square* name, noz in *Cubike* name : al=though the numeratoz bee a *Square*, and the denomi=natoz a *Cube*.

Scholar. That doeth appeare reaſonable, at the firſte ſighte.

Maſter. Then ſeeyng you are ſo readie in lear=nyng : aunſwere me to this queſtion.

A queſtion of a Gonne. A Gonne of ſire inches diameter in the mouthe, doeth ſhotte a bollet of twentiepound weighte : what weighte ſhall that bollette haue, that ſerueth foz a gonne of. 14. inches in the mouthe?

But to helpe you in this queſtion, and in all ſoche like, you ſhall marke well *Euclide* his ſaiyng, in the 18 pzopoſition of his. 12. booke, which is this.

All Globes here together triple that propor=tion, that their diameters doe.

So in this example, the pzopoztion of the *diameters* beyng as. 14. to. 6. Dr as. 7. to. 3. I ſhall triple it, and then haue I the pzopoztion of their Globes.

Wherefoze

of Rootes.

Wherefore I sette the, 3, fractions thus, $\frac{7}{3}$ $\frac{7}{3}$ $\frac{7}{3}$ and thei make $\frac{343}{27}$, that is, 12, $\frac{19}{27}$, And so is the proportion of the Globes, as well in weighte, as in bignesse,

Therfore I must multiplie, 20, that is the weight of the lesser bollette, by the numerator of the propor= tion, and diuide it by the denominator,

And so shall I haue, 254 $\frac{2}{27}$ for the weighte of the greater bollete,

Now prooue you the like woorke, Remembryng that Cubes also, as well as Glo= bes, doe beare triple propor= tion, in comparison of their

$$\begin{array}{r} 343 \\ 20 \\ \hline 6860 \end{array}$$

$$\begin{array}{l} 113 \\ 2412 \\ 6860(254\frac{2}{27} \\ 2777 \\ 22 \end{array}$$

sides, As you learned before by the, 19, proposition, of the, 8, booke of *Euclide*.

A question of .2. Cubes.

A *Cube* of Brasse of, 4, inches square, doeth weighe 7, pounde weighte, what shall a *Cube* of Brasse of, 9, inches square, waie?

Scholar. The proportion of the sides is as $\frac{9}{4}$ whi= che I must set doune thrise, and multiplie them toge= ther, as fractions should bee, And so will it bee thus, $\frac{9}{4}$ $\frac{9}{4}$ $\frac{9}{4}$, that maketh, $\frac{729}{64}$,

Wherefore I multiplie the weighte of the lesser *Cube*, beyng, 7, by, 729, and it maketh, 5103, and that doe I diuide by, 64, and so finde I, 79 $\frac{47}{64}$, whereby I maie knowe, that the weighte of the greater *Cube*, is 79, pounde weighte, and very nighe $\frac{3}{4}$,

Master. These, 2, questions dooe teache you, ra= ther the proportion of *Cubes*, then the vse of the rule : wherfore to make the questiōs more agreable to this rule, I propounde them thus, in backer order,

A bollette of yron of, 7, inches *diameter*, doeth waie 27, pounde weighte : what shall be the *diameter* to that bollette that shall waie, 125, pounde weighte?

Scholar. I praie you aunswer to it your self, that I maie see the apte forme of appliyng soche questions

M.i. to

The extraction

to this rule.

Master. As the *Cubes* are in triple proportion to
the sides, so are the proportions of the sides, to bee
founde by triple diuision : that is to saie, by seking the
Cubike rootes, of the. 2. termes of the proportion.

Wherefore J doe firste set doune the termes of the
proportion of the bollettes, thus : $\frac{125}{27}$. And J see, that
the *Cubike roote* of. 125. is. 5. and the like roote of. 27.
is. 3. whiche nombers J shall set in the roome of the. 2
others, thus : $\frac{5}{3}$ And thei declare the proportion, be=
twene the *diameters* of the. 2. bollettes. Whereof one
that is the lesser, is knowen to be. 7. Therfore J mul=
tiplie that. 7. by. 5. whereof commeth. 35. and that. 35.
doe J diuide by. 3. whiche giueth 11 $\frac{2}{3}$.

Wherfore J saie, that if. 7. inches bee the *diameter*
to a bollette of. 27. pounde weighte, then. 11. inches
and $\frac{2}{3}$ shall be the *diameter* to the bollete of. 125. pounde
weighte.

Scholar. The proofe of this had nede bee certain,
seeyng the woorke is obscure, to the common iudge=
mente.

The proofe.

Master. You saie well. And this is the very or=
der of proofe for it. Multiplie bothe these rootes *Cu=*
bikely. And if their *Cubes* be in soche proportiō as their
waightes bee (that is to saie in this exāple as $\frac{125}{27}$) then
is the woorke good : els not.

Scholar. That must neades bee so. And therefore
will J proue it so in these nombers.

And for that eande, firste J multiplie. 7. *Cubikely*,
and it giueth. 343. Then J multiplie. 11 $\frac{2}{3}$. *Cubikely*,
and it maketh $\frac{42875}{27}$. But now seyng the one nomber
is a fraction, J will for ease tourne the other into a
fraction of thesame denomination : and it will bee $\frac{9261}{27}$
in whiche. 2. fractions, the proportion muste cōsist be=
twene the numeratours. So that thei bothe beeyng
diuided by one common nomber, muste come to this
                                                        fraction

of Rootes.

fraction $\frac{125}{27}$.

And ſo I ſee it will be : for the leſſer beyng diuided by. 343. will yelde 27. And the greater diuided by the ſame. 343. will giue. 125. So that by triall, that woorke is approued good.

Maſter. I will now proue your cunnynge, in a newe queſtion, whiche Braſiers often tymes, haue occaſion to vſe : as thus.

I haue a dice of Braſſe of. 64. vnces of Troye weighte, whoſe ſide is. 3. inches and $\frac{1}{2}$ and would haue an other dice of theſame mettall of. 18. pounde weighte. *A queſtion of weightes.*

My demaunde is : what ſhall be the ſide of the dice?

Scholar. This queſtion muſt firſte bee reduced to one kinde of denomination in the weightes, and then will it be more apte to be aunſwered.

Wherefore I ſhall tourne. 18. pounde into vnces, multipliyng it by. 12. and it will be. 216.

And then I conſider the proportion, that is be=twene thoſe. 2. nombers of weighte. 64. and. 216. and it is certainly. $3\frac{3}{8}$, or $\frac{27}{8}$ out of whiche proportion, I muſt extracte the *Cubike* roote, as I maie eaſily dooe, ſeyng bothe the numerator and the denominator, are *Cubike* nombers.

And ſo is their roote $\frac{3}{2}$: whiche is the proportion of the ſides of the twoo dice.

And ſeyng the ſide of the leſſer die, is knowen to be 3. inches and $\frac{1}{2}$, the other his ſide muſt be in *Seſquialter* proportion to it, that is. $5\frac{1}{4}$: whiche is wroughte alſo thus. I multiplie. $3\frac{1}{2}$ by. 3. and it maketh. $10\frac{1}{2}$ whiche I ſhall diuide by. 2. and there commeth. $5\frac{1}{4}$.

Maſter. Yet one queſtion more I will propounde to giue you occaſion, to vnderſtande the apte confe=rence of maſſes, of diuerſe ſtuffe.

And for that purpoſe, I ſuppoſe this proportion in weighte, to bee betwene maſſes of one biggeneſſe.

D.ii.   That

The extraction

Examples of rates for weightes.

That if I compare Woodde and stone of one quantitie together, the stone shall weighe more then the woodde by $\frac{2}{3}$.

Stoffe.	Weighte.				
Woodde.	60.	1			
Stone.	100	$\frac{3}{5}$	1		
Yron.	150	$\frac{2}{5}$	$\frac{2}{3}$	1	
Brasse.	200	$\frac{3}{10}$	$\frac{1}{2}$	$\frac{3}{4}$	1
Ledde.	280	$\frac{3}{14}$	$\frac{5}{14}$	$\frac{15}{28}$	$\frac{5}{7}$

Likewaies yron to be heuier then stone by $\frac{1}{2}$.

And Brasse to bee heuier then Yron by $\frac{1}{3}$.

Ledde to be heuier then Brasse by $\frac{2}{5}$.

All whiche rates, although thei be taken for examples, and not of truthe, yet thereby maie you learne, how to woorke with true rates, set in a like table.

And now for the vse of this table, take this questiõ.

A question of weighte.

I would haue. 5. weightes of *Cubike* forme, made of these. 5. stuffes.

The weighte of the woodde shall be. 28. pounde.

The stone. 56. pounde.

The yron. 112. pounde.

The Brasse. 224. pounde.

And the Ledde. 448. pounde.

Of all these I haue but the Yron weighte : whose side, or *Cubike* roote is. 12. inches $\frac{2}{3}$.

And my desire is to knowe, of what quantitie the sides of all the other weightes shall bee.

Scholar. The question is pleasaunt : and yet some what harder then the other.

Master. The table will helpe you fully, so that you côferre it well, with that you haue learned before

But bicause I haue litle leiser, to spende moche tyme with you (saue that zeale to your furtheraunce doeth make me partly to forgette my owne businesse) therefore will I leaue this question to your self, to be aunswered at your laisure.

And so in all the rest, I must poste it ouer : and giue an iye to soche maters, that touche me more nighe :

and

of Rootes.

and weighe moze heuily, then all foche weightes, by 20. folde.

Wherfoze, touchyng all the rootes of compounde nombers, you shall at my hand now, haue no pziuate declaration. But foche as you haue learned all redie.

Of compounde rootes.

 F the nomber bee com= pounde, other of *Square nom= bers*, oz of *Cubike nombers*, then accozdyngly as the copofition is, fo shal you draw the roote : and without one of thefe two there can bee no compofition.

Wherefoze to begin with the fmalleft compounde nom= ber in that fozte, whiche is a *Square of fquares*, you shall firfte extracte the fquare roote, as you haue lear= ned befoze. And out of that roote (whiche muft nedes bee a *Square nomber*) you shall extracte his fquare roote alfo : and that roote is the *zenzizenzike roote*, of the firfte *Square of fquares oz zenzizenzike nomber*.

Squares of Squares.

Foz example take. 14641. whofe *Square roote* is 121. and that fame roote is it felf, a *Square nomber* : and hath foz his roote. 11.

```
        2
    14641(2916
    224
```

Wherfoze J maie faie, that. 11. is the *Squared fquare roote*, oz the *zenzizenzike roote* of 14641.

Again 8503056. is a *Square of fquares*, and therfoze a *Square nomber*. And his *Square roote* is 2916. whiche is a *Square nomber* alfo, and hath. 54. foz

```
    4
2916(54
    10
```

```
    341
    499593
    8503056(2916.
    45882
    5
```

D.iii.　　　　　his

The extraction

his roote.

So that. 54. maie well bee called the *zenzizenzike root* of. 8503056.

And so shall you woorke, with all of that name.

Zenzizen=
zizenzikes.

But and if the nomber be compounde, of. 3. *zenzi=
kes*, oz. 3. *Squares*, as a *Square of squared squares*, oz a *zen=
zizenzizenzike* (whiche some men foz shoztnesse, call
zenzizenzenzike). Then shall you drawe firste the
Square roote, and then the *Square roote* of that roote, and
thirdly the *Square roote* of that laste roote.

As foz example. 6561. is a *Square of
squared squares*. And his firste roote is. 81.
whiche is also a *Square nomber*, and hath
9. foz his roote. That. 9. likewaies is a
Square nomber, and hath. 3. foz his roote.

$$6561(81 \\ 16$$

So that the *zenzizenzizenzike roote* of. 6561. is. 3.

And foz these foznes of nombers, I shall not nede
to state foz any moze explication, oz examples : seeyng
the mater is plaine.

Now foz compounde *Cubike nombers*, you shall vn=
derstande the like fozme.

Cubes of
cubes.

If the nomber bee a *Cube of Cubes*, you shall firste ex=
tracte the *Cubike roote*. And bicause that roote is a *Cu=
bike nomber* also, therfoze shall you seke the *Cubike roote*
of it. And that seconde roote shall bee the *Cubicubike
roote* of the firste nomber.

As foz example. 512. is a *Cubike nomber*, oz a *Cube of
Cubes*. And his *Cubike roote* is. 8. whiche. 8. againe is a
Cubike nomber, and hath. 2. foz his roote.

So that. 2. is the *Cubicubike roote* of. 512.

Likewaies. 10077696. is a *Cubicubike nōber*, and
his firste *Cubike roote* is. 216. as you maie easily per=
ceiue by these woozkes : where I haue sette fozthe the
ozder of extraction of his *Cubike roote*, whiche is. 216.
And that. 216. is a *Cubike nomber*, you neade not to
doubte,

of Rootes.

```
        816
10077696(216.
        63
      1323
      7938
      2268
       216
    816696
```

doubte, for that it is one of
the, which you
haue, I dare
saie, in perfecte
memorie: Bi=
cause his roote
is a digite, and
that is. 6.

```
 2816
10077696(21
     6
    12
  1261
  1261
```

By this you maie iudge of *Cubicubes Cubikely, or Cubes of Cubicubes.* that in theim you shall firste seke their *Cubike* roote : And then the *Cubike* roote of that roote. And thirdly the *Cubike* roote of that roote againe. And so haue hou the *Cubicubicubike* roote of that firste nomber. *Cubicubes Cubikely.*

The third waie of composition is, when *Squares* and *Cubes* be compounde together : as *Zenzicubes, Zenzizenzicubes, Zenzicubicubes,* or soche like, as it happeneth diuersely. *The thirde composition.*

In all these you shall as often abate the *Zenzike* roote, as that name is in the composition, and so like= waies of the *Cubike* roote.

So that in a *Zenzicubike,* you shall extracte firste the *Square* roote : and out of that *Square* roote, you shall ex= tracte the *Cubike* roote. *Zēzicubike.*

As. 64. is a *Zenzicubike nomber,* whose *Square* roote is 8. and that. 8. is a *Cubike nomber,* and hath. 2. for his roote.

So. 531441. is a *Zēzizenzicube* : whose firste *Square* roote is. 729. whiche nomber is a *Zenzicube,* & hath for his *Square* roote. 27. And that nō= ber is a *Cube,* and hath for his roote. 3. where= fore I maie iustly saie, *Zenzizenzicube.*

```
 34
729(27
 4
```

```
  144
  430
531441(729
 1444
  1
```

that. 3. is the *Zenzizenzicubike* roote of. 531441.

But as I saied before, that I might not staie long
at

The extraction

at this p͛efente, ſo the vſe of theſe greate nombers is
rare in p͛actiſe : and therefo͛e I will ouerpaſſe them,
fo͛ this tyme.

And yet fo͛ your aied in the meane ſeaſon, I haue
here drawen a table, whiche maie bee called the table
of eaſe : in whiche you haue greate plentie of theſe
nombers, with their rootes in diuerſe kindes.

The table it ſelf iſ ſo manifeſte, that it neadeth no
declaration : if you haue not fo͛gotten, what you lear=
ned befo͛e.

And if you liſte to enlarge this table, you maie ea=
ſily doe it, multiplipng the nombers ſtill by their roo=
tes, whiche bee ſet ouer theim, in the hedde of the ta=
ble. And ſo maie you make it to extende infinitely :
whiche ſhall eaſe you wonderfully, in the extraction
of any kinde of rootes. Fo͛ which at ſome other time
if my leiſure ſerue me better, with quietneſſe, I will
giue you mo͛e ſpecialle rules.

And alſo I councell you, well to examine this ta=
ble, and truſt not to my caſtynge. Fo͛ haſte and
other troubles, maie often times cauſe
erroure in ſupputation.

                                                    ❡ The

In the original 1557 edition of this book, the table replicated across the
following two pages was printed on a single folded pull-out page.
A facsimile of the folded page is available for download at
www.renascentbooks.co.uk

		2	3	4	5	6	7	8	9	10	11
1	Rootes.	2	3	4	5	6	7	8	9	10	11
2	Squares.	4	9	16	25	36	49	64	81	100	121
3	Cubes.	8	27	64	125	216	343	512	729	1000	1331
4	Squares of squares.	16	81	256	625	1296	2401	4096	6561	10000	14641
5	Surfolides.	32	243	1024	3125	7776	16807	32768	59049	100000	161051
6	Squares of Cubes.	64	729	4096	15625	46656	117649	262144	531441	1000000	1771561
7	Seconde Surfolides.	128	2187	16384	78125	279936	823543	2097152	4782969	10000000	19487171
8	Squares of squared fares	256	6561	65536	390625	1679616	5764801	16777216	43046721	100000000	21446888i
9	Cubes of Cubes.	512	19683	262144	1953125	10077696	40353607	134217728	387420489	1000000000	235795691
10	Squares of Surfolides.	1024	59049	1048576	9765625	60466176	282475249	1077741824	3486784401		
11	C.Surfolides.	2048	177147	4194304	48828125	362797056	1977326743	8589934592			
12	Squares of zenzicubes.	4096	531441	16777216	244140625	2176662736					
13	D.Surfolides.	8192	1594323	67108864	1220703125	7059974016					
14	Squares of Bsurfolides.	16384	4782969	268435456	6103515625						
15	Cubes of Surfolides.	32768	14348907	1077741824							
16	Zenzizenzizenzizenzikes.	65536	43046721	4294967296							
17	Esurfolides.	131072	129140163								
18	Squares of Cubicubes.	262144	387420489								
19	Fsurfolides.	524288	1162261467								
20	Zenzizenzisursolides.	1048576	3486784401								
21	Cubes of Bsurfolides.	2097152									
22	Squares of Csurfolides.	4194304									
23	Gsurfolides.	8388608									
24	Zenzizenzizezicubes.	16777216									

		25	26	27
1.	Rootes.	25	26	27
2.	Squares.	625	676	729
3.	Cubes.	15625	17576	19683
4.	Squares of Squares.	390625	456976	531441
5.	Surfolides.	9765625	11881776	14348907
6.	Zenzicubes.	244140625	308915776	387420489
7.	Bsurfolides.	6103515625	9037801776	9075523707

R.i.

be called the table of ease.

12	13	14	15	16	17	18	19	20	21	22	23	24
144	169	196	225	256	289	324	361	400	441	484	529	576
1728	2197	2744	3375	4096	4913	5832	6859	8000	9261	10648	12167	13824
20736	28561	38416	50625	65536	83521	104976	130321	160000	194481	234256	279841	331776
248832	371293	537824	759375	1048576	1419857	1889568	2476099	3200000	4084101	5153632	6436343	7962624
2985984	4826809	7529536	11390625	16777216	24137569	34012224	47045881	64000000	85766121	113379904	148035889	191102976
35831808	62748517	105413504	170859375	268435456	410338673	612220032	893871739	1280000000	1801088541	2494357888	3404825447	4586471424
429981696	815730721	1475789056	2562890625	4294967296	6975757441	11019960576	16983563041					
5159780352	10604499373											

28	29	30	31	32	33	34	35	36	37	38	39	40
784	841	900	961	1024	1089	1156	1225	1296	1369	1444	1521	1600
21952	24389	27000	29791	32768	35937	39304	42875	46656	50653	54872	59319	64000
614656	707281	810000	923521	1048576	1185921	1336336	1500625	1679616	1874161	2085136	2313441	2560000
17210368	20511149	24300000	28629151	33554432	39135393	45435424	52521875	60466176	69343957	79235168	90224199	102400000
481890304	594823321	729000000	887503681	1073741824	1291467969	1544804416	1838265625	2176782336	2565726409	3010936384	3518743761	4096000000
13492928512	17249876309	21870000000	27512614111									

of Cosike nombers.
Of nombers denominate.

Nombers contraƐte.

Hus haue J lightly ouer run the moſte common kindes of nombers *AbſtraƐte.* And now reſteth the treatice of nom=bers *ContraƐte,* o₂ *Denominate.* Of whiche kinde there bee ſome called *nombers deno=minate vulgarely* : and other bee called nombers *denomi=nate Cosikely.* And a thirde ſo₂te there is of nombers *radicalle,* whiche commonly bee called *nombers irratio=nalle* : bicauſe many of theim are ſoche, as can not bee ex₂reſſed, by common nombers *AbſtraƐte,* nother by any certain rationalle nomber. Other men call them mo₂e aptly *Surde nombers.*

And although many menne would not accoumpte them, with nombers *denominate,* yet J maie iuſtly doe it, fo₂ that thei require a reduction to one denomina=tion, if thei haue ſeueralle ſignes of quãtities, as you ſhall heare hereafter. And thoſe nombers neuer goe alone, without ſome other ſigne, and name of rooted quantitie, annexed to theim.

Of the firſt kinde of nombers denominate, whiche are vulgarely denominate, as. 10. ſhillinges. 10. men 20. ſhippes, 100. ſhepe. 1000. yeres, and ſoche like, J will ſpeake nothyng in this treatice. But of the o=ther twoo kindes J will ſomewhat w₂ite, fo₂ youre learnyng and contentation.

Scholar. Sir, J am moche bounde vnto you : And therefo₂e remit all to your owne diſcretion and good will. Truſtynge ſo to applie my ſtudie, and emploie my knowlege, that it ſhall neuer repente you of your curteſie in this behalfe.

Maſter. Then marke well my wo₂des, and you ſhall perceiue, that J will vſe as moche plaineſſe, as J maie, in teachyng : And therfo₂e will beginne with *Cosike* nombers firſt.

S.i. Of

The Arte
Of Cossike nombers.

Ombers *Cossike*, are soche as bee contracte vnto a denomination of some *Cossike* signe as 1. nomber. 1. roote. 1. square 1. Cube. &c.

But as for cōpendiousnesse in the vse of theim, there bee certain figures set for to signifie them : so I thinke it good to expresse vnto you those figures, before wee enter any farther, to thintente we maie procede alwaies in certentie, and knowe the thynges that wee intermedle withall : for thei are the signes of all the arte, that foloweth here to be taught.

And although there be many kindes of irrationall nombers, yet those figures that serue in *Cossike* nōbers, bee the figures also of all irrrtionalle nombers, and therfore being ones well knowen, thei serue in bothe places commodiously.

These therfore be their signes, and significations briefly touched : for their nature is partly declared before.

ꝑ.	Betokeneth nomber absolute : as if it had no signe.
℞.	Signifieth the roote of any nomber.
ʒ.	Represeteth a square nomber.
ᴄᴇ.	Expresseth a Cubike nomber.
ʒʒ.	Is the signe of a square of squares, or Zenzizenzike.
ſʒ.	Standeth for a Surſolide.
ʒᴄᴇ.	Doeth signifie a Zenzicubike, or a square of Cubes.
bſʒ.	Doeth betoken a seconde Surſolide.
ʒʒʒ.	Doeth represent a square of squares squared

ly,

of Cossike nombers.

ly, oz a *zenzizenzizenzike.*

cₑcₑ.   Signifieth a *Cube* of *Cubes.*

ʒ∫ʒ.   Expresseth a *Square* of *Surolides.*

cʒ.   Betokeneth a thirde *Surolide.*

ʒʒcₑ.   Represententh a *Square* of *Squared Cubes*: oz a *Zenzizenzicubike.*

ᴅʒ.   Standeth foz a fourthe *Surolide.*

ʒ ˢᵇʒ.   Is the signe of a *square* of seconde *Surolides.*

cₑ∫ʒ.   Signifieth a *Cube* of *Surolides.*

ʒʒʒʒ.   Betokeneth a *Square* of *squares*, squaredly squared.

ᴇʒ.   Is the fifte *Surolide.*

ʒcₑcₑ.   Expresseth a square of *Cubike Cubes.*

ꜰʒ.   Is the sixte *Surolide.*

ʒʒ∫ʒ.   Doeth represente a square of squared surolides.

cₑᵇʒ.   Standeth foz a *Cube* of seconde *Surolides.*

ʒcʒ.   Is a square of thirde *Surolides.*

gʒ.   Doeth betoken the seuenthe *Surolide.*

ʒʒʒcₑ.   Signifieth a square of squares, of squared Cubes.

And though I maie proceade infinitely in this sozte, yet I thinke it shall be a rare chaunce, that you shall nede this moche : and therfoze this maie suffice. Notwithstandynge, I will anon tell you, how you maie cõtinue these nombers, by progression, as farre as you liste.

And farther you shal vnderstande, that many men doe euer moze call square nombers *zenzikes,* as a shoz ter and apter name, other men call those squares the *firste quantities,* and the *cubes* thei call *seconde quantities,* squares of squares thei call *thirde quantities,* and surso lides *fourthe quantities.* And so namyng them all quan tities (excepte nombers and rootes) thei dooe adde to theim foz a difference, an ozdinall name of nomber, as thei doe goe in ozder succesiuely.

The Arte

As here foloweth in example.

ȝ.	Firſte.	
ꝯ.	Seconde.	
ȝȝ.	Thirde.	
ſȝ.	Fourthe.	
ȝꝯ.	Fifte.	
bſȝ.	Sixte	
ȝȝȝ.	Seuenthe.	*Quantities.*
ꝯꝯ.	Eighte.	
ȝſȝ.	Nineth.	
cſȝ.	Tenthe.	
ȝȝꝯ.	Eleuenthe	
dſȝ.	Twelfthe.	

And ſo forthe, of as many as maie bee reckened.

But althoughe ſomemen accompte this the moꝛe eaſie waie : bicauſe the other names be comberouſe, yet thoſe other names befoꝛe, do expꝛeſſe the qualitie of the nomber, better then theſe later names doe.

Scholar. I thanke you double, ſith you are contente to teache me double names : foꝛ ſo ſhall I be acquainted with bothe foꝛmes, as I ſhall chaunce on them in other mennes bookes.

Therfoꝛe now you maie pꝛoceade to numeration : whiche I thinke it nexte.

Maſter. There be other. 2. ſignes in often vſe, of whiche the firſte is made thus——┼——and betokeneth moꝛe : the other is thus made———and betokeneth leſſe.

And where thei come in any nomber *Coſsike*, oꝛ other, that nomber is called a compounde nomber, bicauſe it conſiſteth of. 2. nombers. And where neither of theim is, the nomber is called vncompounde, although the ſigne be compounde. Foꝛ the compounde ſigne, maketh not a compounde nomber. And now I will pꝛocede to numeration.

Of

of Cofsike nombers.

Of Numeration in nombers

Cofsike, vncompounde.

Master.

Ombers *Cofsike* vncompounde, haue no *Numeration*
difficultie in their numeration : for euer
more the nōber repzefenteth, ſo many of
that *Cofsike* denominatiō (be thei nōbers,
rootes, ſquares, Cubes, ſquares of ſqua=
res, oz any other like) as ther be vnities in that nōber.

So. 6.9. is. 6. nombers : And. 6. ℞ . is. 6. rootes :
20. ℥. is. 20. ſquares : 30. ℭ. betokeneth. 30. Cubes.

Scholar. J ſee it well. For by this nōber. 20. ℥. is
not appoincted any nōber abſolute, of one certaintie,
but onely ſo many *quātities* of that kinde : whiche maie
bee. 80. if. 4. be one ſquare. And if. 9. bee one ſquare,
then 20. ſquares make 180. And if. 25. be one of thoſe
ſquares thereby repzefented, then. 20. ſquares make
500. And as for the ſignes, you taught me thē befoze.

Of Addition.

Master.

His numeration is ſo plaine, that wee   *Addition of*
maie paſſe from it vnto addition : whi=   *like ſignes.*
che is as eaſie alſo, if the quantities be
of one denomination. For then nedeth
no moze, but to adde the nombers to=
gether, and to put that ſame common
Cofsike denomination, to the totall thereof.

Scholar. J take it thus, 20. ℞. added to. 30. ℞.
will make. 50. ℞. And. 12. ℥. added to. 16. ℥. bzyn=
geth forthe. 28. ℥.

Master. As you doe eaſily ſee al the mater of this
addition, ſo maie you as eaſily conceiue, all the wozke   *Subtraction*
of ſubtractiō. For it is wzought as in vulgare nōbers   *of like ſignes*

The Arte

Scholar. Then if J abate. 6. ℞. out of. 10. ℞. there will reſte 4. ℞. And ſo. 9. ℥℥. out of. 25. ℥℥. doeth leaue. 16. ℥℥.

Maſter. This is all for nombers of like ſignes *Coſsike.*

Scholar. What then if J would adde. 10. ℞. to 6. ℥. ? where the ſignes bee vnlike? maie it be doen? ſeyng thei be not of one denominatiõ, nor ſigne *Coſsike.*

Addition of vnlike ſignes.

Maſter. As well as ſhillynges maie bee added with poundes, or penies : and in like forme.

For thei ſhall ſtand ſtill as thei wer, with the ſigne of addition, whiche is this. ——— . & betokeneth more.

So that. 10. ℞. put to. 6. ℥. maketh. 6. ℥. ——— 10. ℞. that is. 6. ℥. more. 10. ℞. or. 6. ℥. and. 10. ℞.

Scholar. And why not. 10. ℞. ——— 6. ℥ ?

Maſter. Bicauſe it is moſte orderly, to ſette the greateſte ſigne *Coſsike,* formoſte in order.

As you ſaie. 20. ſhillynges, and. 6. pennies : rather than. 6. pennies and. 20. ſhillynges.

Scholar. Then J ſe, if. 15. ℞. be added to. 18. ℥℥ it will make. 18. ℥℥. ——— . 15. ℞. An ſo. 12. ſ℥. ioyned with. 20. ℥ ℞. dooe make. 20. ℥ ℞. ——— 12. ſ℥.

Of Subtraction.

Maſter.

Subtractiõ of vnlike ſignes.

 Vbtraction is as eaſie : for it doeth depend onely of the ſigne of abatemente, which is this. ——— . and ſignifieth leſſe, or abatyng. And therefore if J would abate 6. ℞. out of. 10. ℥. J muſt ſette it thus

10. ℥. ——— . 6. ℞ : that is to ſaie. 10. ℥. leſſe. 6. ℞. or abatyng. 6. ℞.

Scholar. Then if J haue 30 ℞. and would abate out of thē. 12. 9. J muſt ſet it thus. 30. ℞ ——— . 12 9. that is. 30. *cubes* ſaue. 12. nombers. And if multiplica-

                                                                                    tion

of Cofsike nombers.

tion and diuiſion, bee as eaſie, thei ſhall neade no
greate ſtudie.

Of Multiplication.

Maſter.

Omewhat moꝛe laboure is there *Multiplica-*
in multiplication and diuiſion, to *tion.*
finde out the newe ſignes as J wil
tell you anon. But foꝛ findyng of
the nombers, the common multi=
plication and diuiſion doeth ſerue.
So that when. 12. ⅔. is multi=
plied by. 6. ℞ . it maketh. 72. ℮ . And if. 24. ℮ . bee
multiplied by. 5. ⅔ . there riſeth. 120. √⅔.

Scholar. This paſſeth my cunnynge, foꝛ the fin=
dyng of the newe ſigne : although the multiplication
of the nombers, be as eaſie as can be.

Maſter. Jf you did well remēber, what you haue
learned befoꝛe : the mater would not ſeme ſo harde.

Doe not you knowe, that a roote multiplied by a
roote, doeth make a ſquare? And a ſquare multiplied
by his roote, doeth bꝛyng foꝛthe a cube?

Scholar. That J knowe right well : and therfoꝛe
a *Square of Squares* multiplied by his roote, will yelde
a *Surſolide.*

Maſter. Then by like reaſon, a *Cube* multiplied
by a *Square,* ſhall make a *Surſolide.*

Scholar. Jn deede it is all one, to multiplie a *cube*
by a *Square,* and a *Square of Squares* by a roote.

Maſter. Then foꝛ a generalle rule, J will ſette
foꝛthe here a pꝛeſidente foꝛ you : whereby you maie
knowe the newe ſigne, in all multiplication oꝛ diuiſi=
on : not onely by ſight very ſpedily, but that you maie
alſo commit it aptly to memoꝛie.

Wherfoꝛe marke wel this table folowing : where
you ſee in the higher rowe, a line of nombers, ſet in
naturall

The Arte

naturall progreſſion : and vnder them you ſee the ſi=
gnes of *coſsike nombers.*

The table of Coſsike ſignes,
and their peculier nombers.

0.	1.	2.	3.	4.	5.	6.
ꝗ.	℥.	℥.	℀.	℥℥.	ſ℥.	℥℀.

7.	8.	9.	10.	11.	12.	13.
bſ℥.	℥℥℥.	℀℀.	℥ſ℥.	cſ℥.	℥℥℀	dſ℥.

This table is largely ſet forthe, in the title of pro=
greſſion, whereunto you maie haue recourſe, if your
nomber be to greate for this table.

By this table maie you reaſily knowe, the ſigne
that ſhall ſerue for your newe ſomme, in multiplica=
tion.

As for example, if I dooe multiple ſquares by roo=
tes : I looke in the table, what nombers ſtande ouer
them bothe, and puttyng thoſe. 2. nombers together,
I ſeke the totall in the ſame line, and vnder it I finde
the newe denomination *coſsike*, whiche I ſhould haue

Scholar. I perceiue ouer. ℥. the nomber of. 1. and
ouer. ℥. the nomber. 2. whiche bothe added toge=
ther make. 3. And bicauſe vnder. 3. I find the figure
or ſigne of. ℀. I muſte take that for the newe deno=
mination.

Maſter. You ſaie truthe.

Scholar. Then if I multiplie. 12. ℥℀. by. 8. ℀.
the ſomme will be. 96. ℀℀. For ouer. ℀. I finde
3. and ouer. ℥℀. ſtandeth. 6. whiche bothe together
doe make. 9. and vnder. 9. I ſee. ℀℀. whiche I take
for the denominator.

And if the ſame rule bee generall, I am cunnynge
inoughe

of Coſsike uombers.

inoughe in it.

Maſter. It is generall, foꝛ multiplication in this kinde.

Of Diuiſion.

Diuiſion.

At foꝛ diuiſion, you muſte abate the one nomber out of the other, to finde a newe denomination.

Therfoꝛe if you would diuide. 96.℀℀ by. 8.℀. the *quotiente* will be. 12.ʒ℀. bi= cauſe that ouer the ſigne of your diuidende, ſtandeth 9. And ouer the diuiſoꝛs ſigne is ſet 3. Wherfoꝛe aba= typng. 3. from. 9. there reſteth. 6. vnder whiche is the ſigne. ʒ℀. that I muſt take, to put to my *quotiente.*

Scholar. Then foꝛ an other triall, if I would di= uide. 260.℮ſʒ. by. 5.ſʒ. the *quotiët* will be. 52.ʒʒ℀. Foꝛ bicauſe that ouer. ℮ſʒ. I finde. 17. and ouer. ſʒ. ſtandeth. 5. then ſubtractyng. 5. frö. 17. there reſteth. 12 vnder whiche in the table I finde. ʒʒ℀.

So diuidyng. 20. ℀. by. 4.℈. the *quotiente* will bee 5.℀ : and ſo of other.

Maſter. But and if you would diuide. 12.℀. by 5. ʒ. that muſt be ſet in foꝛme of fraction, thus. $\frac{12℀}{5ʒ}$.

So. 18.ʒ. by. 7.ℨ℮. maketh. $\frac{18ʒ}{7ℨ℮}$ and. 6.ʒ. by. 2.℀. yeldeth. $\frac{6ʒ}{2℀}$. of whiche fractions, wee will ſpeake e= mongeſte the fractions of *Coſsikes* compoſtde. Foꝛ thei degenerate out of this kinde.

Wherefoꝛe this maie ſuffice bꝛiefly, foꝛ the cuſto= mable woaꝛkes of whole *Coſsike nombers.*

Of Fractions in Coſsike nombers.

Of fraction in nombers Coſsike.

And as foꝛ fractions, the woꝛkyng is like in euery poincte, vnto the woꝛke of nom= bers *Abſtracte* : remembꝛing onely that as thoſe bꝛoken nombers, haue a *Coſsike* de= nomination annered with them, ſo muſt

T.i.   that

The Arte

that denomination followe the rules, now laſte de=
clared.

Wherefore J ſhall not nede to doe any moze, but
to ſet fozthe onely certain examples, of euery kinde of
woozke in them.

Examples of Numeration.

$\frac{2}{5}$ ℞ . Signifieth $\frac{2}{5}$ of a *Roote.*

$\frac{8}{9}$ ✗ . Betokeneth $\frac{8}{9}$ of a *Square.*

$\frac{12}{17}$ ℭ . Repzeſenteth $\frac{12}{17}$ of a *Cube.*

And ſo of all other fozmes of *Coſsike* ſignes : where=
by is intended, that the *Coſsike quantitie,* is diuided in=
to ſo many partes, as the denominatoz containeth,
and there is here repzeſented onely ſo many of them,
as the numeratoz doeth impozte.

Scholar. Hereby J dooe perceiue, that a fraction
Coſsike, maie ſignifie a nomber, and not onely a parte
of an vnitie, as it did in nombers *Abſtracte.*

Foz when J ſaie $\frac{2}{3}$ ✗ . if that *Square* be. 9. then that
fraction ſignifieth. 6. But if the *Square* be. 4. then that
fraction doeth repzeſente. $2\frac{2}{3}$.

Likewaies $\frac{3}{4}$ ℭ . if the *Cube* be. 8. then that fraction
doeth ſignifie. 6. But if the *Cube* be. 27. then that frac=
tion is equalle to. $20\frac{1}{4}$.

Maſter. You doe conſider it well.

Of Addition.

Addition.

Now foz addition, take theſe examples.

$\frac{2}{3}$ ✗ . added to $\frac{3}{4}$ ✗ . doe make $\frac{17}{12}$ ✗ . oz. 1 ✗ $\frac{5}{12}$.

$\frac{5}{7}$ ℭ ioined with $\frac{7}{8}$ ℭ . doe make $\frac{89}{56}$ ℭ . oz 1, ℭ $\frac{33}{56}$.

And in vnlike ſignes.

$\frac{3}{4}$ ✗ . added to $\frac{4}{5}$ ℭ . doe make $\frac{4}{5}$ ℭ . ⊢ $\frac{3}{4}$ ✗ oz els
thus by one common denomi=
natoz.

16. ℭ ⊢ 15. ✗ .
20.

Of

of Cossike nombers.

Of whiche I will speake moze in the *Binomialles*, and therefoze will omitte it, till we come to them.

Scholar. As foz the reste, I see it well : Foz the woozke is all one with fractions *Abstracte*.

And here the denominatiõ oz *Cossike* signe is not va= ried, although here be vsed diuerse multiplications.

Master. And good reason : foz the whole *quotiente* whiche is repzesented by that *Cossike* signe, is not mul= tiplied, but certaine partes of it : and therefoze oughte that *Cossike* signe, to stand vnaltered, as the quantitie repzesented by it, is not multiplied noz altered.

Examples of Subtraction.

$\frac{2}{3}$ ce. abated out of $\frac{3}{4}$ ce. doe leaue $\frac{1}{12}$ ce.

$\frac{5}{9}$ ʒ. out of $\frac{7}{8}$ ʒ. there resteth $\frac{22}{72}$ ʒ.

$\frac{8}{15}$ ʒʒ. subtracted frõ $\frac{4}{5}$ ʒʒ. doe leaue $\frac{20}{75}$ ʒʒ, oz $\frac{4}{15}$ ʒʒ.

And in vnlike signes.

$\frac{3}{8}$ ʒ ce abated frõ $\frac{6}{11}$ cece doe leue $\frac{6}{11}$ cece —— $\frac{3}{8}$ ʒ ce.

$\frac{3}{4}$ ʒ taken out of $\frac{9}{13}$ ce. the reste is $\frac{9}{13}$ ce —— $\frac{3}{4}$ ʒ.

Likewaies as in additiõ, so in this sozte of subtrac= tion, there maie be an other kinde of woozke, whiche I will remit to the treatice of *Binomialles*.

Examples of Multiplication.

$\frac{3}{5}$ ʒ multiplied by $\frac{3}{8}$ ʒ. doe make $\frac{9}{40}$ ʒʒ.

$\frac{5}{7}$ ce. multiplied by $\frac{8}{9}$ ce. bzyngeth fozthe $\frac{40}{63}$ ʒ.

$\frac{12}{19}$ ʒ. multiplied by $\frac{2}{3}$ ʒ. doe yelde $\frac{24}{57}$ ʒʒ, oz $\frac{8}{19}$ ʒʒ.

Here the signes doe alter, as in the multiplication of whole *Cossike* nombers.

T.ii.   Scholar.

The Arte

Scholar. This doeth somewhat trouble me : that the *Coſsike* ſignes ſhould chaunge here, rather then in addition, oʒ ſubtraction : Seyng there was as moche multiplication in any of them bothe, as there is here.

Maſter. Marke the mater well, and you ſhall bee ſone ſatiſſied.

Foʒ in addition and ſubtraction, the multiplicatiõ ſerueth onely foʒ the reduction of the. 2. fractions, vn= to one denomination : And therefoʒe in them, you ne= uer multiplie the numeratoʒs together : but you mul= tiplie croſſe waies, the numeratoʒ of the one, by the denominatoʒ of the other, where as in multiplicatiõ, you vſe no reduction, but doe make a plaine multipli= cation.

And ſo likewaies in diuiſiõ, there is vſed no meane of reduction : and therefoʒe in it the ſignes muſt alter, as befoʒe is declared.

Examples of Diuiſion.

$\frac{6}{7}$ᵹᵹ . diuided by $\frac{6}{11}$ᵹ . doe make in the *quotiente* $\frac{66}{42}$ᵹ . oʒ $\frac{11}{7}$ᵹ .

$\frac{7}{9}$℞ . diuided by $\frac{8}{15}$℞ . doeth yelde $\frac{105}{72}$९ . oʒ els $\frac{35}{24}$.

Foʒ ſeyng J ſhall diuide. ℞. by. ℞. J muſt there= foʒe abate. 3. from. 3. and ſo reſteth nothing, whiche is ſigniſied by this Cipher. o. and that ſtandeth ouer the ſigne of nomber : therefoʒe the fraction, that is as the *quotiente*, muſt be taken as a nomber *Abſtracte*.

Likewaies $\frac{8}{3}$ᵹᵹ . diuided by $\frac{8}{9}$ᵹᵹ . doeth make $\frac{72}{24}$९ . that is to ſaie.3. And ſo $\frac{8}{15}$℞℞. diuided by $\frac{9}{10}$℞. doeth bʒyng foʒthe $\frac{80}{135}$ᵹ℞ , oʒ $\frac{16}{27}$ᵹ℞ .

Scholar. That is ſuſſiciente foʒ diuiſion. Now if you thinke good to ſpeake of pʒogreſſion, J can not but remember you of your pʒomiſe.

Of

of Cossike nombers.
Of Reduction.

Master.

Lthough *Reduction* ſhould go in oꝛder be=   *Reduction.*
foꝛe *Progreſsion*, yet ſeeyng this *Reduction*,
conſiſteth in the onely nombers, and not
in the ſignes : and therefoꝛe agreeth with
vulgare reduction of fractions (as here
you maie ſee befoꝛe in diuerſe examples) therfoꝛe wil
we omitte it, and go in hande with *Progreſsion* : whiche
is moꝛe ſtraunge.

Scholar. J pꝛaie you ſo : Foꝛ J ſee this reduction,
is but to reduce the greater fraction, to a leſſer in nõ=
ber : as J learned long a gone by your other booke.

Of Progreſsion in Cossike ſignes.

Master.

Rogreſsion is thus wꝛoughte : Firſte ſette
doune as many vulgare nõbers, in their
naturall pꝛogreſſion, as you liſte to haue
Cossike ſignes, that by them you maie the
better knolv, the true places of the *Cossike*
ſignes : ſo that you ſet in the firſte place a Cipher, and
vnder it. ℈. And then vnder. 1. ſet. ℞. vnder. 2. put. ⅔
and vnder. 3. wꝛite. ℞. As you ſee in the table fololv=
yng. And by theſe ſhall you ſet, as many as you liſte.

Foꝛ all the vulgare nombers, whiche you haue ſet
in the higher relve, be other compounde nombers, oꝛ
els vncompounde : and if the place, where you would
ſet any *Cossike* ſigne, be noted with a nomber vncom=
pounde, then muſt there be ſet one of the *Surſolides*.

Foꝛ vnder the firſt nõber vncompounde, you muſt
ſet the firſte *Surſolides*, and the ſecond vnder the ſecond
nomber vncõpounde : and the thirde vnder the thirde,

          T.iii.   and

The Arte

and so foxthe.

The nombers vncompoūde, are these in their pro=
greſſion.

5. 7. 11. 13. 17. 19. 23. 29. 31. 37. 41.
43. 47. 53. 59. 61. 67. &c.

Under nethe. 5. muſt you ſet. ſ℥. and vnder. 7. ᵇſ℥
vnder. 11. ᶜſ℥. and vnder. 13. ᵈſ℥. and ſo foorthe, til
you come to. 67. vnder whiche you muſt ſet. ʳſ℥. and
vnder 71 you muſt ſet. ˢſ℥. and ſo as farre as you liſt.

But for any other place, bicauſe the vulgare nom=
ber is compounde, that is ſet (as the peculiare nom=
ber, in the higher rewe) therefore the *Coſsike* ſigne
muſt nedes be compounde, other of. 2. or of. 3. or els of
bothe. And if it be cōpounde of. 2. then ſet doune. ℥. ſo
often tymes as. 2. is in the cōpoſition of that nomber.

As for example : 16. is cōpounde of. 2. fower tymes
(not by addition, but by multiplication, as in ſaiyng,
twiſe. 2. twoo tymes, twiſe.

Scholar. J perceiue twiſe. 2. to bee. 4. and twiſe
that to be. 8. and twiſe that to make. 16.

Maſter. So maie you worke backewarde, in ſai=
yng. 16. diuided by. 2. maketh. 8. that is ones : then. 8.
by. 2. yeldeth. 4. that is twiſe. Again. 4. by. 2. maketh
2. that is thriſe : and. 2. for himſelf, is the fourth : wher=
fore vnder. 16. J muſt ſet doune. ℥℥℥℥.

And ſo vnder. 32. J muſte ſette. 5. ℥. in one thus.
℥℥℥℥.

And vnder. 64. J ſhall ſette it. 6. tymes, thus.
℥℥℥℥℥℥. Bicauſe. 64. is made of. 6. multipli=
cations by. 2.

Scholar. Here by J ſee, that vnder. 8. J muſte put
3. tymes that ſigne : and vnder. 4. twiſe theſame.

Maſter. So muſt you in deede.

And now for other places, if their nombers bee cō=
pounde

of Cossike nombers.

pounde of. 3. onely, then must you set doune the signe of *Cube*, as oftentymes as. 3. is multiplied, to make that nomber.

As for example. 27. is compounde onely of. 3. and not of. 2. (for of all other compounde nombers herein then of soche as be cōpounde of. 2. oʒ. 3. we take no re= garde.) And. 3. multiplied thrise, doeth make. 27. in saipng. 3. tymes. 3. thrise. And therefore vnder. 27. J shall set this signe of. ℭ. three times, thus. ℭℭℭ. whiche betokeneth a *Cube of Cubes Cubikely.*

But and if the nomber bee compounde, bothe of. 2. and. 3. then for euery tyme that. 2. is multiplied, to that composition, J shall sette. ʒ. and for euery tyme that. 3. is multiplied, J shall set. ℭ. remembʒyng still to set. ʒ. before. ℭ. and not after hpm.

As for example. Vnder. 24. J shall set. ʒʒʒℭ. bicause that. 2.2.2.3. that is to saie. 2. tymes. 2. twise thrise, doeth make. 24. Or by resolution, thus. 24. di= uided by. 2. giueth. 12. For that firste. 2. set. ʒ. Again 12. diuided by. 2. yeldeth. 6 : for this seconde. 2. set. ʒ. also. Then diuide. 6. by. 2. and it maketh. 3. For the. 2. J must set. ʒ. and for. 3. J must put. ℭ. and so all to= gether maketh. ʒʒʒℭ. in the. 24. place.

Likewaies vnder. 36. J must sette. ʒʒℭℭ. bi= cause that. 2.2.3.3. doeth make it, that is. 2. tymes. 2. thʒise, thʒise. And by resolution, thus. 36. diuided by 2. giueth. 18. For that. 2. J set. ʒ. Againe. 18. diuided by. 2. maketh. 9. For that. 2. J sette doune againe. ʒ. Thirdly, for bicause. 9. can not bee diuided by. 2. but by. 3.3. tymes : therefore J muste sette doune, for those twoo. 3. twise. ℭ. ⁊ so the whole signe is. ʒʒℭℭ.

Now if the nomber of the place, oʒ peculiare nom= ber, bee compounde of one of theim twoo, with some other nomber vncōpounde, then must we ioyne their signes together.

As. 10. is compounde of. 2. and. 5. therefore must J
                                                              set

The Aarte

set vnder. 10. the signe that is in the fifth place, whi=
che is. ∫ℨ. and before it I muste set the signe of. ℨ. for
2. So must that signe be. ℨ∫ℨ.

Likewaies, bicause. 15. is compounde of. 3. and. 5.
I shall ioine together the signe of. ᴔ. and of. ∫ℨ. and
make it. ᴔ∫ℨ.

Scholar. So I vnderstande it now, that I cannot
misse it. Saue that for lacke of vse, and throughe for=
getfulnesse, when I heare the name of composition
in nombers, I doe mistake it sometimes for addition,
els here can be no erroure. For when I doe consider,
that. 20. is compounde of. 2,2,5. that is twise. 2. and. 5
(sith. 2. tymes. 2. maketh. 4 : and. 5. tymes. 4. maketh
20.) I maie sone consider, to set. ℨ. twise before. ∫ℨ.
and then it will be. ℨℨ∫ℨ. to be put in the. 20. place.

Likewaies in the. 21. place, I set. ᴔ ᵇℨ. seyng 21
is compounde of. 3. and. 7. and. ᴔ. is the signe to the
thirde place, as. ᵇℨ. serueth for the. 7. place.

Master. What shall you set in the. 84. place?

Scholar. 84. is compounde of. 2,2,3,7. therefore
his signe must be. ℨℨᴔ ᵇℨ.

Master. Now I see, you are cunnyng inough in
this, and therefore take here this table, for a patrone :
and then will we procede to the worke of *Cossike* nom=
bers compounde.

The table for progression Cossike,
whiche maie increase it self infinitely,
without any difficultie.

0.	1.	2.	3.	4.	5.	6.	7.	8.	9.	10.	11.

12.	13.	14.	15.	16.	17.	18.	19.	20.

21.	22.	23.	24.	25.	26.	27.	28.

29.	30.	31.	32.	33.	34.	35.	36.

37.	38.	39.	40.	41.	42.	43.	44.

45.	46.	47.	48.	49.	50.	51.

52.	53.	54.	55.	56.	57.	58.

59.	60.	61.	62.	63.	64.	65.

66.	67.	68.	69.	70.	71.	72.	73.

74.	75.	76.	77.	78.	79.	80.

In this table, ₿. cℓ. and. ſℬ. are the groundes:
of all the reſte aboue them. For of theſe
thre, all thoſe other bee made.

U.i.   Df

The Arte
Of Coſsike nombers compounde.

Oſsike nombers compounde, are made by addition of. 2. oꝛ moꝛe ſimple *Coſ=ſike* nombers together :
As, 6.℥. ┼ .5.℥. oꝛ.
12. ℰ. ┼ 4.℥. ┼ 3.℈. and ſo foꝛꝛthe in diuerſe foꝛmes, whiche be infinite. Howbeit foꝛ bꝛiefneſſe, we maie compꝛehende, vnder the ſame name (bicauſe of the like woꝛke) all other *reſidualles Coſsike*, whiche be made by ſubtraction : as. 3.ℰ. ——— .4.℥. And al=ſo thoſe that bee made by addition and ſubtraction, bothe together : As. 9.℥℥. ┼ .4.℥. ——— 6.℥. In whoſe numeration is no hardneſſe.

Scholar. Then your rules maie be the ſhoꝛter.

Of Numeration.

Maſter.

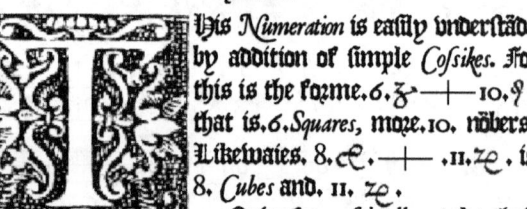

His *Numeration* is eaſily vnderſtãde by addition of ſimple *Coſsikes.* Foꝛ this is the foꝛme.6.℥ ┼ 10.℈. that is.6.*Squares*, moꝛe.10. nõbers. Likewaies, 8.ℰ. ┼ .11.℥. is 8. *Cubes* and. 11. ℥.

Now foꝛ *reſidualles*, take theſe examples. 9.℥℥. ——— .12. ℰ, whiche is. 9. *Squares of Squares*, ſaue. 12. *Cubes.* Alſo. 4.ſ℥. ——— .15. ℥. that is. 4. *ſurſolides*, abatyng. 15. *ſquares.*

And foꝛ bothe together, this is the foꝛme. 10. ℥℥ ┼ .6. ℰ. ——— 30. ℥. whiche ſignifieth 10. *Squares of Squares*, and. 6. *Cubes*, abatyng. 30. *rootes.*

Scholar. This is plaine. Foꝛ ſo maie J vnder=ſtande of all other. As. 9.℥ℰ. ——— .3.℥. ┼ 8.℈. that is. 9. *Squares of Cubes*, leſſe 3. *Squarss*, moꝛe. 8. nõbers.

Maſter.

of Cossike nombers.

Master. It were moze ozderly, to kepe the signes of moze and lesse in ozder, then to followe the ozder of the *cossike* signes : bicause that addition, is ozderly placed befoze subtraction. So were it better to set theim thus. 9. ʒ cℰ. —— 8. 9. ——3. ʒ. Hobeit in deede all is one in these kinde of nombers, but not so in other *Surde nombers*, where the ozder foloweth of necessitie, as shall be declared in their place moze largely.

Of Addition.

IN addition, you must haue consideration of the *Cossike* signes : foz noe other nomber, maie bee added into one, then soche as appertain to one signe *Cossike.*

As in bulgare denominations, you doe not adde the nöbers of shillynges, to the nombers of pennies : but you ioine shillynges to shillinges, and pennies to pennies : & poundes to poundes, so in *Cossike* nombers, *Cubes* muste bee ioined to *Cubes, Squares* to *Squares,* and generally, like to like.

Scholar. If this be al, I cã marke it well inough.

Master. There is somewhat moze to be considered, that if there bee any signe in the one nomber, whiche is not in the other, that seueralle signe with his nomber, muste bee sette doune with his figure of —— .oz. —— . as it standeth there.

And farther, touchyng those twoo signes. —— .
—— . whiche bee the figures of moze and lesse, you must giue regarde, whether thei bee like oz vnlike, in those nombers that must be added : Foz if thei be like in nombers, of one denomination, then muste thei so remain as thei be. But if thei be vnlike, euermoze abate the smaller nomber of theim, that followe those

U.ii.   vnlike

The Arte

vnlike ſignes, out of the greater : and ſette doune the
reſte, with the ſigne of the greater nomber.

Scholar. By examples, I ſhall better conceiue
thoſe rules.

Maſter. Take theſe examples.

10, ℥ .	—+—	,12, 9 ,	10, ℥ .	——— ,12, 9 ,
4, ℥ .	—+—	,8, 9 ,	4, ℥ .	——— ,8, 9 ,
14, ℥ .	—+—	,20, 9 ,	14, ℥ .	——— ,20 9 ,

10, ℥ .	———	,8, 9 ,	10, ℥ .	—+— ,8, 9 ,
4, ℥ .	———	,12, 9 ,	4, ℥ .	—+— ,12, 9 ,
14, ℥ .	———	,20, 9 ,	14, ℥ .	—+— ,20, 9 ,

10, ℥ .	—+—	,12, 9 ,	10, ℥ .	——— ,12, 9 ,
4, ℥ .	———	,8, 9 ,	4, ℥ .	—+— ,8, 9 ,
14, ℥ .	—+—	,4, 9 ,	14, ℥ .	——— ,4, 9 ,

10, ℥ .	—+—	,8, 9 ,	10, ℥ .	——— ,8, 9 ,
4, ℥ .	———	,12, 9 ,	4, ℥ .	—+— ,12, 9 ,
14, ℥ .	———	,4, 9 ,	14, ℥ .	—+— ,4, 9 ,

Here haue I varied one example diuerſly, to the in
tente you maie marke the vſe of your rules in theim.
And for the reaſon of thoſe rules, you ſhall marke
those

fo Cofsike nombers.

thofe examples well.

For where in the firſte example, bothe ſignes are ——— , it muſt nedes be, that after the addition of the firſte nombers, the feconde muſte bee added with the ſigne. ——— .

In the feconde example, where bothe the ſignes be ——— , bicaufe there wanteth. 21. $\wp$. of the firſt. 10. $\mathcal{Z}$. Therfore is it reafon, that bothe thofe wantes ſhould be fette doune with the ſigne of. ——— : and fo in the thirde and fourthe examples.

In the fifth example, the feconde fomme is not ful= ly. 4. $\mathcal{Z}$ .but there wanteth of it. 8. $\wp$. and therefore if you put donne the. 4. $\mathcal{Z}$. fully, you muſt abate. 8. out of the. 12. $\wp$. in the higher fomme : and fo of the other examples.

But for more practife, and better declaration of the vfe of them, here are other exaples, of more varietie.

20. $\mathcal{Z}$ cℓ . ——— . 9. $\mathcal{Z}$. ——— . 120. $\mathcal{Z}$.
15. $\mathcal{Z}$ cℓ . ——— . 5. $\mathcal{Z}$. ——— 16 . $\mathcal{Z}$.
_____
35. $\mathcal{Z}$ cℓ . ——— . 14. $\mathcal{Z}$. ——— . 104. $\mathcal{Z}$.

16. $\mathcal{Z}\mathcal{Z}$. ——— . 28 . $\wp$. ——— . 16 . $\mathcal{Z}$.
12. $\mathcal{Z}\mathcal{Z}$. ——— . 12. $\mathcal{Z}$. ——— . 19 . $\wp$.
_____
28. $\mathcal{Z}\mathcal{Z}$. ——— . 9 . $\wp$. ——— . 4 . $\mathcal{Z}$.

In the firſte example of thefe. 2. you fee. 120. $\mathcal{Z}$. with the ſigne of leſſe, to bee added with. 16 . $\mathcal{Z}$. with the ſigne of more : and therefore, feeyng the ſignes of one Cofsike denomination difagree, J dooe fubtracte the leſſer, out of the greater : and that. 104. whiche remaineth, J doe fet doune with the ſigne of

U. iii. leſſe

The Arte

leſſe, bicauſe the remainer is of that nomber, that
bare that ſigne.

And in the ſeconde exãple, the placyng of the ſigne
———+——— befoze ———— maketh nombers to bee ſette be=
foze ſquares : and ſo the like denominations, dooe not
ſtande one ouer an other. Yet is the woozke dooen as
if thei did ſtande eche ouer his like.

Scholar. I pzaie you lette me trie my cunnynge,
with an example oz twoo.

$$17. \sqrt{3}\sqrt{3}. \;+\; .10. \text{cc} . \;-\; .2. \text{ze} .$$
$$16. \sqrt{3}\text{cc}. \;+\; .12. \sqrt{3} . \;-\; .6. \text{ze} .$$

$$16. \sqrt{3}\text{cc}. \;+\; 17. \sqrt{3}\sqrt{3} . \;+\; 10. \text{cc} .$$
$$+\; 12. \sqrt{3}. \;-\; .8. \text{ze} .$$

I ſet the example, as nombers came to my mynde :
but I had almoſte ſet my ſelf on grounde : ſaue that I
called to remembzaunce, the compariſon that you
made, to vulgare denominations of poundes, ſhillin=
ges, and pennies : and ſo was inſtructed to place eue=
ry ſeueralle denomination ſeuerally. And to ſette the
greateſte denominatiõ firſt, ⅋ eche other in his ozder.

Now will I pzoue an other example, oz twoo.

$$3 . \sqrt{3} . \;+\; .4. \text{cc} . \;-\; .20. \text{q} .$$
$$20. \text{cc} . \;-\; .8. \sqrt{3} . \;-\; .16. \text{q} .$$

$$3 . \sqrt{3} . \;+\; .24. \text{cc} . \;-\; .8. \sqrt{3} . \;-\; .36. \text{q} .$$

$$13. \sqrt{3}\text{cc}. \;+\; .8. \text{cc} . \;-\; .4. \text{ze} .$$
$$7. \sqrt{3}\text{cc} . \;-\; .6. \text{cc} . \;-\; .7. \text{q} .$$

$$20. \sqrt{3}\text{cc}. \;+\; .2. \text{cc} . \;-\; .4. \text{ze} . \;-\; .7. \text{q} .$$
$$6. \sqrt{3} .$$

of Cossike nombers.

6. ℨ . —|— .10. ℥ . ——— .8. ♀ .
4. ℨ . —|— . 17. ♀ . ——— .7. ℥ .

---

10. —|— .3. ℥ . —|— .9. ♀ .

4. ℨℭ . —|— .5. ℨ . —|— .6. ℥ .
8. ℭ . ——— .8. ℨ . ——— 10 ℥ .

---

4. ℨℭ . —|— .8. ℭ . ——— .3.ℨ . ——— .4. ℥ .

Master. You haue doen well : And for proofe of your worke, you maie in this arte not onely proue it, by the contrary kynde, as you did in nöbers *Abstracte*, but alſo by the reſolution of all thoſe *Cossike* nombers into nöbers *Abstracte*, takyng any nomber for a roote and then the *Squares* and *Cubes*. &c. accordyngly. As here in this table, you maie briefly ſee, but more largely in the table at the eande of nombers figuralle.

A table for trialle by reſolution,
of any woorke in this arte.

℥	ℨ	ℭ	ℨℨ	ſℨ	ℨℭ	bℨ
2	4	8	16	32	64	128
3	9	27	81	243	729	2187
4	16	64	256	1024	4096	46384

ℨℨℨ	ℭℭ	ℨſℨ
256	121	1024
6561	19683	59049
65536	262144	1048576

And if this table in any parte, ſeme to ſhorte or to litle :

The Arte

litle : you maie haue recourſe to the table, at the ende
of figuralle nombers, whiche therfoꝛe is made large
and generalle : ſo that it maie well be called the frute=
full table, oꝛ table of eaſe.

But now foꝛ triall of the laſte example :
firſte there is. 4. ƶ c℮ : foꝛ whoſe roote J take
2. and therefoꝛe thoſe. 4. ƶ c℮. make. 256.   256.
whiche J ſette doune in nomber *Abſtracte*.   20.
Perte is. 5. ſquares, whiche accoꝛdyng to that   12.
roote, muſt nedes be. 20. and that. 20. J ſette   288.
doune alſo : and then. 6. rootes, whiche make
12. And all thei yelde. 288. and that is all the
firſte ſomme.

Then foꝛ the ſeconde ſomme, J ſee firſte. 8.
Cubes, whiche make. 64. to bee added. Then   32.
foloweth. 8. ſquares leſſe, that is. 32. to bee a=   20.
bated, and alſo. 6. rootes leſſe, that is. 20. alſo   52.
to bee abated : So muſt J abate. 52. (foꝛ theim
bothe) out of. 64. and then there reſteth but   64.
12. whiche added vnto 288. of the firſt ſomme   52.
doe yelde. 300.   12.

Now if the totall agree with this, then is
the woꝛke good.   288.

Foꝛ triall whereof, J reſolue. 4. ƶ c℮. in=   12.
to nomber *Abſtracte*, and thei will make. 256.   300.
then. 8. c℮. maketh. 64 : whiche bothe yelde
320. Then foloweth in theſame ſomme. 3. ƶ   256.
and. 4. ᴢ℮. to be abated. The. 3. ƶ. make. 12.   64.
and the 4. rootes yelde. 8. whiche together do   320.
amounte to. 20. and that muſt bee abated frõ
theſaid ſomme of 320 and then there remaineth one=
ly 300. agreable to the foꝛmer ſomme aboue the line.

Scholar. This pꝛoofe J like well : And J perceiue
that if J would woꝛke the like, takyng foꝛ the roote
3, oꝛ any other nomber, the pꝛoofe will ſuccede a like.

Maſter. Now to make an eande of Addition, bi=
cauſe

of Cofsike nombers.

cauſe you ſhall the better remember the rules of it, J
will giue you them in this bꝛiefe foꝛme.

> *In greateneſſe like and ſignes alſo,*
> *Adde like to like there nedes no mo:*
> *And where the greateneſſe diſagree,*
> *Place eche by other ſeuerally.*
> *with ſigne of eche, as doeth require,*
> *But if the ſignes vnlike appere:*
> *Then from the more abate the leſſe,*
> *The greater his ſigne with the exceſſe.*
> *will make the ſomme,*
> *Of that addition.*
> *The proofe is by reſoluyng,*
> *Eche nomber into his reckenyng.*

This leſſon doeth containe the foꝛmer rules onely
in bꝛief, and therefoꝛe neadeth no declaration : but the
greateneſſe doeth betoken the *Cofsike* denomination,
and ſignes betoken ſpecially, ——|— and. ——— . the
ſignes of moꝛe and leſſe, and no other ſignes.

Scholar. This bꝛief leſſon will helpe memoꝛie
moche : and ſhall ſuffice foꝛ the rules of Addition.

Of Subtraction.

Maſter.

Hen foꝛ ſubtraction, this ſhall you
marke in eſpeciall : that when your
nombers are ſette doune, after the
cōmon maner, firſte the totall, and *1. Rule.*
then the deduction : you ſhall conſi=
der well, whether the ſignes bee
——|— oꝛ.——— . Foꝛ in the de=
duction, if you haue ——|— then muſt that be ſubtrac=
ted from the like aboue.

And if that ſomme in the deduction, that hath the *2. Rule.*

The Arte

ſigne ——┼—— bee greater then the nomber of the like
quantitie ouer hym, with the like ſigne ——┼—— , then
abate the higher out of the lougher, and wꝛite the
reſte with this ſigne ———— .

3. Rule.   But if the like quantitie in the totall, haue the
ſigne ———— , then adde bothe nombers together and
ſet them vnder the line with that ſigne ———— .

4. Rule.   And if the ſeconde ſomme (that is the deduction oꝛ
abatement) with any nomber, haue this ſigne of leſſe
———— , it muſt be accoumpted foꝛ moꝛe, and muſt be
added to the like nomber ouer it, excepte the ouer nõ=
ber haue the ſigne of leſſe alſo : Foꝛ then muſt you a=
bate the leſſer, out of the greater, and ſette doune the
reſte, with the ſigne of the greater nomber : whiche
thei haue at this conferẽce : I meane to regarde what
the ſigne of the ſeconde ſomme is by eſtimation, and
not by wꝛityng, foꝛ thei are contrary.

Scholar. I ſee good reaſon in this : Foꝛ in any a=
batemente, the moꝛe is abated, the leſſe by ſo moche
ſhall remain : and the leſſe is abated, the moꝛe doeth
remain by ſo moche.

5. Rule.   Maſter. Yet one thyng moꝛe is to bee marked,
that if there be ſome denominations, in the one ſome
that are not in the other, you ſhall marke in whiche
ſomme thei bee. Foꝛ if thei bee in the firſte, then ſhall
thei kepe ſtill their owne ſigne. And if thei bee in the
ſeconde ſomme, whiche is the deduction, then ſhall
thei chaunge their ſigne to the contrary : But where
ſoeuer thei be, thei muſt be ſet in the remainer.

Scholar. I can better vnderſtande you, then re=
membꝛe thoſe rules.

Maſter. Then take this bꝛief leſſon, apter to bee
remembꝛed, then to bee vnderſtande, but by the let=
ter befoꝛe, and by the examples folowyng. But me=
moꝛie liketh well ſoche aide.

                                                   ◖A

of Cofsike nombers.

A brief rule of Subtraction.

1.　　　*When fignes and greatneffe bothe agree,*
　　　Your woorke procedeth forthe commonly.

2.　　　*But if thabatemente greater bee,*
　　　Thexceffe fhall chaunge his figne therby.

3.　　　*And where the fignes doe diffagree,*
　　　The higher figne muft reft duely:
　　　And though the batemente be the greater,
　　　The refte ftill ioyneth bothe fommes together.

4.　　　*If quantities doe diffagree,*
　　　Place them with fignes all feuerallie:
　　　The totall kepeth the figne he had,
　　　The batemente ftill, to chaunge is glad.

Scholar. Now fome examples, will lighten thefe rules well.

Mafter. I will propounde the like, as I did in addition, to the intète you maie iudge the likeneffe, and diuerfities in bothe woozkes.

$$10. \; \bar{3} \cdot \;\; +\!\!-\!\!- \; .12. \; \mathcal{9} \cdot \qquad 10. \; \bar{3} \cdot \;\; +\!\!-\!\!- \; 8. \; \mathcal{9} \cdot$$
$$4 . \; \bar{3} \cdot \;\; +\!\!-\!\!- \; .8. \; \mathcal{9} \cdot \qquad 4 . \; \bar{3} \cdot \;\; +\!\!-\!\!- \; 12. \; \mathcal{9} \cdot$$

$$6 . \; \bar{3} \cdot \;\; +\!\!-\!\!- \; .4. \; \mathcal{9} \cdot \qquad 6 . \; \bar{3} \cdot \;\; -\!\!-\!\!- \; 4. \; \mathcal{9} \cdot$$

$$10. \; \bar{3} \cdot \;\; -\!\!-\!\!- \; .12. \; \mathcal{9} \cdot \qquad 10. \; \bar{3} \cdot \;\; -\!\!-\!\!- \; .8. \; \mathcal{9} \cdot$$
$$4 . \; \bar{3} \cdot \;\; -\!\!-\!\!- \; .8. \; \mathcal{9} \cdot \qquad 4 . \; \bar{3} \cdot \;\; -\!\!-\!\!- \; .12. \; \mathcal{9} \cdot$$

$$6 . \; \bar{3} \cdot \;\; -\!\!-\!\!- \; .4. \qquad 6 . \; \bar{3} \cdot \;\; +\!\!-\!\!- \; .4.$$

The Arte

10. ʒ . ——+—— .12. ♀ .	10. ʒ . ——+—— .8. ♀ .
4 . ʒ . ———— .8. ♀ .	4 . ʒ . ———— .12. ♀ .
6 . ʒ . ——+—— .20. ♀ .	6 . ʒ . ———— .20. ♀ .

10. ʒ . ———— .12. ♀ .	10. ʒ . ———— .8. ♀ .
4 . ʒ . ——+—— .8. ♀ .	4 . ʒ . ——+—— .12. ♀ .
6 . ʒ . ———— .20. ♀ .	6 . ʒ . ———— .20. ♀ .

The firste and thirde examples be very plaine : and in the seconde where. 12. should bee abated out of. 8. there is. 4. to fewe : and therefore I abate the higher, out of the lougher, and I set doune. 4. with the signe of wantyng, or abatemente.

In the fourthe example : bicause the higher nomber is the lesser, I doe subtracte him out of the nether, and sette doune the reste. 4. with a contrary signe of ——+——.

But in the. 4. later examples, where the signes do disagree, the nöbers that followe the signes, are not subtracted one from an other, but are added together : and thei take still the higher signe. Bicause in value, the signe of abatemente is contrary, to that it appeareth to bee.

And for your exercise, to make you full prompte in this arte, I haue set forthe more examples.

6. ℞ . ——+—— .120. ♀ .	8. ʒ℞ .
9. ℞ . ———— .40. ♀ .	9. ʒ℞ . ———— 89. ♀ .
160. ♀ . ———— .3. ℞ .	89. ♀ . ———— .1. ʒ℞ .

3. ʒ .

of Cossike nombers.

$$3. \mathcal{Z}. \quad +\!\!\!\!-\!\!\!\!- \quad 18. \mathcal{Z} . \qquad \mid \qquad 18. \mathcal{Z} . \quad +\!\!\!\!-\!\!\!\!- \quad .3. \mathcal{Z} .$$
$$12. \mathcal{Z} \quad -\!\!\!\!-\!\!\!\!- \quad .3. \mathcal{Z} . \qquad \mid \qquad 12. \mathcal{Z} . \quad -\!\!\!\!-\!\!\!\!- \quad 3. \mathcal{Z} .$$

$$6. \mathcal{Z} . \quad +\!\!\!\!-\!\!\!\!- \quad .6. \mathcal{Z} . \qquad \mid \qquad 6. \mathcal{Z} . \quad +\!\!\!\!-\!\!\!\!- \quad .6. \mathcal{Z} .$$

$$3. \mathcal{Z} . \quad +\!\!\!\!-\!\!\!\!- \quad .18. \mathcal{Z} . \quad -\!\!\!\!-\!\!\!\!- \quad .10. 9 .$$
$$12. \mathcal{Z} . \quad +\!\!\!\!-\!\!\!\!- \quad .8. 9 .$$

$$3. \mathcal{Z} . \quad +\!\!\!\!-\!\!\!\!- \quad .6. \mathcal{Z} . \quad -\!\!\!\!-\!\!\!\!- \quad .18. 9 .$$

$$4. \sqrt{\mathcal{Z}} . \quad +\!\!\!\!-\!\!\!\!- \quad .10. \mathcal{C} . \quad -\!\!\!\!-\!\!\!\!- \quad .6. \mathcal{Z} .$$
$$5. \mathcal{Z}\mathcal{Z} \quad +\!\!\!\!-\!\!\!\!- \quad .12. \mathcal{Z} . \quad +\!\!\!\!-\!\!\!\!- \quad .3. 9 .$$

$$4. \sqrt{\mathcal{Z}} \quad +\!\!\!\!-\!\!\!\!- 10. \mathcal{C} . \quad -\!\!\!\!- 5. \mathcal{Z}\mathcal{Z} . \quad -\!\!\!\!- 18. \mathcal{Z} . \quad -\!\!\!\!- 3 9 .$$

Here in the firſte example, where I would abate 9
℮. out of. 6. ℮ . I maie eaſily perceiue, that there are
3. ℮. to fewe. And therefoꝛe doe I ſette doune. 3. ℮.
with this ſigne ———, whiche ſignifieth wante oꝛ a=
batement : and the. 2. nombers that followe the vn=
like ſignes, I ſet doune bothe added into one : and put
therto the ſigne of the totall oꝛ ouermoſte ſomme.

In the ſeconde example, there is the like wooꝛke :
Foꝛ in abatyng. 9. out of. 8. I finde. 1. to fewe : that. 1.
doe I ſet doune with his denomination of. $\mathcal{Z}$ ℮ : and
the ſigne. ———.

And the nomber 89 that foloweth the ſigne ———
in the ſeconde ſomme, ſtandeth in foꝛce as —┼—, foꝛ
the leſſe is abated, the moꝛe muſt remain : therfoꝛe in
the remainer, I ſet not the ſigne of moꝛe, befoꝛe that
nomber of. 89. but I put it in the firſte place of the
ſomme : whiche place of it ſelf, ſignifieth ſtill moꝛe.

Ӿ.iii.   And

The Aarte

And bicaufe ouer that nomber 89. there are no nō=
bers in the totall, therefore J mufte putte doune that
fomme as it is, without addyng to it, oz abatyng frō
it, in it felf.

Scholar. Thofe. 2. examples might be fet thus, as
J thinke, bicaufe the places doe fo require.

$$6.c\!\ell. \quad +\!\!\!-\!\!\!- \quad .120.\,\text{ℊ}.$$
$$9.c\!\ell. \quad -\!\!\!-\!\!\!- \quad .40.\,\text{ℊ}.$$

$$-\!\!\!- \quad .3.c\!\ell. \quad +\!\!\!-\!\!\!- \quad .160.\,\text{ℊ}.$$

$$8.\text{ℨ}c\!\ell.$$
$$9.\text{ℨ}c\!\ell. \quad -\!\!\!-\!\!\!- \quad .8.\,\text{ℊ}.$$

$$-\!\!\!- \quad .1.\text{ℨ}c\!\ell. \quad +\!\!\!-\!\!\!- \quad .8.\,\text{ℊ}.$$

Mafter. Remember your felf well, and marke
the remainer how it is written.

Scholar. J fee my owne ouerfighte : For no nom=
ber maie begin, with figne of leffe : and therfore muft
their places be altered of neceffitie, and fet in ozder as
thei were before.

Mafter. Then for all the refte of the examples, oz
any other like, J fhall not neade to giue you any far=
ther inftruction : fith that by thefe former, you maie
iudge of all other.

Proofe.   And for the examination of your wozke, the trialle
by refolution doeth ferue here, as well as els where :
rememberyng onely (as the ozder of fubtraction maie
admonifhe you) that the fomme of the totalle, whiche
is the firfte fomme, muft counteruaile the other bothe
fommes : that is of the deduction, and of the remainer.

So to trie the firfte example, takyng. 3. for a roote :
6.cℓ. make. 162. whiche J put to. 120. and it yeldeth
282. Then in the feconde fomme. 9.cℓ. are. 243.
whereof. 40. muft bee abated for the figne ————, fo
                                                    is

of Coſsike nombers.

is that ſomme. 203. Again in the remainer. 3. ℞. are 81. whiche muſt bee abated out of. 160. and ſo reſteth 79. whiche with. 203. doe make. 282. agreable with the firſte ſomme.

Scholar. This doe J well vnderſtande, and praie you to procede to multiplication.

Of Multiplication.

Maſter.

JN multiplication, there is no diffi= cultie, ſo that you dooe well marke the ſignes ——+—— and ———, whi= che beyng bothe like, will haue the ſigne ——+—— ſette in the totalle and beyng vnlike, thei will haue in the totalle the ſigne ———.

And likewaies in diuiſion ——+—— diuided by ——— or côtrary waies ——— by ——+—— will alwaies haue in the totalle ——— : but ——+—— diuided by ——+——, or ——— by ———, will make alwaie ——+——.

Whiche rule for ready remembraunce, J haue gi= uen you here in meter.

> Who that will multiplie,
> Or yet diuide trulie:
> Shall like ſtill to haue more,
> And miſlike leſſe in ſtore.
> Their quantities doe kepe ſoche rate,
> That .M. doeth adde: and .D. abate.

Scholar. So meane you, that like ſignes multi= plied together, doe make more, or ——+—— : And vnlike ſines multiplied together, doe yelde leſſe, or ———.

Maſter. So is the rule. But to go forward now : of the nexte difficultie, as touchyng Coſsike quantities that chaunge their denomination, here is no more to bee

The Arte

bee ſaied, then was taught in multiplication of nom=
bers *Coſſike* vncompounde, and in the table ſet foꝛthe
foꝛ the chaunge of their names.

    Scholar. I vnderſtande, that in multiplication
(that is .℈.) their figures muſt bee added. And in .D.
(oꝛ diuiſion) thei muſte bee abated. Therefoꝛe a fewe
examples ſhall ſuffice foꝛ the reſte.

    Maſter. Take theſe foꝛ a pꝛeſidente, of all that
wooꝛke : by whiche you maie iudge of all other like.

$$10.\mathcal{C}\!\!\!\mathcal{C}. \;+\; .9.\mathcal{Z}. \;+\; .20.\mathcal{Z}\!\!\!e.$$
$$5.\mathcal{Z}. \;+\; .7.\mathcal{Z}\!\!\!e. \;-\; .8.\wp.$$

$$-\;80.\mathcal{C}\!\!\!\mathcal{C}. \;-\; .72.\mathcal{Z}. \;-\; .160.\mathcal{Z}\!\!\!e.$$
$$70.\mathcal{Z}\mathcal{Z}. \;+\; .63.\mathcal{C}\!\!\!\mathcal{C}. \;+\; .140.\mathcal{Z}.$$
$$59\mathcal{Z} \;+\; .45\mathcal{Z}\mathcal{Z} \;+\; 100\,\mathcal{C}\!\!\!\mathcal{C}.$$

$$59\mathcal{Z} \;+\; 115\mathcal{Z}\mathcal{Z} \;+\; 83\,\mathcal{C}\!\!\!\mathcal{C} \;+\; 68\mathcal{Z} \;-\; 160\,\mathcal{C}\!\!\!\mathcal{C}$$

$$15.\mathcal{Z}\mathcal{C}\!\!\!\mathcal{C}. \;-\; .12.\mathcal{Z}.$$
$$14.\mathcal{Z}. \;+\; 2.\mathcal{Z}\!\!\!e. \;-\; .5.\wp.$$

$$-\;.75.\mathcal{Z}\mathcal{C}\!\!\!\mathcal{C}. \;+\; 60.\mathcal{Z}.$$
$$30.b\!/\!\mathcal{Z} \;-\; 24.\mathcal{C}\!\!\!\mathcal{C}.$$
$$210.\mathcal{Z}\mathcal{Z}\mathcal{Z} \;-\; 168.\mathcal{Z}\mathcal{Z}.$$

$$210.\mathcal{Z}\mathcal{Z}\mathcal{Z} \;+\; 30.b\!/\!\mathcal{Z} \;-\; 75\mathcal{Z}\mathcal{C}\!\!\!\mathcal{C} \;-\; 168\mathcal{Z}\mathcal{Z}$$
$$-\;24\mathcal{C}\!\!\!\mathcal{C}. \;+\; .60.\mathcal{Z}.$$

    Scholar. I perceiue, that theſe woꝛkes doe appere
moꝛe hard, then thei bee in deede, and that bicauſe of
their ſtraunge foꝛmes : but by vſe I truſte to bee ac=
quainted with them well inough and therfoꝛe I will
begin with moꝛe eaſie examples. As theſe bee, that
folowe

of Cossike nombers.

folloe here.

$$18. \; \mathcal{Z} \cdot \; \longrightarrow \!\!+\!\!\longrightarrow \; .20 \; \varphi .$$
$$15. \; \mathcal{Z} \cdot \; \longrightarrow\!\!\longrightarrow \; .4. \; \varphi .$$

$$\longrightarrow 72. \mathcal{Z} \cdot \longrightarrow \; .80. \varphi .$$
$$270. \mathcal{C} \cdot \; \longrightarrow \!\!+\!\!\longrightarrow 300 \, \mathcal{Z} .$$

$$270. \mathcal{C} \cdot \; \longrightarrow \!\!+\!\!\longrightarrow 300. \mathcal{Z} \cdot \; \longrightarrow 72. \mathcal{Z} \cdot \; \longrightarrow .80. \varphi .$$

$$16. \mathcal{Z} \cdot \; \longrightarrow \!\!+\!\!\longrightarrow \; .14. \mathcal{Z} \cdot$$
$$8. \mathcal{C} \cdot \; \longrightarrow \; .7 . \varphi .$$

$$\longrightarrow 112. \mathcal{Z} \cdot \; \longrightarrow 98. \mathcal{Z} \cdot$$
$$128. \sqrt{\mathcal{Z}} \; \longrightarrow \!\!+\!\!\longrightarrow 112 \; \mathcal{Z} \mathcal{Z} \cdot$$

$$128. \sqrt{\mathcal{Z}} \; \longrightarrow \!\!+\!\!\longrightarrow 112. \mathcal{Z} \mathcal{Z} \cdot \longrightarrow 112. \mathcal{Z} \cdot \; \longrightarrow 98. \mathcal{Z} \cdot$$

And this J fee farther noe, that thefe woozkes feme moze difficulte to looke on, then thei be in pzactife, if a manne giue good hede to the fignes, and the quantities.

Mafter. Befoze we go any farther, J will fhee you fomewhat of the reafon, why the fignes ought to chaunge. And that by twoo plaine woozkes, in nombers *Abftracte*. As here foloeth.

Where you fee that when J had multiplied. 16 ——+— 12 by 20 it made. 320 ——+— 240 that is in all. 560.

But bicaufe the multipli= are ought not to be fo moche by 4 therfoze it is reafon, that

16. ——+—	.12.
20 ————	.4.
64 ————	.48.
320 ——+—	.240
560 ————	121
that is	448.

J fhall multiplie the higher fomme by. 4. and abate that out of the fozmer totall.

P.i.　　　Whiche

The Arte

Whiche thyng you see here doen by. ——— .64.
——— .48. whiche bothe make. 112. to bee deducted
out of 560. and so remaineth 448. The iuste somme
that commeth of that multiplication.

    Scholar. This I vnderstande well : and
maie proue it in this sorte. 16. ——— .12.
maketh. 28 : and. 20. ——— .4. is. 16.
Then if I multiplie. 28. by. 16. it will yelde
448. as the woorke here declareth.

    And hereby maie I iudge, of *Cossike* nom=
bers likewaies.

28.
16.
168.
28.
442

    Master. Yet one example more will I propound
bicause I would put you out of all doubte. Wherfore
marke this forme of woorke.

    Here you maie see, that if
the firste somme of 24 ——— 3
wer multiplied by 15 it would
make. 360. ——— .45. that is
315. But it ought not to bee so
moche, but lesse by. 2. tymes
24 ——— 3. that is. 48 ——— 6 :
bicause the multiplier doeth wante. 2. of. 15.

24. ———	.3.
15. ———	.2.
——— 48 ———	.6.
360 ———	.45.
366 ———	93.
that is. 273.	

    And so abatyng. 42. or. 48. ——— .6. out of. 315.
there resteth. 273. whiche is the iuste totall, when. 21
is multiplied by. 13. wherby the multiplication is de=
clared to bee good.

    And for bicause that ——— multiplied ——— with
doeth make ——— : marke here, that you maie not a=
bate fully. 48. but 48. ——— 6.

    Then seeyng in abatemente, the signes in figure
are contrary to their owne estimation and force : ther=
fore that. 48. must be made ———, and the ——— be=
fore. 6. tourned into ——— .

    Scholar. I see it well, it must nedes be so.

    For if thei were set, to bee subtracted, then should
thei stande so. 48. ——— .6 : whiche declareth that 42
should

of Coſsike nombers.

ſhoulð bee abateð.

But when theſame nombers, are ſet emongeſte o=
ther to be aððeð : as it is here in woʒkyng of multipli=
cation, then muſt thei be wʒitten thus.———— 48 ——|— 6
ðeclaryng that if you abate. 48. you muſte aððe. 6. a=
gain, bicauſe you abateð. 6. moʒe then you ought.

Maſter. You vnðerſtanð it well. Wherfoʒe here
will wee make an eanðe of multiplication : ſith there *The profe of*
reſteth nothyng but the pʒoofe of it : which maie bee *multiplac=*
wʒought by reſolution, of all the *Coſsike* nombers, in= *tion.*
to nombers *Abſtraſte,* as in other kinðes befoʒe. One=
ly conſiðeryng that the reſolutions of the firſt and ſe=
conðe ſommes, muſt be aððeð together.

Anð therfoʒe if you liſte to pʒoue the firſte example
takyng. 2. foʒ the roote, you ſhall finðe the firſte ſome
80.——|—36 ——|— 40. that is. 156. Anð the ſeconðe
ſomme is. 20. ——|— .14.———.8. that is. 26. The
thirðe ſomme is. 1600. ——|— .1840. ——— .664.
——|— 272.——— .320. whiche maketh. 4056. Anð
ſo ðoeth. 156. multiplieð by. 26.

Scholar. This maie J pʒoue at any tyme : ſo that
you ſhall not neðe to ſtaie aboute it.

Of Diuiſion.

Maſter.

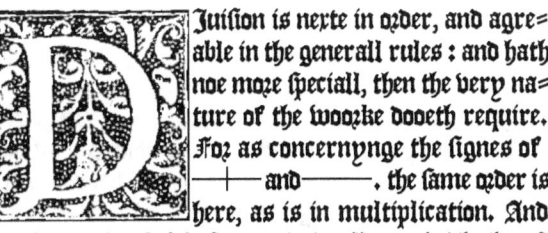

Iuiſion is nexte in oʒðer, anð agre=
able in the generall rules : anð hath
noe moʒe ſpeciall, then the very na=
ture of the wooʒke ðooeth require.
Foʒ as concernynge the ſignes of
——|— anð———. the ſame oʒðer is
here, as is in multiplication. Anð
touchyng the *Coſsike* ſignes, it is all one with that J
ſaieð in ðiuiſion of nombers *Coſsike* vncompounðe.

Scholar. Then a fewe examples maie ſupplie the
P.ii.   ðeclaration

The Arte

declaration of the vſe of the rules, with the pꝛactike
wooꝛke.

Maſter. Take theſe foꝛ your purpoſe.

An erample of the firſte woꝛke.

60.
12. ȝ ȝ ———— .16. cℰ. ———— 80. ȝ. (2. ȝ.
6. ȝ ———— .8. ℤℯ.

The remouyng of the diuiſoꝛ,
foꝛ the ſeconde wooꝛke.

66.
12. ȝ ȝ ———— 16. cℰ. ———— 80. ȝ. (2 ȝ ———— 10 ℤℯ
6. ȝ. ———— .8. ℤℯ.

The pꝛoofe in nombers *Abſtracte*,
accoumptyng .2. foꝛ roote.

3   480.
192. ———— 608. ———— .320. (8.
24. ———— 16.

480
192 ———— 608 ———— 320. (8 ———— 20.
24. ———— 16.

| The ſame woꝛke in vulgare foꝛme. | Here J haue not onely parted the wooꝛke, foꝛ your eaſe in vn= derſtanding : but J haue alſo put againſt it, the declaration of the ſame, by reſoluyng the *Coſsike* nöbers, into nombers *Abſtracte*. |

3
1120 (28.
440

And finally, J haue putte one erample of theſame
nombers,

of Cossike nombers.

nombers, after the vulgare forme, all whiche, 3. agree
together : and vouche one an other,

Scholar. Yet I praie you woorke, one example
more,

Master. Here is an other.

**¶The firste extraction
of the diuisor,**

$$4\emptyset . \mathfrak{z} c\ell . \qquad 48 . \mathfrak{z} \mathfrak{z} \qquad 20 c\ell \qquad 24 . \mathfrak{z}\ell . \quad (8 . \mathfrak{z} \mathfrak{z}$$
$$5 . \mathfrak{z} . \qquad 6 . 9 .$$

**¶The remouynge for=
ward of the diuisor,**

$$4\emptyset . \mathfrak{z} c\ell . \qquad 48 \mathfrak{z} \mathfrak{z} \qquad 20 c\ell \qquad 24 c\ell \quad (8 \mathfrak{z} \mathfrak{z} \qquad 4 \mathfrak{z}\ell$$
$$5 . \mathfrak{z} \qquad . 9 .$$

**¶The comprobation of thesame by resolu=
tion, accoumptyng still. 2. for a roote,**

$$2560 \qquad 768 \qquad 160 \qquad 48 .$$
$$2\emptyset \qquad 6 . \qquad\qquad (128 .$$

¶The settyng forward of the diuisor,

$$2560 \qquad 768 \qquad 160 \qquad 48 .$$
$$2\emptyset \qquad . 6 . \qquad 128 \qquad 8 .$$

Scholar. Yet ones again, I praie you worke the
like,

For although I perswade my self, that I perceiue
the woorke : yet would I see more confirmation of it,
before I would be to constante in my persuasion,

Master. Good aduisemēte is euer sure : but if you
doubte, your councelloure is not farre absente,

Scholar. I maie iustly reioice thereof : But for e=
uery mater to require aied, and neuer to trauell my
owne witte, it might seme mere dastardlinesse. And

The Arte

ſo were it plaine babiſheneſſe, to couet euery moꝛſell, to be chawed befoꝛe hande, and put into my mouthe.

Ɱaſter. Then take this other example, in one platte complete : But with a caueat, to beware of to moche confidence, while you ſeme to flee doubtefulle daſterdlineſſe.

16.

14ȝcꝛ ——— 30ȝȝ ——— 16ȝ ——— 6ſȝ ——— 6cꝛ (7cꝛ ——— 8ᴢℯ ——— 3ȝ
2. cꝛ. ——— .2.ᴢℯ.
         2.cꝛ. ——— 2.ᴢℯ.   2cꝛ ——— 2ᴢℯ

Scholar. Now haue J, that J looked foꝛ.

Ɱaſter. Softe, lette vs trie this wooꝛke, as wee haue doen the other : befoꝛe we goe from it.

Scholar. J pꝛaie you let me doe it.

Ɱaſter. With a good will.

64	16
14	30
256	480
64	
896	

Scholar. J kepe ſtill the old roote 2. Then is the.ȝcꝛ. 64 : whiche be= ing multipled by. 14. maketh. 896. And ſo. 30.ȝȝ. doe yelde. 480. And 16. ſquares make. 64. All thei toge= ther yelde. 1440.

The reſte of the nombers, muſt be abated, bicauſe of the ſignes. ——— . and thei make

32	8
6	6
192.	48

240. Foꝛ euery.ſȝ. is. 32. and then. 6. times that, that makeh 192. whereunto J put. 48. foꝛ 6. *Cubes* : and ſo haue J. 240. to be aba= ted out of. 1440. and then remaineth. 1200. foꝛ the diuidende. The diuiſoꝛ is but. 20. ſith. 2. cꝛ. are. 16. and. 2. rootes make. 4,

| 896 |
| 480 |
| 64 |
| 1440 |

| 1440 |
| 240 |
| 1200 |

1200 (60
   2 0

If J diuide now. 1200. by. 20. the *quotiente* will be. 60. agreably to the foꝛmer *quotiente*. Foꝛ 7. cꝛ . make. 56
And

of Cossike nombers.

And. 8. rootes yelde. 16. that is. 72. From whiche I
must abate. 3. ʒ. that is. 12. And then it is iuste. 60.

Master. This is well doen.

Scholar. Yea sure, I am perfecte inought, in this
feate of diuision, I trowe.

Master. You doe well to doubt.

Scholar. I thinke my self sure without doubte :
As by one oz twoo examples, I will declare.

And first I take this nöber 322 bſʒ —|— 115 ʒ cℓ
——— 42. cℓ —|— 69. ʒ. —|—30. ℀. to be diuided
by. 14. ʒ. —|— 5. ℀. wherefore I sette them doune
thus.

$$322\,bſʒ —|— 115\,ʒ\,cℓ——42\,cℓ—|—69\,ʒ—|—30\,℀ \quad (23\,ſʒ —|— 3\,℀$$

14. ʒ. —|— 5. ℀.	14 ʒ —|— 5. ℀.
322 bſʒ —|— 115. ʒ cℓ.	42 cℓ —|— 15 ʒ.

And finde the firste *quotiente* to bee. 23. ſʒ. by
whiche I multiplie the diuisoz, and it taketh awaie
all the nombers ouer it : Wherefore I set the diuisoz
fozward, & finde 3 ℀. foz the *quotiente*, whiche I mul=
tiplie into the diuisoz, & it maketh 42 cℓ —|— 15 ʒ.
wherby I am at a staie. Foz although I see in the di=
uidende, the like nombers, yet the signe of ——— de=
clareth, that it is not possible, to abate this newe nö=
ber thens :seyng ——— 42. cℓ. is lesse then naughte.

Master. Therefoze consider it, in chosyng your
quotiente : and giue your *quotiente* the like signe.

Scholar. But then riseth an other doubte. Foz
there will be ———. 15. ʒ. whiche disagreeth in signe
from the nomber ouer it.

Master. Yet maie you subtracte it well inoughe,
if you haue not fozgotten, your rules of subtraction.

Scholar. Now I dooe better remember my self :
that by good reason, I must leaue as a remainer, not
onely the whole nomber ouer it, which is. 69. ʒ.
but

The Arte

but I muſt adde therto.15. ʒ. moʒe.

So ſhall I cancell the. 69. and ſet ouer it. 84. And then doe I remoue the diuiſoʒ foʒward, ſettyng 14 ʒ vnder. 84. ʒ. and the reſte in oʒder, whereby I per= ceiue, that the newe *quotiente* will be.——— .6. ꝗ.

84.

$$\begin{array}{c} \cancel{322.6}ʒ \ + \ \cancel{115}ʒ\,cE \ —— \ 4^2 cE \ + \ \cancel{69}ʒ \ + \ 30\,cE \ (23 \int ʒ \ —— \ 3\,zE \ —— \ 6\,ꝗ \\ 14.ʒ. \ + \ .5.\,zE \quad | \quad —— 14ʒ. \ + \ .5.\,zE. \\ \hline \cancel{322}\,6ʒ \ + \ \cancel{115}ʒ\,cE \ | \ —— \ 4^2\,cE \ —— \ 15ʒ. \\ \qquad\qquad\qquad\qquad\qquad\quad 84.ʒ. \ + \ 30\,zE. \\ \qquad\qquad\qquad\qquad\qquad\quad \overline{14.ʒ. \ + \ 5.\,zE.} \end{array}$$

Whiche *quotiente* I doe multiplie into the diuiſoʒ, and it doeth make. 84. ʒ. ——— .30. zE. agreable to the ſomme ouer it. And ſo there remaineth nothyng.

Maſter. You haue dooen well. But in choſynge your diuidende, and the diuiſoʒ, your lucke was bet= ter then your cunnyng.

Scholar. That ſhall I pʒoue againe, by an other example, takyn alſo at all aduentures.

I would diuide this ſomme.

16. ʒ cE. ——— .20. ʒ. ——— .12. zE. ——— .8. ꝗ. by 4. ʒ. ——— .2. zE. And therfoʒe I ſet theim doune in oʒder thus.

$$16.ʒ\,cE \ + \ 20\,ʒ \ + \ 12\,zE \ —— \ 8\,ꝗ. \qquad (4.\,ʒ\,ʒ$$
$$4.ʒ. \ + \ 2.\,zE.$$

And firſte I ſee, that. 4. is contained in. 16. fower tymes : and ſo maie I finde. 2. in any other nombers there. 4. tymes. Wherfoʒe I ſet. 4. in the *quotiente.*

And bicauſe the. 4. in the diuiſoʒ are. ʒ. and the 16 to bee diuided, are. ʒ cE. accoʒdyng to the foʒmer ru= les, I finde the newe denomination *Coſsike* to be. ʒ ʒ whiche

of Cossike nombers.

whiche J set in the *quotient* with 4 and so is it. 4.ʒꝛ

Then saie J. 4.ʒʒ. multiplied by. 4.ʒ. do make 16 ʒ ₡. and therfore cleareth and consumeth al that some ouer it. Then farther saie J. 4. ʒʒ. multiplied by. 2. ℞ . doe yelde. 8.∫ʒ : But J see noe soche deno‐
mination in the diuidende.

Master. Then maie you perceiue, that you haue missed.

Scholar. Why sir, J thinke J ought to doe as you did : that is to multiplie the *quotiente* into euery parte of the diuisor.

Master. That is true : but J wil detecte the faulte vnto you. And that is this.

That all nombers *Cossikes* compounde, can not bee diuided orderly, by diuisors compounde. And those that can bee diuided, will not receiue any other diui‐
sor of thesame kinde, but one of. 2. nombers, by mul‐
tiplication of whiche, it was made : and so the other of those. 2. shall be the *quotiente* : As it came to passe in all those. 3. examples, which J set forthe. And therfore it is loste laboure, to goe aboute to diuide theim in that sorte.

Scholar. Then are there but fewe nombers of *Cossikes* compounde, that maie be diuided.

Master. So many men saie. But J saie thereto, that though many of them can not be diuided, by like nombers *Cossikes* copounde, yet are there many thou‐
sandes, that maie be so diuided.

And again J saie, that all sortes of theim, maie bee diuided, by an *Abstract* nomber. And also any of them maie be diuided, by conuersion into a fraction : And so maie your example be set thus.

16.ʒ ₡. ——— .20.ʒ. ——— 12.℞ . ——— .8.ꝗ.
_____
4.ʒ. ——— .2.℞ .

Z.i.   And

The Arte

And in all other cafes like, fette the diuidende ouer a line, and the diuifor vnder thefame line, and fo is your diuifion eanded : and this is the reddiefte waie, and the mofte indifference, in all foche nombers.

Scholar. That is fone learned. And therfore nea= deth no moare examples.

It is like in nombers *Abftracte*, when the greater nomber, doeth diuide the leffer. As. 6. diuided by. 11. maketh $\frac{6}{11}$.

Mafter. Somewhat like it is. Howbeit here is a woorke moare like thereunto, as when we fhould di= uide the leffer *Cofsike* nomber, by the greater, for then we muft fet them in that forme. So. 6.ʒ. diuided by 7.℞. fhall be fet thus : $\frac{6ʒ}{7℞}$. And. 20. ℞. diuided by 5.√ʒ. muft ftande in this maner : $\frac{20℞}{5√ʒ}$.

Scholar. Why? 20. maie be diuided by. 5.

Mafter. But. ℞. can not be diuided by. √ʒ. And in *Cofsike* nombers, the chief regard is to be had, to the *Cofsike* fignes.

Scholar. Then, as for any other forme, of regu= lare diuifion, here is none.

Mafter. Noe, excepte your diuifor, bee a nomber *Abftracte* : Or at the leafte, if it haue one onely *Cofsike* figne, and be vncompounde, that figne muft be other equalle, or leffer then the leafte *Cofsike* figne, in the di= uidende.

For fo. 60.ʒ℞. ——|—— 48℞. ——|—— 18.ʒ. maie bee diuided by any nomber, hauyng one of thefe. 3. fi= gnes *Cofsike*.ʒ.℞.9.

Scholar. I vnderftand it well. For.ʒ. is the lafte figne in the diuidende : And. ℞. and.9. are not onely leffe then it, but alfo.9. leaueth the nomber, as if it were a nomber *Abftracte*.

So if I would diuide your nomber, affigned by 40.ʒ. the *quotiente* would bee thus.

60.ʒ.

of Cossike nombers.

$$60.\maltese\, c\!e + 48\, c\!e + 18\maltese\, (1\tfrac{1}{2}\maltese\,\maltese + 1\tfrac{1}{5}\,\maltese + \tfrac{9}{20}\,\wp.$$
$$40.\maltese.\qquad\qquad 40.\maltese.\qquad\qquad 40.\maltese.$$

Master. Befoze we eande this wozke of diuiſion, I will admoniſhe you, of one eaſie aied, in the diuiſiō of diuerſe nombers. And that is, to conſider, whether your diuidende, doe omit any Coſſike denominations, betwene them, whiche it hath. Foz if it doe, you muſt yet ſupplie their roomes, with ſignes and Cyphers. As by example, you ſhall vnderſtande.

I require to haue this nomber. 8. $c\!e$. $+$ 64. $\wp$. diuided by. 2 $\maltese$. $+$ .4. $\wp$.

Scholar. That will I doe quickely. Foz I ſee. 4. will be the firſte quotiente and his denomination will be. $\maltese$. ſith. $c\!e$. diuided by. $\maltese$. doe make. $\maltese$.

But firſte I ſette doune the nombers ozderly. And then I multiplie the diui- ſoz by the quotiente, & there riſeth. 8. $c\!e$. $+$ 16. $\maltese$.

$$8.\,c\!e + 64.\wp.\,(4.\maltese.$$
$$2.\maltese + .4.\wp.$$
$$\overline{\,8.\,c\!e + 16.\maltese.\,}$$

Master. Stande you now amaſed, foz all your greate confidence? You ſee that you can not finde any $\maltese$. in the diuidende. Therfoze ſet doune the nomber as I told you befoze, in this ſozte.

$$\qquad\qquad —\ 16\maltese.$$
$$8.\,c\!e + .0.\maltese. + .0.\maltese + 46.\wp.\,(4.\maltese.$$
$$2.\maltese + .4.\wp.$$
$$\overline{\,8.\,c\!e + .16.\wp.\,}$$

And then I take the ſame quotient that you did, and I finde the remainder to be. $—$ .16. $\maltese$. Wherefoze I doe again ſette fozward the diuiſoz : And finde the quotiente to bee$—$8. $\maltese$. by whiche I multiple the

The Arte

diuiſoʒ, and it maketh. 16.ʒ. —— .32. ℥ ſo that a=
batyng the. 16.ʒ. the reſte, that is, —— .32. ℥.
ſhall be the remainer with the ſigne —+— by the rule
of ſubtraction.

$$-16\,ʒ\; -\!\!\!+\!\!\!- \; 32\,℥.$$
$$8\,℈\; -\!\!\!+\!\!\!-\; 0\,ʒ\; -\!\!\!+\!\!\!-\; 0\,℥\; -\!\!\!+\!\!\!-\; 64\,9\quad (4\,ʒ\; -\!\!\!-\; 8\,℥\; -\!\!\!+\!\!\!-\; 16\,9.$$
$$2\,℥\; -\!\!\!+\!\!\!-\; 4.9.$$
$$2\,℥\; -\!\!\!+\!\!\!-\; 4.9.$$

⸿Then vnder that remainer, I remoue the diuiſoʒ,
and finde the newe *quotiente* to bee —+— 16.9. And
ſo is the nomber clerely conſumed.

Scholar. If I foʒgette any parte of this, I am de=
ceiued to foule.

Maſter. Then haue you learned this parte, well
inough, foʒ this tyme. And therfoʒe will we go foʒth
vnto fractions, whiche partly were omitted befoʒe,
and partly are compounde of them ſelf.

Of fractions, and their numeration.

Ractions of this kinde appere ſim=
ple : and yet are ſcante ſo to bee iud=
ged : as $\frac{48}{3℥}$ betokeneth 4.ʒ. to bee
diuided by. 3. ℥. Likewaies this
fractiõ $\frac{12\,ʒ}{58\,℥}$ doeth impoʒt that 12 ʒ
muſte bee diuided by. 5. ʒ ℥. But
$\frac{10\,ʒ}{19}$ betokeneth. 10.ʒ. to bee parted
into. 19. poʒtions.

And here ſhall you note, the doubtfull foʒme, that
many menne in this arte vſe, whiche wʒite that laſte
fraction thus $\frac{10}{19}\,ʒ$. where as this fractiõ doeth repʒe=
ſent $\frac{10}{19}$ of a ſquare : and not 10.ʒ to be diuided by. 19.

Scholar. Bicauſe you ſaie, that ſome doe ſo vſe it,
and

of *Cossike* nombers.

and I would gladly excuse all good writers : I maie
saie for them that as in vulgare nombers, when. 10.
should be diuided by 19. And is set thus $\frac{10}{19}$ it doeth im=
porte bothe that. 10. is diuided into. 19. and also that
euery portion of those. 19. is $\frac{10}{19}$ of an vnitie : so that if
10.℔. should be parted emongest. 19. men, euery man
should haue $\frac{10}{19}$ of.1.℔.

Master. Your wordes haue so moche apperaūce
that thei maie persuade hym, that is not very precise
in termes, especially seyng there is no other *quotiente*
there, but thesame nomber. But as the somme of
10.℔. beyng diuided by. 19. is farre more then $\frac{10}{19}$ of an
vnitie : So. 10. ᷓ. to bee diuided by. 19. differ moche
from $\frac{10}{19}$ of a square. For the one is 19. tymes so moche
as the other. And therfore oughte to haue a distincte
forme in writyng.

Scholar. Then you would haue me to write thē
so, that $\frac{10}{19}$ of a Square, should haue the signe againſt
the line, as here is set $\frac{10}{19}$ᷓ : and when I would repre=
sent. 10. ᷓ. diuided by. 19. I shall write it thus $\frac{10ᷓ}{19}$,
with the signe aboue the line.

Master. You maie see their agremente, and their
difference by resolution, in this maner $\frac{10ᷓ}{19}$ will make
$\frac{40}{19}$ accoumptynge. 2. for a roote, and $\frac{10}{19}$ᷓ maketh $\frac{10}{19}$ of
4. or $\frac{40}{19}$ of. 1.

Again, accoumptyng. 3. for the roote, then $\frac{10ᷓ}{19}$ yel=
deth $\frac{90}{19}$: and $\frac{10}{19}$ᷓ maketh $\frac{90}{19}$ of an vnitie : so thei appere
to bee equall in valewe by reduction.

But now maie you see, that the one doeth betoken
the firste nōber, whiche is to be diuided : and the other
doeth signifie the *quotiente* of the diuision : and so are
thei distincte in office and nature. But bicause by re=
solutiō, the one tourneth into the other, therfore ma=
ny men accoumpt them as one. Howbeit, we stand to
longe aboute this, consideryg the erroure, is not al=
waies daungerous.

Z.iii.   But

The Arte

But their ouerſighte is moꝛe daungerous, whiche miſplace the ſigne, when it ſhould bee ſette vnder the line : as a greate clerke doeth (except I ſhall foꝛ his ex= cuſe, impute the faulte to the pꝛinter) foꝛ he meaning to diuide. 3. by. 7. ℥·℥· wꝛiteth it thus. $\frac{3}{7}$℥·℥· where he ſhould wꝛite it thus. $\frac{3}{7℥·℥·}$: and again, myndyng to diuide. 7. by. 3. ℥·℥· he wꝛiteth it thus $\frac{7}{3}$℥·℥· where he ſhould wꝛite. $\frac{7}{3℥·℥·}$·

Scholar. This faulte is manifeſte, and detecteth the firſte negligence : Foꝛ $\frac{7}{3}$℥·℥· doeth make in nom= ber, after the foꝛmer reſolution. $\frac{112}{3}$ and. $\frac{7}{3℥·℥·}$ dooeth make. $\frac{7}{48}$·

Maſter. Well, ſeyng you perceiue the faulte, we will ſtande no longer aboute it. Therfoꝛe to pꝛocede diſtinctly and certainly, whether that fraction be com pounde, oꝛ ſimple, where the numeratoꝛ is a *Coſſike* nomber, and the denominatoꝛ, a nomber abſolute, yet maie you boldly thinke, that fraction to bee com= pounde, whoſe numeratoꝛ is a nomber *Coſſike* and the denominatoꝛ an other *Coſſike* of vnlike ſigne : as . $\frac{3ꝯ}{5℥}$· and $\frac{58ꝯ}{12ᴣ}$·

Yet as in nombers Abſtracte, it maie ſeme moſte aptly to bee called a fraction, when the numeratoꝛ, is leſſer, then the denomintoꝛ, ſo in nombers *Coſſike*, moſte aptly the ſigne of the denominatoꝛ, ſhould bee the greater. Yet bothe foꝛmes come in vſe.

And foꝛbicauſe eaſineſſe in woꝛkyng, doeth often= times bꝛing certaintie with it befoꝛe we take in hãde the addition of fractions, I thinke it good to ſpeake ſomewhat of Reduction, to an other denomination. So that you foꝛgette not, that any. 2. nombers *Coſſike* compounde, with a line betwene them, maie be called a fraction. As thus. $\frac{5ꝯ \,+\, 8ᴣ \,-\, 6ᵹ}{3℥ \,+\, 12ᵹ}$ that is, 5·ꝯ· —+— ·8·ᴣ· ——— ·6·ᵹ· to bee diuided by 3·℥·

Examples of Numeration.

3.ʒ. ——+—— .12.ꝗ. and ſo of other like.

Of Reduction of fractions.

Ractions *Coſsike*, not onely in their nombers, but alſo in their ſignes maie be reduced to other valutions, and namely to their leaſte termes, and yet continue ſtill in one propoꝛtion, betwene the numeratoꝛ, and denominatoꝛ.

So $\frac{28\,ʒʒ}{36\,c\!e}$ maie bee reduced to $\frac{7\,\mathcal{Z}}{9\,ꝗ}$: foꝛ ſo high as. $c\!e$ is aboue. ꝗ. that is in the thirde place from it : So is ʒʒ. in the thirde place aboue. $\mathcal{Z}$.

Againe. $\frac{27/ʒ}{39\,ʒ}$. by reduction doeth make $\frac{9\,c\!e}{13\,ꝗ}$: And ſo

12ʒ$c\!e$ ——+—— 15ʒ ——— 9$c\!e$ will bee by reduction.

———————

21ʒʒ ——+—— 6$c\!e$ 4$c\!e$ ——+—— $\mathcal{Z}$ ——— 3ꝗ

—————————————————

                7$\mathcal{Z}$ ——+—— 2ꝗ

And ſo in all other fractions, where the nombers bee commenſurable.

But if any one nomber, bee incōmenſurable with the other, then can there be made no reduction in the nombers. Yet in the ſignes *Coſsike*, there maie be a reduction, other to greater, oꝛ to ſmaller ſignes : Foꝛ thoſe ſignes be euer commenſurable.

And there is no exception, but thei maie bee reduced to ſmaller quantities, excepte any one quantitie of theim bee. ꝗ. that is a number. Foꝛ that can bee no ſmaller. And therfoꝛe none other maie be altered, ſith euery one muſt be abated alike.

And looke how moche, the ſmalleſte quantitie of that fraction, is aboue a number, ſo moche maie thei all bee abated : foꝛ thei are neuer reduced to the ſmalleſte, till one of them be a nomber.

Scholar. And why maie not this reduction, ſerue foꝛ whole *Coſsike* nombers?

Maſter. Bicauſe the whole nomber, doeth not cō
                                 ſiſt

The Arte

ſiſt of a proportion, as the fraction doeth, and ſo maie bee exprꝛeſſed in diuerſe termes : but it importeth one ſomme certaine, whiche maie nother bee increaſed, noꝛ decreaſed, but it will chaunge his valewe, and al= ter his office.

And if J ſaie : a foote is $\frac{2}{6}$ of a yarde, J maie ſaie as truely, increaſyng bothe nombers, in the like proporꝛ= tion, a foote is $\frac{4}{12}$ of a yarde : oꝛ in leſſer termes : a foote is $\frac{1}{3}$ of a yarde.

But when J ſaie in whole nomber, a yarde is. 3. foote, oꝛ a foote is. 12. ynches, J ſaie truely : and if J doe increaſe oꝛ abate any of thoſe nombers, my woꝛ= des will be falſe.

So although in this nomber. 8.$\sqrt{3}$. ——— .6.℞. ——— 10.$\mathcal{3}$. by reaſon of bothe nombers and ſignes, there might bee a reduction, yet bicauſe it is a whole nöber, it ſhould therby bee abated moche : as here you maie ſee. 4.℞. —— 3.℞. ———.5.𝟡. whiche by re= ſolution into vulgare nombers, 2. beyng ſette as the roote, doeth make. 32 —— 6. ———.5. that is. 33. and the other nomber befoꝛe, doeth yelde by the like reſo= lution. 256 —— 48. ———.40. that is. 264. and is 8. tymes ſo moche as the other.

Scholar. J perceiue now good reaſon, why redu= ction ſerueth foꝛ fractions onely. And if there bee noe moꝛe difficultie in it, then you haue declared. J can woꝛke it eaſily.

Reduction in ſignes onely Foꝛ other the reduction cöſiſteth in the ſignes *Coſ*= *ſike* onely, as $\frac{10\mathcal{3}}{13\mathcal{℞}}$ where the nombers bee vncommen= ſurable, and therfoꝛe can not bee altered to any leſſer termes. But the ſignes *Coſſike* maie bee abated by. 3. denominations : ſeyng the ſmalleſte of them, is ſo ma= ny in oꝛder aboue. 𝟡. And therfoꝛe it maie be reduced to $\frac{10\mathcal{3}}{13\mathcal{𝟡}}$.

Reduction in nöbers onely Other els ſecondarily, the reduction conſiſteth in the nombers onely, when the nombers be communi= cante.

of Cossike nombers.

cante. And the signes *Cossike* bee all redy at the leaste : as when one of theim is. ϙ. So $\frac{16 c\!\!\!c}{12\,ϙ}$ will bee reduced to. $\frac{4 c\!\!\!c}{3 ϙ}$.

Reduction in signes and nomberses.

Or els thirdly, the reduction maie bee wroughte, bothe in signes, and also in nombers. When all the signes be aboue. ϙ. and the nombers be communicant So $\frac{50/з}{35з\cdot c\!\!\!c}$ maie be reduced well vnto. $\frac{10\,ϙ}{7\,z\!\!e}$.

An other reduction.

Master. Yet one forme of reduction more, I will showe you, where not onely the like woorke maie be, but also the nomber maie be broughte from his com= position, to a more simplicitie, by abatyng some of his partes.

As this number $\frac{6 c\!\!\!c \;—\!\!\!+\!\!\!—\; 18 з}{8 з\cdot c\!\!\!c \;—\!\!\!+\!\!\!—\; 24/з}$ maie bee reduced,

firste by his nombers to $\frac{3 c\!\!\!c \;—\!\!\!+\!\!\!—\; 9 з}{4 з\cdot c\!\!\!c \;—\!\!\!+\!\!\!—\; 12/з}$.

Secondarily, by his signes it maie be altered thus. $\frac{3 z\!\!e \;—\!\!\!+\!\!\!—\; 9 ϙ}{4 з\cdot з \;—\!\!\!+\!\!\!—\; 12 c\!\!\!c}$.

Thirdely, by abatynge the nombers, that followe signe of compositiõ (that is —+—) it maie be brought to. $\frac{3 ϙ}{4 з\cdot з}$. or. $\frac{3 ϙ}{4 c\!\!\!c}$. whiche fractions, kepe the self same proportion, that the firste fraction did.

Likewaies with the signe of ——— . nombers resi= dualles, maie bee reduced. As. $\frac{6 c\!\!\!c \;—\!\!\!—\; 18 з}{8 з\cdot c\!\!\!c \;—\!\!\!—\; 24/з}$. will bee reduced, as the other was to $\frac{3 ϙ}{4 c\!\!\!c}$.

Scholar. This is vnto me a marueilouse mater, that those. 2. contrary nombers, should be reduced to one fraction.

Master. The like happeneth in vulgare nom= bers. For. $\frac{18 —+— 6}{24 —+— 8}$. will bee reduced to $\frac{3}{4}$. For firste it maketh $\frac{24}{32}$ and then $\frac{3}{4}$. So likewaies $\frac{18 —— 6}{24 —— 8}$ will make firste $\frac{12}{16}$ and then $\frac{3}{4}$.

And the reason of it, doeth depende of the. 19. pro= position, of the fifth booke of *Euclide*, where it is writ= ten thus.

Aa.j.　　　　If

The Arte

If the proportion of the abatemente vnto a=
batemente be, as the whole is in proportion to the
whole. Then shall the residue bee in like propor=
tion to the residue, as the whole is to the whole.

That is in the laste example. As. 18. is vnto. 24. so
is 6 vnto 8. Therfoze shall 12 be to 16. as 18. is to 24.

And foz to exercise you the better, loe, here are one
oz twoo examples moze, of the like reduction.

$$\frac{7\text{Œ}}{8\sqrt{\text{з}}} \text{---} \frac{14\text{з}}{16\text{з з}} \text{ maketh } \frac{7\text{Œ}}{8\sqrt{\text{з}}} \text{ oz } \frac{7\text{я}}{8\text{з}}. \text{ Again } \frac{192\text{я}}{28\text{з}} \text{------} \frac{48\text{ze}}{7\text{Œ}}$$

yeldeth $\frac{192\text{я}}{28\text{з}}$ oz $\frac{48\text{я}}{7\text{з}}$.

But this muste you farther marke, that in *Cossike*
nombers, not onely the nombers, but also the *Cossike*
signes must bee, accozdyng to *Euclides* propofition.

Scholar. What doe I see.

Foz in the laste example: As. я. is to. з. so. ze. is
to. Œ.

And in the nexte example befoze: As. Œ. is to. √з.
so is. з. to. з з.

Likewaies in the other examples, as Œ is to з Œ
so is. з. to. √з.

Al this is good and reasonable.

Master. Now doe you see, bothe the maner of re=
duction, and also some reason foz it. Therfoze I will
pzocede, to declare the woozke of Addition.

Of Addition and Subtraction.

IN Addition there is nothyne moare,
then you haue learned befoze : Foz as
foz the multiplications of the denomi=
natozs together, and then crosse waies
with the numeratoz of thother, is iuste
agreable with the reductions of Ab=
stracte fractions, to bzyng theim to one common de=
nominatoz

of Cofsike nombers.

nominatoʒ.

And then the numeratoʒs added together, dooe make the newe numeratoʒ in addition.

And likewaies the leſſer numeratoʒ, ſubtracted frõ the other, doeth make the numeratoʒ in ſubtraction : wherfoʒe a fewe examples maie ſuffice.

Examples of Addition.

$$54.\textrm{z}. \;\underline{}\!\!+\!\!\underline{}\; .28.\textrm{ce}.$$
$$\frac{6}{7}.\textrm{z}. \quad \textrm{to} \quad \frac{4}{9}.\textrm{ce}.$$
$$63.$$

$$40.\textrm{z}. \;\underline{}\!\!+\!\!\underline{}\; .42.\textrm{ze}.$$
$$\frac{5}{6}.\textrm{z}. \quad \textrm{to} \quad \frac{7}{8}.\textrm{ze}.$$
$$48.$$

That is in ſmal= | $20.\textrm{z}. \;\underline{}\!\!+\!\!\underline{}\; .21.\textrm{ze}.$
ler termes. | $24.$

Here you ſee how the. 2. fractions be ſette betwene 2. lines : and vnder the nethermoſte line, is ſette the newe denominatoʒ : and ouer the higher line, are ſet the. 2. newe numeratoʒs ioyned in one.

The firſte of them, can not be reduced to any ſmal= ler termes, bicauſe the nombers be not all. 3. commẽ= ſurable : ſ the denominatoʒ, alſo is a nomber Abſtract.

The ſeconde hath alſo a nomber Abſtracte foʒ his denominatoʒ, and therfoʒe there can be noe reduction in ſignes : but the nombers all. 3. beyng comenſura= ble, ſ diuiſible by. 2. maie be reduced, as there you ſee.

More examples of Addition.

$$16.\textrm{fz}. \;\underline{}\!\!+\!\!\underline{}\; .4.\textrm{ce}.$$
$$12.\textrm{fz}. \;\underline{}\!\!+\!\!\underline{}\; .9.\textrm{ce}. \quad \textrm{to} \quad 4.\textrm{fz}. \;\underline{}\!\!\!\!\underline{}\; .5.\textrm{ce}.$$
$$20.\textrm{z ce}. \qquad\qquad 20.\textrm{z ce}.$$
$$20.\textrm{z ce}.$$

Aa.ii.   That

The Arte

That is in smal= $4.\mathfrak{z}. \quad + \quad .1.\wp.$
ler termes. $\overline{\qquad\qquad}$
                         $5.\mathfrak{ce}.$

Here is noe multiplication wroughte, bicause the
denominators are like.

Another Example of Addition.

$$5.\mathfrak{z}\mathfrak{ce}. \quad + \quad .20.\mathfrak{ce}. \quad \text{———} \quad .3.\sqrt{\mathfrak{z}}.$$

$$5.\mathfrak{z}\mathfrak{ce}. \quad + \quad .3.\sqrt{\mathfrak{z}}. \qquad\qquad 20.\mathfrak{ce}. \quad \text{———} \quad 6.\sqrt{\mathfrak{z}}.$$
$$\overline{6.\mathfrak{ce}\mathfrak{ce}.} \qquad \textit{to} \qquad \overline{6.\mathfrak{ce}\mathfrak{ce}.}$$

$$6.\mathfrak{ce}\mathfrak{ce}.$$

That is in les= $5.\mathfrak{ce}. \quad + \quad .20.\wp. \quad \text{———} \quad .3.\mathfrak{z}.$
ser termes. $\overline{\qquad\qquad\qquad\qquad}$
                       $6.\mathfrak{z}\mathfrak{ce}.$

Here is noe multiplication, nor reduction to one
common denominator : sith thei bee one all ready : no=
ther can the nombers be reduced, to any other leſſer :
but the quantities onely be reduced as you see.

Scholar. I praie you let me proue.

Another Example.

$$80.b\sqrt{\mathfrak{z}} \quad + \quad 90\,\mathfrak{z}\,\mathfrak{ce} \quad + \quad 60\,\mathfrak{z}\,\mathfrak{ce} \quad \text{———} \quad 30\sqrt{\mathfrak{z}}.$$

$$8.\mathfrak{ce} \quad + \quad 9.\mathfrak{z}. \qquad\qquad 6.\mathfrak{ce}. \quad \text{———} \quad .3.\mathfrak{z}.$$
$$\overline{10.\mathfrak{ce}.} \qquad \textit{to} \qquad \overline{10.\mathfrak{z}\mathfrak{z}.}$$

$$110.b\sqrt{\mathfrak{z}}. \qquad\qquad\qquad\qquad \text{That is}$$

Master. Marke your worke well, before you re=
duce it.

Scholar. I see my faulte : I haue sette. 2. nombers
seuerally, with one signe *Cosike* : by reason I did not
foresee, that. $\mathfrak{ce}$. multiplied with. $\mathfrak{ce}$. doeth make the
like

of Coſsike nombers.

like quantitie, as. $\partial\cdot\partial$. multiplied by. ∂. Therefore it ſhould be thus.

$$80.\text{\textit{bſ}}\partial. \quad + \quad .150.\partial.\text{\textit{c}}. \quad - \quad .30.\textit{ſ}\partial.$$
$$\overline{110.\text{\textit{bſ}}\partial.}$$

Whiche maie bee reduced, by meane of the nom= bers, to this ſomme.

$$8.\text{\textit{bſ}}\partial. \quad + \quad .15.\partial.\text{\textit{c}}. \quad - \quad .3.\textit{ſ}\partial.$$
$$\overline{11.\text{\textit{bſ}}\partial.}$$

And now conſideryng the *Coſsike* ſignes, and wor= kyng as J haue marked you to dooe : That is to abate the leaſte ſigne, out of theim all : bicauſe. $\textit{ſ}\partial$. is here the leaſte, J abate it out of. $\text{\textit{bſ}}\partial$. and there reſteth. ∂. and ſo doing with the other ſigne. $\partial\text{\textit{c}}$. there remai= neth. $\text{\textit{c}}$ & then $\textit{ſ}\partial$ out of $\textit{ſ}\partial$ doeth leaue. 0. or nöber : So will the fraction bee thus : $\dfrac{8\partial + 15\text{\textit{c}} - 3\text{\textit{0}}}{11\partial}$ by reduction in ſignes and nombers alſo.

Maſter. Seyng you haue ſo well marked the re= duction of the ſignes (whiche followeth the forme, taught before in diuiſion) J thinke it not nedefull, to ſtaie any longer aboute this.

Wherfore we will goe forward to ſubtraction, af= ter that J haue admoniſhed you of fractions, in appe= raunce ſimple, whiche in deede by addition, bee come compounde. As this $\frac{2}{3}\text{\textit{c}}$. added to $\frac{3}{4}\partial$. maie firſte be added by the common ſigne of addition, thus.

$\frac{2}{3}\text{\textit{c}}. \quad + \quad .\frac{3}{4}\partial$. whiche by reduction, vnto one deno= mination, wil be thus written. $\dfrac{8\text{\textit{c}} + 9\partial}{12}$

But as this is eaſie inough to vnderſtand, ſo maie it helpe often times, for ſpedie worke, as well in addi= tiö, as in ſubtractiö, by the onely addyng of the ſigne.

As if J would ſubtracte this fraction $\frac{5}{7}\partial\partial$. out of   *Subtraction.*
Aa.iii. $\frac{9}{10}\partial\text{\textit{c}}$.

The Arte

$\frac{9}{10}$ ʒℛ . I maie wꝛite it thus. $\frac{9}{10}$ ʒℛ . —— $\frac{5}{7}$ ʒʒ .
And ſo is the Subtraction wꝛoughte.

Yet maie you reduce theim, to one denomination, if you will, after theſame foꝛme, as you did in additi= on. And then will it bee. $\frac{63\text{ʒℛ} — 50\text{ʒʒ}}{70}$ whiche can not bee reduced to any ſmaller termes, bicauſe the nombers are not commenſurable : and one of theim (that is to ſaie, the denominatoꝛ) is a nomber *Abſtract*.

Scholar. I ſee in this, there is no difference from Addition, but in the ſignes. ——|—— and. ———— . wher= foꝛe I will pꝛoue an other example, by your leaue.

I would ſubtracte $\frac{3}{4}$ √ʒ . out of. $\frac{4}{5}$ ʒʒ . and it will bee at the firſte $\frac{4}{5}$ ʒʒ . —— $\frac{3}{4}$ √ʒ . And by reduction $\frac{16\text{ʒʒ} —— 15\,\sqrt{ʒ}}{20}$

Maſter. Your woorke is well doen, accoꝛdyng to your firſte meanyng : But as the numeratoꝛ of this laſte reduction doeth declare, it can not bee well, that 15. √ʒ . maie bee abated out of. 16. ʒʒ . Foꝛ the grea= ter abſolutely, can not well be abated out of the leſ= ſer : and therfoꝛe you might rather haue abated $\frac{4}{5}$ ʒʒ out of. $\frac{3}{4}$ √ʒ .

Scholar. I ſee it well now : foꝛ the √ʒ . is alwaies double oꝛ triple, oꝛ yet moꝛe tymes greater, then the ʒʒ . Bicauſe the √ʒ . commeth by multiplication of the ʒʒ by his firſte roote.

Maſter. Yet here in is diſcretion to be vſed, foꝛ in fractions, ſometyme the nomber of the greater ſigne maie be the leſſer. As foꝛ example $\frac{3}{16}$ √ʒ is leſſer then $\frac{3}{4}$ ʒʒ . as by reſolution you maie pꝛoue, accomptyng 2. foꝛ the common roote.

Scholar. 2. beyng the roote. 32. is the. √ʒ . and his $\frac{3}{16}$ maketh. 6. then. $\frac{3}{4}$ ʒʒ . beeyng. 12. dooeth appere double to it : and therefoꝛe greater by moche.

If I doe by the like reſolutiõ, pꝛoue the other frac= tions befoꝛe, $\frac{3}{4}$ √ʒ . will bee. 24 : and $\frac{4}{5}$ ʒʒ . will bee 12 $\frac{4}{5}$: whiche is leſſer moche.

So,

of Cossike nombers.

So, I perceiue the greatnesse and smalnesse of the fractions, must be considered, as well in the nombers as in the Cossike signes. And farther, if their fractions be nigh of one greatnesse, oz the fraction of the lesser signe the greater, then can not the subtraction, ap= peare reasonable.

Master. That is true, if those. 2. fractions stande alone : els beyng partes of other nombers, it maie ap= peare reasonable inough. As in this example of com= pounde fractions. $\frac{4}{10}$ cℓ. ——— $\frac{3}{4}$ ℥. maie bee abated out $\frac{3}{5}$ cℓ. ——— $\frac{3}{5}$ ℞. and yet in the abatemente af= ter ——— not onely the nomber $\frac{3}{4}$ is greater, then $\frac{3}{5}$ in the other, but also, the Cossike signe. ℥. is greater then the other Cossike signe. ℞.

Scholar. I consider it to be so : and yet $\frac{3}{5}$ cℓ. doeth so moche excede $\frac{4}{10}$ cℓ. that it supplieth sufficiently the other defaulte : els could it not be well doen.

But foz this woozke, I must craue your helpe : bi= cause I haue not seen the like.

Master. You maie doe in this, as I saied befoze, generally foz all subtractions.

Set doune bothe nombers in due ozder, so that the abatemente dooe folowe in ozder : and putte betwene them the signe of subtraction : as thus.

$$\frac{3}{5} cℓ ——— \frac{3}{5}℞. ——— \frac{4}{10} cℓ. ——— \frac{3}{4}℥.$$

Howbeit, if you will firste reduce euery cōpounde fraction, into one fraction, it will seme moze apte. As thus. $\frac{3}{5}$ cℓ. ———. $\frac{3}{5}$ ℞. beyng reduced by additiō will make $\frac{15cℓ ——— 15℞}{25}$. and by farther reduction of nombers. $\frac{3cℓ ——— 3℞}{5}$. Likewaies $\frac{4}{10}$. cℓ. ——— $\frac{3}{4}$℥. will make by the firste addition. $\frac{16cℓ ——— 30℥}{40}$. and by farther reduction $\frac{8cℓ ——— 15℥}{20}$.

Now ioyne theim together, with the signe of sub= traction, and thei will stande thus.

$$\frac{3cℓ ——— 3℞}{5} ——— \frac{8cℓ ——— 15℥}{20}.$$

Scholar.

The Arte

Scholar. This doeth appeare verie straunge vn=
to me : but by vse I shall finde it moze familiare : See=
yng I see the reason of this wozke, to agree with the
wozke of common fractions.

But foz pzoofe of it, I will resolue eche wozke, in=
to nombers absolute, accoumptyng. 2. foz a roote.

Master. So shall you finde it true : But foz easie
woozke, take rather. 10. foz the roote.

Scholar. I thanke you foz your aide.

Then if. 10. be the roote, the square will be. 100.
and the *Cube.* 1000. Now $\frac{3}{5}$ ℞. that is $\frac{3}{5}$ of. 1000.
is. 600. And $\frac{3}{5}$ of. 10. whiche is the roote, will bee. 6.
whiche bothe put together, doe make. 606. and that
is the greater nomber.

Then foz the lesser $\frac{4}{10}$ ℞. are in this example. 400
Foz the Cube beeyng. 1000. his $\frac{1}{10}$ is. 100. Againe
the square beyng. 100. $\frac{3}{4}$ ℥. must nedes bee
75. whiche beeyng put vnto. 400. dooeth
make. 475.

Then doe I abate. 475. out of. 606. and
there will reste. 131. How now?

606.

475.

131.

Master. I perceiue you staie, as beeyng astonis=
shed, bicause in the former wozke, there is not lefte a
remainer : But the. 2. firste sommes onely altered by
reduction, and ioyned together, with the signe of sub=
traction : wherein if you had continued your wozke,
you should haue founde thesame nombers.

Foz. 3. ℞. must nedes bee. 3000. seyng. 1. ℞. is a
1000. And also. 3 ℥. are. 30 : whiche bothe added to
gether, make. 3030. Diuide them by. 5. (as the deno=
minatoz would) and it will be. 606. as the valewe of
the firste fraction.

Then come to the later nomber : and you maie sone
thinke that. 8. ℞. are. 8000. And. 15. Squares are
1500. adde theim together, and thei will make
9500. whiche must bee diuided by. 20. (as the deno=
minatoz

of Cossike nombers.

minatoz impozteth) and there will a=
mounte. 475. the valewe of the leſſer
fraction : whiche nombers appeare the
ſame, that were befoze : and thereby
the woozke is good.

$$\begin{array}{r} xx \\ 9500 \ (475 \\ 2220 \end{array}$$

But if you will bzyng it to a remainer, doe thus.
Reduce theſe. 2. newe fractions, into one denomina=
tion.

Scholar. That can J doe, by multipliyng the nu=
meratozs together : that is. 20. by. 5. and thereof com=
meth. 100. whiche ſhall be the common numeratoz :
then muſt J multiplie in croſſe waies, the numeratoz
of the firſte, by the denominatoz of the ſeconde, and
contrarily.

So foz the firſte numeratoz
J woozke thus. And thereby
dooeth amounte (as you ſee)
60. ce. ——|—— 60. ze. And foz
the ſeconde numeratoz, J multiplie. 8. ce ——|—— 15 ʒ
by. 5. and there doeth riſe. 40. ce ——|—— .75. ʒ . eche
of theim hauyng one common numeratoz. 100.

$$\begin{array}{c} 3.\,ce.\ ——|——\ .3.\,ze.\\ 20.\\ \hline 60.\,ce\ ——|——\ 60.\,ze. \end{array}$$

Wherfoze, ſeyng bothe nombers, haue one deno=
minatoz, J ſhall abate the leſſer numeratoz out of the
greater, as here in example is ſet fozthe : and then the

$$\begin{array}{c} 60.\,ce.\ ——|——\ .60.\,ze.\\ 40\,ce.\ ——|——\ .75.\,ʒ.\\ \hline 20.\,ce.\ ——|——\ 60.\,ze.\ ———\ .75.\,ʒ. \end{array}$$

remainer will bee (as you ſee). 20. ce. ——|—— 60. ze.
——— 75. ʒ. vnto whiche J muſte adde the common
denominatoz. 100. and it will be thus.

$$\begin{array}{c} 20.\,ce.\ ——|——\ .60.\,ze.\ ———\ .75.\,ʒ.\\ \hline 100. \end{array}$$

The Arte

Now prone whether this remainer, doe not agree to thother remainer befoze, in your trial: which was 131

Scholar. 20 $c\!$ do make. 20000. & 60 $\mathcal{Z}$. yelde 600 : those 2 sommes I must adde together, bicause of the signe. ———. and it will be. 20600. then. 75. ჳ. are. 7500. whiche I must abate from the fozmer somme of. 20600. and there will remaine. 13100. foz the numeratoz, and 100. foz the denominatoz, thus $\frac{13100}{100}$.

$$\begin{array}{r} 20600. \\ 7500 \\ \hline 13100 \end{array}$$

Maſter. And what doe you thinke of it?

Scholar. By that I learned in the vulgare fracti= tions, I knowe that it is iuſte. 131. and ſo doeth it a= gree pzecisely, with the fozmer pzoofe.

Maſter. Well yet foz moare exactneſſe in this wozke, I will farther reduce that fractiõ, by diuiding the numeratoz by the denominatoz: wherfoze. 20. $c\!\!\!\!\!\!\!\!\!\!$ diuided by. 100. doeth yelde. $\frac{1}{5}$ $c\!\!\!\!\!\!\!\!\!$. And. 60. $\mathcal{Z}$. diui= ded by. 100. doeth make $\frac{3}{5}$ $\mathcal{Z}$. And laſtly. 75. ჳ. di= uided by. 100. will yelde $\frac{3}{4}$ ჳ. ſo is theſame fraction ſo reduced $\frac{1}{5}$ $c\!\!\!\!\!\!\!\!$ ——— $\frac{3}{5}$ $\mathcal{Z}$. ——— $\frac{3}{4}$ ჳ. And now trie what that is, by the fozmer pzoofe.

Scholar. I maie ſone perceiue, that $\frac{1}{5}$ $c\!\!\!\!\!\!\!\!$. is. 200. when the Cube is. 1000 : And ſo $\frac{3}{5}$ $\mathcal{Z}$. is. 6. whiche I muſt adde together, and it will be. 206. Then $\frac{3}{4}$ ჳ. is 75. whiche if I dooe abate from. 206. there will re= main. 131. agreably as befoze. And ſo is this wozke fully examined.

Maſter. Yet will I pzopounde one oz two exam= ples moze, partly to pzactiſe your memozie, and part= ly to admoniſhe you, if you happen to ſee any ſoche miſſe wzoughte, in ſome other bokes (as I haue doen) how you maie amende the erroure, and not ſtaie at it.

Firſte take this example. I would ſubtracte.

$$48.\mathcal{P}.\qquad\qquad\qquad 48\mathcal{P}.$$

——————— out of ———

$$12.\mathcal{Z}.\ ———\ 3.ჳ.\qquad\qquad 7.ჳ.$$

Scholar.

of Coſsike nombers.

Scholar. J muſt firſt multiplie the denominatoꝛs together, and ſo it will make, as here is ſette foorthe 84.℀.——— 21.ӡ ӡ .

Then J multiplie the nu=	12. 𝓏℮ ——— .3. ӡ .
meratoꝛ of the firſte, by the	7. ӡ .
denominatoꝛ of the ſeconde,	——————————————
and it will bꝛyng	84.℀ ——— 21. ӡ ӡ .

48. 9 .
7. ӡ .
———
336. ӡ .

foorthe. 336. ӡ : whiche is the numeratoꝛ foꝛ the abatement.

Afterward J multiplie the numeratoꝛ of the ſeconde,	12. 𝓏℮ ——— 3 ӡ .

by the denominatoꝛ of the firſte, and it will make	48. 9 .
————————————
576. 𝓏℮ . ——— 144. ӡ .	96
48
————————————
576. 𝓏℮ . ——— 144. ӡ .

Now if J ſubtracte that 336. ӡ . out of. 576. 𝓏℮ . ——— 144. ӡ . it will bee 576. 𝓏℮ ——— 480. ӡ .foꝛ the abatemēte that ſhould be ſubtracted now, is ſette after the ſigne ——— with the foꝛmer ſomme of .144.

Finally, to make the remainer complete, as that laſte nomber is the numeratoꝛ, ſo vnto it J muſt adde the common denominatoꝛ. 84.℀ . ——— .21. ӡ ӡ . and it will bee. $\frac{576\,𝓏℮ —— 480\,ӡ}{84\,℀ == 21\,ӡӡ}$. that is in leſſer termes $\frac{192\,9 —— 160\,𝓏℮}{28\,ӡ === 7\,℀}$.

Maſter. Now pꝛoue your cunnyng in this ſome, $\frac{48\,9}{12\,𝓏℮ —— 3\,ӡ}$, ſubtractyng it out of. $\frac{232\,𝓏℮ —+— 576\,9}{84\,ӡ —— 21\,℀}$.

Scholar. Firſte J muſt reduce theim, to one com= mon denominatoꝛ : by multipliyng bothe denomina=

84. ӡ . ——— .21.℀ .
12. 𝓏℮ . ——— .3 . ӡ .
————————————————
1008.℀ . ——— .252. ӡ ӡ .
63.√ ӡ . ——— .252. ӡ ӡ .
————————————————————
63.√ ӡ . —+— .1008.℀ . ——— .504. ӡ ӡ .

The Arte

toʒs together. And ſo wil it be. 63.ſℨ —+— 1008 ℭℛ
——— 504.ℨℨ. as by ſpeciall wooʒke, J haue here
pʒoued.

Then doe J multiplie the numeratoʒ of the totall,
by the denominatoʒ of the abatemente, as here alſo J
haue perticularly ſet foʒthe in wooʒke, foʒ my owne
eaſe, and auoidyng of erroure : And ſo J finde it to be
1056.ℨ. —+— .6912.ℛℯ. ———696. ℭℛ. whiche
ſhall bee the numeratoʒ of the totalle.

$$
\begin{array}{l}
232.\,\mathcal{R}e. \;—+—\; .576.\, \\
12.\,\mathcal{R}e \;—+—\; .3.\,\mathfrak{z}. \\
\hline
2784\,\mathfrak{z}. \;—+—\; .6912.\,\mathcal{R}e. \\
———696.\,\mathcal{C}\!\mathcal{R}. \;———\; 1728.\,\mathfrak{z}. \\
\hline
1056.\,\mathfrak{z}. \;—+—\; .6912.\,\mathcal{R}e. \;———\; .696.\,\mathcal{C}\!\mathcal{R}.
\end{array}
$$

Then doe J multiplie the numeratoʒ of the abate=
mēte, by the denominatoʒ of the totalle (whiche thing
is eaſily dooen, bicauſe the one nomber, is a nomber
Abſtracte) and ſo haue J foʒ the numeratoʒ of the aba=
temente. 4032.ℨ ——— 1008.ℭℛ.

And ſeyng theſe two nombers, haue one common
denominatoʒ, J ſhall abate the leſſer numeratoʒ, out

$$
\begin{array}{l}
1056.\,\mathfrak{z}. \;—+—\; 6912.\,\mathcal{R}e. \;———\; .696.\,\mathcal{C}\!\mathcal{R}. \\
4032.\,\mathfrak{z} \;——————————\; .1008.\,\mathcal{C}\!\mathcal{R}. \\
\hline
6912.\,\mathcal{R}e. \;—+—\; .312.\,\mathcal{C}\!\mathcal{R}. \;———\; .2976.\,\mathfrak{z}.
\end{array}
$$

of the greater, & ſo wil there be left foʒ the numeratoʒ
of the remainer 6912 ℛℯ —+— 312 ℭℛ ——— 2976 ℨ.
vnto whiche, J ſhall adde the common denominatoʒ,
and then will it be.

$$
\begin{array}{l}
6912.\,\mathcal{R}e. \;—+—\; .312.\,\mathcal{C}\!\mathcal{R}. \;———\; .2976.\,\mathfrak{z}. \\
\hline
63.\,\!\smallint\!\mathfrak{z}. \;—+—\; .1008.\,\mathcal{C}\!\mathcal{R}. \;———\; .504.\,\mathfrak{z}\mathfrak{z}.
\end{array}
$$

That

fo Coſsike nombers.

That is in leſſer termes.

$$2304.\wp. \longrightarrow .104.\chi. \longrightarrow .992.\mathcal{Z}e.$$
$$21.\chi\chi \longrightarrow .336.\chi. \longrightarrow .168.ce.$$

Maſter. You haue wꝛought it well. And hereby I coniecture, that you are experte inough in ſubtrac=
tion. Wherfoꝛe now we will goe in hand, with mul=
tiplication and diuiſion.

Of Multiplication.

AND firſte, concernyng multiplica=
tion, here is no moꝛe to bee ſaied,
then hath been taughte befoꝛe.

For the nombers ſhall bee mul=
tiplied, as common fractions are
wonte to bee : that is to ſaie, nume=
ratoꝛ, by numeratoꝛ, and denomi=
natoꝛ, by denominatoꝛ.

Multipli= cation.

And foꝛ the chaunge of their denominations *Coſ=
ſike*, the rules giuen befoꝛe ſhall ſuffice : ſo that a fewe
examples ſhall ſufficiently inſtruct you, in the woꝛke
of it.   As this foꝛ the firſte.

$$20.\chi. \longrightarrow 19.\mathcal{Z}e.$$
$$6.ce \longrightarrow 3.\wp.$$
$$120.\int\chi. \longrightarrow 114.\chi\chi.$$
$$\longrightarrow .60.\chi. \longrightarrow 57.\mathcal{Z}e.$$
$$120.\int\chi \longrightarrow 114.\chi\chi \longrightarrow 60.\chi \longrightarrow 57\,\mathcal{Z}e.$$

Where I dooe multiplie.

$$20.\chi. \longrightarrow 19.\mathcal{Z}e. \quad by \quad 6.ce. \longrightarrow .3.\wp.$$
$$31.ce. \qquad\qquad\qquad 4.\chi.$$

And firſte I ſhall multiplie, numeratoꝛ by nume=

Bb.iii.   ratoꝛ :

The Arte

rato꞉ : where. 20. ꝫ. multiplied by. 6. ℀. dooe make 120. ſꝫ. as the former table of multiplication, for chaunge of *Cofsike* ſignes doeth declare. And ſo in all the reſte, there is no difficultie, if you remember that, that you haue learned before.

Scholar. I perceiue it well. And ſo the whole newe numerator will bee. 120. ſꝫ. —— 114. ꝫꝫ. —— .60. ꝫ. —— .57. ℥. And the denominator will be. 124. ſꝫ.

So will the whole fraction bee.

$$\frac{120.\,\mathit{ſꝫ} \;+\; 114.\,\mathit{ꝫꝫ} \;-\; 60.\,\mathit{ꝫ} \;+\; 57.\,\mathit{℥}}{124.\,\mathit{ſꝫ}}$$

That is not to bee reduced to ſmaller termes of nom= bers, bicauſe thei be vncommenſurable, but in *Cofsike* ſignes, it mighte bee broughte to one leſſer, as.

$$\frac{120.\,\mathit{ꝫꝫ} \;+\; 114.\,\mathit{℀} \;-\; 60.\,\mathit{℥} \;+\; 57.\,\mathit{ꝗ}}{124.\,\mathit{ꝫꝫ}}$$

Now will I proue an other nomber, as fortune doeth offer it to mynde. That is $\frac{32\,℀ \;-\; 28\,\mathit{ſꝫ}}{21\,ꝫ \;-\; 5\,℥}$. to bee multiplied by. $\frac{12\,ꝫ \;+\; 9\,ꝗ}{36\,℥}$.

An Abſurde nomber ex= preſſeth leſſe then naught

Maſter. It appeareth that you take theim, at all aduentures. For your firſte nomber, ſemeth to be an *Abſurde* nomber. Seyng his numerator, is leſſe then naughte, in appearaunce. And then maie it not bee diuided by any nomber : and moche leſſe by ſo greate a denominator.

Scholar. It is eaſie to ſee, now that I am admo= niſhed thereof. For it is not poſſible, that any *Surſolide* nomber, can bee leſſe then fower tymes ſo moche, as the *Cube* of theſame nature. Seeyng euery *Surſolide* is made, by multiplicyng the *Cube* by the *ſquare* of the like *Roote*, but leſſe then. 4. is there no *Square*. And there= fore euery *Surſolide*, doeth excede his *Cube* fower times at the leaſte.

So

of Cossike nombers.

So that. 32.℞. ——— 8.√℥. were nothyng, and so is an *Absurde* nōber. And therfoze. 32.℞ ——— 28.√℥. is moche lesse then nothyng, and is therby an *Absurde* nomber also.

Master. Yet maie your example serue, to teache and pzactise multiplication by, as well as any other.

And farthermoze, J will tell you by this occasion, that J spake to you, moze after the opinion of the cō= mon nomber of artes men, then after my owne iud= mente.

Scholar. J might thinke so, by termynge of your sentence : but yet was your saiyng true.

Master. Yet maie that fraction stand well, if you take a bzokē nomber *Abstracte* foz the roote. Although in whole nombers, it bee an *Absurde* nomber.

Scholar. That will J pzoue, by settyng. $\frac{3}{4}$. foz a

$\frac{3}{4}$ The *Roote.*
$\frac{9}{16}$ The *Square.*
$\frac{27}{64}$ The *Cube.*
$\frac{81}{256}$ The *zenzizenzike.*
$\frac{243}{1024}$ The *Sursolide.*

Roote. Then will the *Square* be $\frac{9}{16}$. and the *Cube.* $\frac{27}{64}$. Also the *Square of squares* will bee. $\frac{81}{256}$. And the *Sursolide* $\frac{243}{1024}$.

And now to pzoue by reso= lution, how my nomber will rise, J take. 32. ℞. that is. $\frac{864}{64}$, oz 13. $\frac{1}{2}$. whiche J note as the firste somme. Then J take likewaies. 28.√℥. whiche yeldeth $\frac{6804}{1024}$, that is. $6\frac{165}{256}$. And now J see that J maie abate it very well, out of. 13$\frac{1}{2}$.

Master. So maie you see, that as in whole nom= bers, euer moare the greater *Cossike* signes, will haue the greateste nombers : So in fractions resolued by *Cossike* signes, the greatest fraction, aunswereth to the leaste signe : and the leaste fractiō, agreeth to the grea= teste signe.

The reason of it is this. That the moare any frac= tion is multiplied by a fraction, the lesser it waxeth. Foz as whole nombers by multiplication, maie in= crease infinitely : so fractions by multiplication, maie

decrease

The Arte

decreaſe infinitely.

But befoze wee paſſe from multiplication, J will pzoue you with one example moare. J would haue $\frac{19\,\mathbb{c}\!-\!\!-\!\!3\,\mathbb{z}\!-\!\!-\!\!5\,\mathbb{q}}{7\,\mathbb{z}\!-\!\!-\!\!6\,\mathbb{q}}$ multiplied by $\frac{24\,\mathbb{c}}{9\,\mathbb{z}}$——. $\mathbb{z}$.

Scholar. J am troubled with the multiplier. Foz J knowe not what to make of it?

Maſter. You doubte (J thinke) of the numeratiõ of it, bicauſe you had not the like example befoze : foz it is a mixte number of a fraction, and a whole nom= ber. But ſeyng the ſigne of abatemente, is ſet againſt the whole fraction, and nother againſte the numera= toz, noz denominatoz, therfoze muſt that 4. $\mathbb{z}$. be vn= derſtande, to be abated out of the full fraction.

Scholar. Now J perceiue the mater. Foz there might be. 3. diuerſe fozmes, to place that abatemente. As here J haue ſet them. $\frac{24\,\mathbb{c}\!-\!\!-\!\!4\,\mathbb{z}}{9\,\mathbb{z}}$ & $\frac{24\,\mathbb{c}}{9\,\mathbb{z}\!-\!\!-\!\!4\,\mathbb{z}}$.

And as it was ſet by you, $\frac{24\,\mathbb{c}}{9\,\mathbb{z}}$.——4. $\mathbb{z}$. whiche J will reſolue into abſolute nombers, to ſee their dif= ference the better. And ſo, taking 3. foz the roote, theſe will be their. 3. fozmes.

The firſte.   Foz the firſte $\frac{648\!-\!\!-\!\!12}{81}$. oz els $\frac{636}{81}$ that is $\frac{212}{27}$.

Foz the ſeconde $\frac{648}{81\!-\!\!-\!\!12}$ oz els $\frac{648}{69}$ that is $\frac{216}{23}$.

And foz the thirde nomber, whiche is our ſpecialle nomber. $\frac{648}{81}$——12. that is. 8.——.12. and is an *Abſurde* nomber. Foz it betokeneth leſſe then naught by. 4.

Maſter. Jf you would haue it no *Abſurde* nom= ber, you muſt increaſe the pzopoztion of the fraction, by augmentyng the numeratoz, oz abatyng the deno= minatoz, oz els thirdly, by abatyng the nomber, after the ſigne of abatement. As $\frac{40\,\mathbb{c}}{9\,\mathbb{z}}$——4. $\mathbb{z}$: oz els ſecondarily, thus. $\frac{24\,\mathbb{c}}{4\,\mathbb{z}}$.——.4. $\mathbb{z}$. oz thirdely $\frac{24\,\mathbb{c}}{9\,\mathbb{z}}$——2. $\mathbb{z}$.

Howbeit foz examples ſake, you maie woozke, as well with *Abſurde* nombers, as with any other.

                                                      But

of Cofsike nombers.

But for you eafe, I will fhewe you the woozke of this example, in twoo formes.

Firft, you fhall multiplie the firfte whole nomber, by the fraction of the feconde nomber, that is.

$$\frac{19\text{\ss}———3\text{\ss}———5\text{\ss}}{7\text{\ss}———6\text{\ss}} \text{ by } \frac{24\text{\ss}}{9\text{\ss}}, \text{ and it will bee.}$$

$$\frac{456.\text{\ss}\text{\ss}. \quad+\quad 72.\text{\ss}\text{\ss}. \quad—— \quad .120.\text{\ss}.}{63.\text{\ss}\text{\ss}. \quad—— \quad .54.\text{\ss}.}$$

As here in woozke you maie fee it plaine.

$$\begin{array}{l}
19.\text{\ss}. \quad+\quad .3.\text{\ss}. \quad—— \quad .5.\text{\ss}. \\
24.\text{\ss}. \\
\hline
456.\text{\ss}\text{\ss} \quad+\quad .72.\text{\ss}\text{\ss}. \quad—— \quad .120.\text{\ss}.
\end{array}$$

$$\begin{array}{l}
7.\text{\ss}. \quad—— \quad .6.\text{\ss}. \\
9.\text{\ss}. \\
\hline
63.\text{\ss}\text{\ss} \quad—— \quad 54.\text{\ss}.
\end{array}$$

That is in lefſer termes, bothe of nombers, and of fignes *Cofsike.*

$$\frac{152.\text{\ss}\text{\ss}. \quad+\quad .24.\text{\ss}. \quad—— \quad .40.\text{\ss}.}{21.\text{\ss}. \quad—— \quad .18.\text{\ss}.}$$

And this is the firfte parte of your fomme.

Then for the nexte parte, multiplie your firfte nõ= ber, that is $\frac{19\text{\ss}——3\text{\ss}——5\text{\ss}}{7\text{\ss}——6\text{\ss}}$ by the abatement of the feconde nomber, that is by. —— .4. $\text{\ss}$. and it will be.

$$\frac{20.\text{\ss}. \quad—— \quad .76.\text{\ss}\text{\ss}. \quad—— \quad .12.\text{\ss}.}{7.\text{\ss}. \quad—— \quad .6.\text{\ss}.}$$

The Arte

As by this woo2ke you maie see.

$$19. c\!\!\!C. \;\rule[0.4ex]{1em}{0.4pt}\!\!\!+\;\; .3. \mathcal{Z}. \;\rule{2em}{0.4pt}\; .5. \mathcal{P}.$$
$$\rule{2em}{0.4pt}\; .4. \mathcal{Z}.$$
$$20. \mathcal{Z}. \;\rule{2em}{0.4pt}\; .76. \mathcal{Z}\mathcal{Z}. \;\rule{2em}{0.4pt}\; .12. \mathcal{Z}.$$

whiche being reduced to the denomination of the fo2=
mer nomber, will be tripled (fith that denominato2 is
triple to this) and fo will it be $\dfrac{60\mathcal{Z}\;\rule{1.5em}{0.4pt}\;228\,\mathcal{Z}\mathcal{Z}\;\rule{1.5em}{0.4pt}\;36\,\mathcal{Z}}{21\,\mathcal{Z}\;\rule{1.5em}{0.4pt}\;18\mathcal{P}}$
Now adde thofe two nombers together, by puttyng
their bothe numerato2s in one, and it will be.

$$20. \mathcal{Z}. \;\rule{2em}{0.4pt}\; .76. \mathcal{Z}\mathcal{Z}. \;\rule{2em}{0.4pt}\; .12. \mathcal{Z}.$$
$$21. \mathcal{Z}. \;\rule{2em}{0.4pt}\; .18. \mathcal{P}.$$

As here appeareth in woo2ke.

$$152. \mathcal{Z}\mathcal{Z}. \;\rule[0.4ex]{1em}{0.4pt}\!\!\!+\;\; .24. \mathcal{Z}. \;\rule{2em}{0.4pt}\; .40. \mathcal{Z}.$$
$$60. \mathcal{Z}. \;\rule{2em}{0.4pt}\; .228. \mathcal{Z}\mathcal{Z}. \;\rule{2em}{0.4pt}\; .36. \mathcal{Z}.$$
$$20. \mathcal{Z}. \;\rule{2em}{0.4pt}\; .76. \mathcal{Z}\mathcal{Z}. \;\rule{2em}{0.4pt}\; .12. \mathcal{Z}.$$

whiche will not bee reduced to any fmaller fraction,
bicaufe the nombers be incommenfurable and one of
the *Coffike* fignes is. $\mathcal{P}$. And fo is that the fomme of the
multiplication.

An other waie you maie woo2ke it, and all foche
like, by reducynge the multiplier, into one vnifozme
fraction. As here in. $\dfrac{24 c\!\!\!C}{9\mathcal{Z}}$ $\rule{2em}{0.4pt}$ 4. $\mathcal{Z}$. you fhall mul=
tiplie $\rule{2em}{0.4pt}$ 4 $\mathcal{Z}$. by. 9. $\mathcal{Z}$. whiche is the fo2mer de=
nominato2, and it will be $\rule{2em}{0.4pt}$ .36. $c\!\!\!C$. Then putte
that to. 24. $c\!\!\!C$. ouer the line, and fet the common de=
nominato2. 9. $\mathcal{Z}$, vnder the line, and it will be in one
fraction reduced $\dfrac{24 c\!\!\!C \;\rule{1.5em}{0.4pt}\; 36 c\!\!\!C}{9\mathcal{Z}}$.

Scholar. Here J maie fee at the firfte vewe, that
this fraction is an *Abfurde* nomber : fo2 the abatement
after the figne $\rule{2em}{0.4pt}$, is greater then the nomber be=
fo2e

of Cossike nombers.

foze it.

Master. That was cõfessed befoze. But yet maie you wozke the erample by it.

Scholar. That is true : and so will the numera=toes, beeyng multiplied together, make eractely, 60.℞.——— 228.ℨ℞.——— .36.ℨℨ. As here in erample of woozke, J haue set it, foz my owne ease and certentie.

$$
\begin{array}{l}
19.℞.\;+\!\!-\;.3.℥.\;————\;.5.\Phi. \\
24.℞.\;————\;.36.℞. \\
\hline
456.ℨ℞.\;+\!\!-\;72.ℨℨ.\;————\;.120.℞. \\
————\;684.ℨ℞.\;————\;108ℨℨ.\;+\!\!-\;.180.℞. \\
\hline
60.℞.\;————\;.228.ℨ℞.\;————\;.36ℨℨ.
\end{array}
$$

And that is the newe numeratoz.

And then foz the seconde nomber, if the firste deno=minatoz. 7ℨ———6.Φ. be multiplied by the seconde denominatoz, 9.ℨ. it is easily seen, that thei will make. 63ℨℨ———54ℨ. whiche shall be the newe denominatoz.

And so the intere fraction shall bee.

$$
\frac{60.℞.————.228.ℨ℞.————.36.ℨℨ.}{63.ℨℨ————.54.ℨ.}
$$

That is in the smalleste nombers and figures Cossike.
$$
\frac{20℥————76ℨℨ————12ℨ}{21ℨ————18\Phi}
$$: whiche somme, dooeth in all thynges fully agree, with the foemer nomber that you wzought.

Master. Proue theim bothe by resolution : And then shall you knowe, the reason of their agremente.

Scholar. J see that the woozke of the denomina=toes, doeth agree. Wherfoze J will take. 3. foz a roote to pzoue how the woozke of the numeratoes wil agree

And so foz. 19.℞. J shall haue. 513. And foz. 3.℥.

The Arte

I ſhall haue. 9. to be added to. 513. And ſo haue I. 522
out of whiche ſomme I muſt abate. 5.
And then remaineth. 517. to bee multi=   648
plied by. 24. ℞. that is by. 648. And the   5 17
totalle will bee (as here in woꝛke appea=   4536
reth). 335016. whiche ſomme muſt be a=   648
bated to a ſmaller nomber, in like rate as   3240
the other was reduced, firſte by partition   335016
into. 3. And then will it be. 111672. And
again, it muſt bee diuided by. 9. foꝛ that is the quan=
titie of a ſquare, by whiche the foꝛmer reduction, was
wꝛoughte foꝛ the *Coſſike* ſignes : and then will it bee.
12408. And that is the firſte parte of the firſt woꝛke.
Then foꝛ the ſeconde parte of that woorke, I ſhall
multiplie the firſte nombers, that is 517 by the abate=
mente of the fraction, that is by —— 4℞, oꝛ. —— 12.
(ſith. 3. is the roote) and thereof will come —— 6204.
whiche ſomme I muſt triple, as I did his equalle (that
is. 20.℞. —— 76.ℨℨ. —— 12.ℨ.) And ſo wil
it bee —— 18612. Now ſhall I adde this ſomme,
with the firſte parte, whiche was. 12408. and it will
bee. 12408. —— 18612. that is. 6204. leſſe then
nothyng : and is the numeratoꝛ of the firſte woꝛke.

Wherfoꝛe I pꝛocede to the ſeconde woorke, where
the numeratoꝛ of the fraction, beeyng reduced to the
common denominatoꝛ, is. 24. ℞. —— 36. ℞. whi=
che is —— 12. ℞. and in nombers reſolute (keping
3. ſtill as a roote) it is —— 324. by whiche if I mul=
tiplie. 517. it will yelde. 167508. And that ſomme
beyng abated, by diuiſion into. 3. and. 9. as the other
was, oꝛ els diuided by. 27. whiche is all one, it giueth
6204. as the foꝛmer woorke did.

Maſter. Thus I ſee, you are experte inoughe in
multiplication : Wherfoꝛe I will ſhewe you now the
oꝛder and foꝛme of diuiſion.

Of

of Cossike nombers.
Of Diuision.

Here is noe speciall rule to be giuen, for the woorke of Diuision, other then soche as are all ready taughte in other workes of diuisiõ before. Wherfore I will by one or 2. examples, shewe you the woorke of it.

The firste example of Diuision.

$$\frac{14.\text{℀} \ + \ 9.\text{ʒ}.}{15.\text{ᵱ}.} \quad \text{to be diuided by} \quad \frac{5.\text{ʒ}. \ + \ 2\,\text{℥}.}{3.\text{℥}.}$$

doeth yelde. $\frac{42.\text{ʒʒ}. \ + \ 27.\text{℀}.}{75.\text{ʒ}. \ + \ .30.\text{℥}.}$ that is in a lesser fraction, by bothe reductions of nombers & signes.

$$\frac{14.\text{℀}. \ + \ 9.\text{ʒ}.}{25.\text{℥}. \ + \ 10.\text{ᵱ}.}$$

An other example.

$$\frac{12.\sqrt{\text{ʒ}}. \ - \ 16.\text{ʒ}.}{2.\text{℀}. \ - \ .5.\text{℥}.} \quad \text{diuided by} \quad \frac{19.\text{ʒ}. \ - \ .3.\text{ᵱ}.}{4.\text{ʒ}. \ - \ .5.\text{ᵱ}.}$$

doeth make.

$$\frac{48.\text{ᵇ}\text{ʒ}. \ + \ 60.\sqrt{\text{ʒ}}. \ - \ 64.\text{ʒʒ} \ - \ 80\text{ʒ}.}{38.\sqrt{\text{ʒ}}. \ + \ 15.\text{℥}. \ - \ 101.\text{℀}.}$$

whose nombers bee incommensurable, and therefore maie not bee reduced, but by abatyng one denomination Cossike. And so will it be.

$$\frac{48.\text{ʒ}\,\text{℀}. \ + \ 60.\text{ʒʒ} \ - \ 64.\text{℀} \ - \ 80.\text{℥}.}{38.\text{ʒʒ}. \ + \ .15.\text{ᵱ}. \ - \ .101.\text{ʒ}.}$$

Cc.iii.   Scholar.

The Arte

Scholar. I see that you multiplie crosse waies (as in vulgare fractions) the numerator of the one nomber, by the denominator of the other. And so is diuision of noe difficultie, to hym that remembreth the former rules.

Of the golden rule.

Master.

He golden rule, that is the rule of proportion, should folowe now, by the commō order. But seyng there is no difficultie in it, nother any other forme of woorke, then is in vulgare nombers, I will not staie any tyme aboute it. Saue that for your pleasure, I haue set here certaine examples, as wel in whole nombers *Cossike*, as in broken.

$$32.\gamma. \diagdown 4.\sqrt{\gamma}. \qquad 250.\mathcal{C}. \diagdown 20.\mathcal{Z}.$$
$$6\mathcal{C}. \diagup \tfrac{3}{4}\gamma\mathcal{C}. \qquad 26.\gamma. \diagup 2\tfrac{2}{25}\mathcal{C}.$$

$$5.\gamma.+.3.\mathcal{Z}. \diagdown 4.\mathcal{C}.-.5.\wp.$$
$$15.\mathcal{C}-.9.\wp. \diagup \dfrac{60\gamma\cdot\mathcal{C}-111\mathcal{C}+45\wp}{5\gamma+3\mathcal{Z}}$$

$$\dfrac{3\mathcal{C}+5\gamma}{12\mathcal{Z}} \diagdown 14.\mathcal{Z}.+.4.\wp.$$
$$61\gamma-7\wp \diagup \dfrac{10248\gamma\gamma+2928\mathcal{C}-1170\gamma-336\mathcal{Z}}{3\mathcal{C}+5\gamma}$$

Scholar. These fewe examples, dooe sufficiently teache the forme of the whole rule. So that here neadeth noe farther explication.

Wherfore, if in this arte, there be any forme of extraction of rootes, I praie you to procede therto.

Of

of Cossike nombers.
Of extraction of rootes.

Maſter.

S in nombers *Abſtracte*, euery nomber is not a rooted nomber, but ſome certaine onely emongeſt theim, ſo in nombers *Coſsike*, all nombers haue not rootes : but ſoche onely emongeſt ſimple *Coſsike* nom= bers are rooted, whoſe nomber hath a roote, agrea= ble to the figure of his denomination.

So that. 16 cℰ. is not a Square nomber, nother hath any roote. Foꝛ although. 16. bee a ſquare nom= ber, and hath. 4. foꝛ his roote, yet the denomination (whiche is. cℰ.) hath noe ſquare roote : but. 16. ℨ. is a ſquare nomber : and hath. 4.ℤℯ, foꝛ his roote.

Likewaies. 8. cℰ. is a *Cubike* nomber, and his roote is. 2.ℤℯ : but. 8. ℨ. hath noe roote. Foꝛ bicauſe. 8. hath no ſquare roote, agreable to the ſigne. ℨ. nother is it a *Cubike* nomber, although it haue a *Cubike* roote, bi= cauſe the roote is diſagreable from the ſigne. ℨ.

Scholar. J perceiue that in theſe nombers, as wel as in all other, the roote beeyng multiplied by it ſelf, will make the nomber, whoſe roote it is. And there= foꝛe can no nomber be called ſquare, oꝛ *Cubike*, oꝛ any waies els a rooted nomber, excepte the roote of the nomber agree with his ſigne : Whereby J perceiue well, that. 32.ſℨ. is a rooted nomber, foꝛbicauſe that 32. hath a *Surſolide* roote, agreable to the ſigne. So likewaies. 125. cℰ. is a rooted nomber, ſeyng 5. is the *Cubike* roote of. 125. But. 27. ℨ. is no rooted nöber.

Maſter. Thus you vnderſtande ſufficiently, the iudgemente of rooted nombers, and their knowlege, in ſimple *Coſsike* nöbers, that be vtterly vncopöunde.

Wherfoꝛe, foꝛ extraction of their rootes, take this bꝛief oꝛder.

                                                          Extracte

The Arte

Extracte the roote of your nomber, as if it were abſolute, and put to it. *ℨℯ* . foz the denomination.

So. 27. *Cubes* hath foz his roote. 3. *ℨℯ* .

And. 49. *ℨ·* . hath. 7. *ℨℯ* . foz his roote.

Again, the roote of. 216. *cℰ* . is. 6. *ℨℯ* .

Scholar. This I perceiue. And by like reaſon, the roote of. 243. *ſℨ·* . is. 3. *ℨℯ* . But why dooe you name nöbers *Coſsike* vtterly vncompounde? Foz as I vnderſtande, that there bee nombers compounde, in their ſignes, ſo I ſee that thei maie haue rootes alſo.

As. 16. *ℨ·ℨ·* . hath foz his roote. 2. *ℨℯ* . And likewaies. 64. *ℨ·cℰ* . hath. 2. *ℨℯ* . foz his roote.

Maſter. And dooe you not ſee, that thoſe compounde nombers, maie haue moare rootes then one? Sith. 16. *ℨ·ℨ·* . hath foz his ſquare roote. 4. *ℨ·* . as wel as it hath. 2. *ℨℯ* . foz his *zenzizenzike* roote.

So. 4. *ℨ·ℨ·* . hath foz his Square roote. 2. *ℨ·* . And hath no *zenzizenzike* *ℨℯ* agreable to his whole ſigne.

Likewaies. 9. *ℨ·cℰ* . hath no *zenzicubike* roote, acco?ding to his whole ſigne : but it hath a ſquare roote agreable to parte of the ſigne, and that is. 3. *cℰ* .

Scholar. I ſee that alſo. And ſo hath. 8. *ℨ·cℰ* . noe *zenzicubike* roote, but a *Cubike* roote : which is. 2. *ℨ·* .

Maſter. Therfoze in cöpoüde ſignes, if the ſigne maie haue ſoche a roote, as the nomber will yelde, it is a rooted nomber, els not.

Whereby you maie perceiue, that if any nomber cöpounde in ſigne, haue a roote agreable to his whole ſigne, then maie it haue alſo, as many rootes, as ther be partes in that compounde ſigne.

So. 4096. *ℨ·ℨ·cℰ* . hath not onely a *zenzizenzicubike* roote, which is. 2. *ℨℯ* : but it hath a *Square* roote, that is. 64. *ℨ·cℰ* . And alſo it hath a *Cubike* roote, that is. 16. *ℨ·ℨ·* : Farther moze, it hath a *zenzizenzike* roote, whiche is. 8. *cℰ* . And fourthly, it hath a *zenzicubike* roote, that is. 4. *ℨ·* .

And

of Cossike nombers.

And so shall you iudge, of all other like.

Scholar. This shall suffice, as I will practise the mater, at moare leiser. But and if the nombers bee compounde, with signes of addition, is there then any speciall order for their rootes? As in this example. $81.\mathfrak{z}\mathfrak{z}$. ———— $27.c\!\!\mathcal{C}$. where I haue made eche parte to be a rooted nomber.

Master. In deede. $81\mathfrak{z}\mathfrak{z}$. hath bothe a Square roote, and also a *zenzizenzike* roote. But. $27c\!\!\mathcal{C}$. hath none of those twoo rootes, although it haue a *Cubike* roote, whiche the other nomber wanteth. And therfore is not that whole nomber, a rooted nomber.

But to the intente, that you maie be the more certein of rooted nombers, I will tell you certein notes, how it maie bee knowen, whether your number be a rooted nomber.

Firste, if the nomber annexed to the greatest signe of that compounde *Cossike* nomber, bee not a rooted nomber, the whole nomber can not be a rooted nöber

Secondarily, if the nomber that is ioyned with the leaste *Cossike* signe, be not a rooted nomber, the whole nomber can not be a rooted nomber.

And eche of these bothe rootes (if thei haue any) are partes of the whole roote, for the compounde *Cossike* nomber.

Thirdly, if the nomber be a rooted nomber, euery parte of it, that is not a rooted nomber, is a meane nomber, betwene the greateste and the leaste.

Fourthly, if. $\mathcal{Z}\!\!e$. bee any denomination in it, then is. $\mathcal{J}$. an other denomination in it also.

Fiftly, and generally, all rooted nöbers, other are specially framed, by orderly multiplication, or els are nombers equalle to some one rooted nomber *Abstract*.

Now specially framed are soche, as are made by multiplicatiö of one nomber by it self, and litle or nothyng altered from that very forme.

           Do.i.   Example

The Arte

Exãple of. 529 ʒ cℰ ——— 184 ʒ ʒ ——— 16 ʒ whiche is a *Square* nomber, made by multiplication of. 23. cℰ ——— 4. ze . by it ſelf. This nomber maie haue his Roote o2derly extracted thus.

529. ʒ cℰ ——— 184 ʒ ʒ ——— 16 ʒ (23 cℰ + 4 ze
23   46. cℰ.

In the firſte nomber, I finde the *Square* roote to bee 23. And fo2 his denomination, I take halfe the *Coſsike* ſigne ʒ cℰ , and that is. cℰ . Fo2 as. cℰ . multiplied by cℰ. doeth make. ʒ cℰ. So in diuiſion by. 2. and in ex=traction of *Square* rootes, I ſhall take the. cℰ . fo2 the halfe of ʒ cℰ and the denomination of his roote : and ſo ſet it doune in the *quotiente*.

Then I ſhall double the nomber *Abſtracte* of that *quotiente* (kepyng his *Coſsike* ſigne vnaltered) and that double ſhall I ſet euermo2e vnder the nexte nomber, toward the righte hande. As here, you ſee, I haue ſet 46 (whiche is the double of 23) with his ſigne cℰ. vn=der the ſeconde nomber. And there I perceiue, I maie haue it. 4. tymes, if I doe diuide (as I ought) 184. by 46. And that. 4. I ſette in the *quotiente*, with the ſigne ——— , and the denomination. ze : ſeyng. ʒ ʒ. diui=ded by. cℰ. doeth yelde. ze .

Laſte of all, I muſte multplie that parte of the *quo=tiente*. 4. ze . by it ſelf, and it will yelde. 16. ʒ. whiche beyng ſubtracted alſo (as it ſhould) leaueth nothyng remainyng of the ſquare nomber.

This o2der muſt you kepe in all ſquare nombers, how greate ſo euer thei be. As in this ſeconde exãple.

————— 90 ʒ ʒ .
25 ʒ cℰ ——— 80 ſ ʒ —— 26 ʒ ʒ ——— 144 cℰ ——— 81 ʒ (5 cℰ + 8 ʒ — 9 ze
5. cℰ.   10 cℰ ——— 64 ʒ ʒ .
——— 10 cℰ ——— 16 ʒ . ——— 9. ze .

The

of Cossike nombers.

The roote of the firſt nomber is. 5 ℀ , whiche I ſet in a *quotiente.*

Then doe I double that. 5, and it maketh. 10, to be ſette vnder. 8. with his denomination, whiche is. ℀. like to the roote.

That. 10. ℀. maie be founde in. 80. ſʒ. 8. times, ⁊ therfore I ſet. 8. in the *quotiente,* with the ſigne ——⊦— and the denomination. ʒ. And then dooe I multiplie that. 8. ʒ. ſquaredly, whiche giueth. ——⊦— 64. ʒʒ. to be ſubtracted out of —— 26. ʒʒ. and ſo remai= neth —— 90. ʒʒ.

After this I double all the *quotiente* again, where= of commeth ——⊦— 10. ℀. ——⊦— 16. ʒ. And bicauſe there is a remainer, ouer the nomber that I wrought laſte, I muſt ſet. 10. ℀. vnder the remainer, and the other nomber in order, as you ſee it ſet.

Then ſeke I how often tymes maie. 10. ℀. diuide 90. ʒ ʒ, and I finde the *quotiente* to be —— 9. ℃. And likewaies ——⊦— 16 ʒ. multiplied by —— 9 ℃ doeth make —— 144. ℀. equalle to the ſomme o= uer it : And ſo ſubtracteth it cleane. Wherfore to ende that woorke, I multiplie the laſte *quotiente,* by it ſelf ſquare, and it yeldeth. ——⊦— 81. ʒ. whiche is to bee ſubtracted out of the like ſomme, in the ſquare nom= ber : and ſo reſteth nothyng. Wherefore I iuſtly af= firme, that the firſte nomber is a ſquare nomber, and hath for his roote. 5. ℀. ——⊦— 8. ʒ. —— 9. ℃.

Scholar. That maie I ſone proue, if I multiplie

$$5℀. \quad ——⊦—8ʒ. \quad .9. ℃.$$
$$5℀. \quad ——⊦—8ʒ. \quad .9. ℃.$$

$$25ʒ℀ \quad ——⊦— 40ſʒ \quad —— 45\ ʒʒ$$
$$——⊦— 40ſʒ \quad ——⊦—64ʒʒ$$
$$81.ʒ. \quad —— 72. ℀ \quad —— 45\ ʒʒ$$
$$—— 72. ℀.$$

$$25ʒ℀ ——⊦— 80.ſʒ —— 26ʒʒ —— 144℀ ——⊦—81ʒ$$

The Arte

that roote by it ſelf, as here J haue doen it. VVherby
J haue not onely confirmed it to be a ſquare nomber :
but alſo J haue eſpied, that you vſed the nomber not
ſo plainly ſet doune, as the particulare multiplicati=
on did make it : but rather as a reaſonable reduction
would expreſſe it. J meane in the. $\frac{z}{z}$. where the
particulare multiplication hath ——+ 64. $\frac{z}{z}$. and
—————— 90 $\frac{z}{z}$. For whiche twoo nombers you ſette
one, that reſulteth of the bothe, that is ——— 26 $\frac{z}{z}$

    Maſter. But if you would take the nöber in that
ſorte, the woorke would be not onely all one : but alſo
ſomewhat plainer to bee perceiued of a learner. And
therefore for your pleaſure, J will ſet forthe here, the
example of that woorke. And loe, here it is.

$$25\frac{z}{} \,\text{\textcent} + 80 / \frac{z}{} + 64\frac{z}{z} - 90\frac{z}{z} - 144\frac{z}{} + 81\frac{z}{} \; (5\,\text{\textcent} + 8\frac{z}{} - 9\,\text{\textcent}$$
$$5.\text{\textcent}. \quad\quad 10\,\text{\textcent} + 64\frac{z}{z} \quad\quad 10.\text{\textcent}. + 16.$$

    Scholar. By comparynge theſe bothe formes of
woorke together, J dooe better vnderſtande, the rea=
ſon of the firſte woorke.

    Maſter. One example moare of this kinde of ex=
traction of rootes, will J ſet doune, that maie be a ge=
neralle patrone, for all the varieties, in this ſorte of
rooted nombers. And if you examine it diligently,
and marke it well, you ſhall neade fewe other exam=
ples, for this kinde of ſquare nombers.

The *Square* nomber, with the
woorke of extraction
of his roote fo=
loweth
here.

The

In the original 1557 edition of this book, the example replicated across the
following two pages was printed on a single folded pull-out page.
A facsimile of the folded page is available for download at
www.renascentbooks.co.uk

$$
\begin{array}{l}
\text{—24} \quad \text{+12} \quad 10 \\
48 \quad +36 \quad 6 \quad 8 \quad 26
\end{array}
$$

36ʒ.fʒ. + 60 cece —— 23ʒ.ʒ.ʒ. — 4 bfʒ. + 22ʒ.ce — 12fʒ. + 35ʒ.ʒ. ——
6.fʒ.   12.fʒ. + 25ʒ.ʒ.ʒ.
         12.fʒ. + 10ʒ.ʒ. + 16ʒ.ce
            12 fʒ + 10ʒ.ʒ. — 8 ce. + 9ʒ.ʒ.
               12.fʒ. + 10ʒ.ʒ. — 8 ce. +
                   12 fʒ. + 10ʒ.ʒ. ——

The Roote.

6.fʒ. + 5.ʒ.ʒ. — 4.ce. + 3.ʒ. — .2.ze. + .1.9.

The proofe by Multiplication.

6.fʒ. + .5.ʒ.ʒ. · .4.ce. + .3.ʒ. — .2.ze. + .1.9.
6.fʒ. + .5.ʒ.ʒ. · .4.ce. + .3.ʒ. — .2.ze. + .1.9.

——————————————————————

36ʒ.fʒ. + 30 cece —— 24ʒ.ʒ.ʒ. + 18bfʒ. —— 12ʒ.ce + 6fʒ.
     + 30.cece + 25ʒ.ʒ.ʒ. — 20bfʒ. + 15ʒ.ce — 10fʒ. + 5ʒ.ʒ.
       — 24ʒ.ʒ.ʒ. — 20bfʒ. + 16ʒ.ce — 12fʒ. + 8ʒ.ʒ. ——
            + 18 bfʒ. — 15ʒ.ce + 12fʒ. + 9ʒ.ʒ. ——
                — 12ʒ.ce — 10fʒ. + 8ʒ.ʒ. ——
                     + 6 fʒ. + 5 ʒ.ʒ. ——

——————————————————————

36ʒ.fʒ. + 60 cece — 23ʒ.ʒ.ʒ. — 4bfʒ. + 22ʒ.ce + 32fʒ. + 35ʒ.ʒ.

Scholar. It maie appeare easily, that this example serueth for many other, it doeth contain so diuersly.

And in this nomber also, as well as in the other, I see that many numbers be omitted, by reduction sirthe orders of nombers. For in the. 2. firste orders, and in the. 5. laste, is no varietie of the signes

Wherfore to see the varietie of woorke, I will sette doune the numbers, as thei rise in particulare experimente of my cunnyng. As here foloweth.

                                                             30

36ʒ.fʒ. + 60 cece + 25ʒ.ʒ.ʒ. — 48ʒ.ʒ.ʒ. + 36bfʒ. — 40bfʒ. + 46ʒ.ce
6.fʒ.   12.fʒ.   12.fʒ. + 10ʒ.ʒ.
         60 cece + 25ʒ.ʒ.ʒ.   48 ʒ.ʒ.ʒ. —————————— 40bfʒ. + 16ʒ.ce
                                     12.fʒ. + 10ʒ.ʒ. — .8. ce.
                                     36bfʒ. + ——————————— 30ʒ.ce

of extraction of his roote.

$$\overset{8}{20\text{cc}} + \overset{6}{10\,\text{z}} - 4\,\text{ze} + 1\,\text{?}.$$

$$6.\text{z}. + 4.\text{z}.$$
$$.8.\text{cc} + 6\,\text{z}. - 4\,\text{ze} + 1.\text{?}$$

$$.4.\text{cc}.$$
$$.6.\text{cc} + .3.\text{z}.$$
$$.6.\text{cc} + .4.\text{z} - 2\,\text{ze}$$
$$4.\text{cc}. + 3.\text{z}. - 2\,\text{ze} + 1\,\text{?}.$$
$$\overline{20\text{cc} + 10\,\text{z} - 4\,\text{ze} + 1\,\text{?}.}$$

many varieties of sigues *Coſsike*, multiplied ſo

: namely in the thirde, fourthe, fifte, and
$+$ and $-$.
multiplication, and in it will J make an

$$- 24\,\text{z}\,\text{cc} + 12\,\text{ſz} - \overset{20}{44\,\text{ſz}} + \overset{\overset{10}{26}}{35\,\text{z}\,\text{z}} - \overset{8}{20\text{cc}} + \overset{6}{10\,\text{z}} - 4\,\text{ze} + 1\,\text{?}.$$

$$- 24\,\text{ſz} - 9\,\text{z}\,\text{z}$$
$$12\,\text{ſz} \quad\qquad + \quad 10\,\text{z}\,\text{z}. - 8.\text{cc}. - .6\,\text{z}$$
$$- 24\,\text{z}\,\text{cc} \quad\qquad - 20\,\text{ſz}. - 16\,\text{z}\,\text{z} - 12\text{cc} + 4\,\text{z}.$$
$$12.\text{ſz}. \quad + \quad 10.\text{z}\,\text{z} - .8.\text{cc} + .6.\text{z} - 4.\text{ze}. + 1.\text{?}.$$

Do.iii.   Where

of Cossike nombers.

Where for myne owne eafe, and aied of memorie, I haue fet vnder euery doublyng of the *quotiente* : And the fomme that amounteth, by the multiplication of thefame, into the newe *quotiente*, with the Square of thefame newe *quotiente*.

Whereby I perceiue that the nombers, doe not go in foche order, that euery odde place, maketh a newe roote, as it doeth in nombers *Abstracte*. But fometime I muft take. 2. places nexte together, and at an other tyme, I fhall fkippe. 2. or. 3. places.

Mafter. You marke it well. And yet that is a good and true rule, that fome menne teache : that in thefe *Cossike* nombers, as well as in other *Abstracte* nombers, you fhall marke euery odde place, and vnder eche of them to finde a Square roote. But that is to be vnderftande, when the nombers are fette, in their brefefte and exactefte order.

Thefe fewe examples maie fuffice, for a declaratiõ of extractyng the roote of Square nombers, made by multiplication. And now touchyng thofe nombers, that bee equalle to fome rooted nomber, and namely foche as be equalle to a fquare nomber, I will teache you how their roote maie be extracted.

The rootes of nõbers equal to be fqures.

But firfte you fhall marke, that a Square beeyng compared, as equalle to rootes and nombers, the roo= tes maie bee coupled with the nombers onely, in. 3. formes. That is. $\mathcal{Z}$. —— . φ (whiche is all one with φ . —— . $\mathcal{Z}$) or els thus. φ . —— . $\mathcal{Z}$. Or thirdly, $\mathcal{Z}$ —— φ . And for eche of thefe. 3. fortes, there is fome varietie, in the extraction of the roote. And in them all moche agremente.

For the firft forme, where $\mathcal{Z}$ —— φ is equalle to $\mathcal{Z}$ take thefe exãples : $\mathcal{Z}$ is equall to. 4. $\mathcal{Z}$ —— 21 φ or. 1 $\mathcal{Z}$. is equalle to 35. φ . —— 2. $\mathcal{Z}$. Likewaies 1 $\mathcal{Z}$ is equalle to. 10 $\mathcal{Z}$. —— 75. φ . or. 1. $\mathcal{Z}$. is equalle to 103. φ . —— .8. $\mathcal{Z}$.

The firfte forme.

Dd.iiij.   In

The Arte

In all thefe eꝛäples, and other foche like, you muſt
firſt confider the nomber anneꝛed with the ſigne. ℞ .
(whiche is the middell quantitie) and the halfe of it
ſhall you note, foꝛ with it ſhal you woꝛke twiſe. Firſt
you ſhall multiplie halfe of that nomber by it ſelf, and
this is the firſte woꝛke, and to it ſhall you adde the o=
ther whole nomber, that is ioyned with. ℥. And thei
will euer moꝛe make a ſquare nomber : out of whiche
ſquare you ſhall eꝛtracte the roote. And to that roote
ſhall you adde halfe the nomber, that was anneꝛed
with the ſigne of. ℞ . (whiche was the nomber that I
bade you to marke). And this is the feconde woꝛke.
The totall that commeth of this addition, is the roote
of the compounde *Coſsike* nomber.

An example   Eꝛample of the firſte. 4. ℞ . —— . 21. ℥ . halfe the
nomber anneꝛed with. ℞ . is. 2. whoſe Square is. 4.
that ſhall I put to. 21. and there riſeth. 25. beeyng a
ſquare nomber, and hauyng. 5. foꝛ his roote. To that
5. I ioyne halfe the nomber anneꝛed with. ℞ . and it
maketh. 7. whiche is the nomber that I ſeke foꝛ : and
is the roote to. 4. ℞ . —— 21. ℥ .

The proofe.   Foꝛ triall whereof take. 4. rootes, that is. 28. and
putte to it. 21. and thereof commeth. 49. whiche is a
ſquare nomber, and hath. 7. foꝛ his roote.

An other
example.   Scholar. Then can I doe the like with the fecond
eꝛampl. 35. ℥ . —— . 2. ℞ . And firſte the halfe of. 2. is
1. and the Square of it is. 1. whiche I put to. 35. and it
maketh. 36. a Square nomber : whoſe roote is. 6. To
that. 6. if I adde. 1. that was the halfe befoꝛe reſerued,
it will make. 7. whiche is the roote that I doe ſeke.

The proofe.   The pꝛoofe is this : 2. rootes maketh. 14. and. 35. gi=
ueth. 49. whoſe roote is. 7.

The thirde
example.   Likewaies foꝛ the thirde eꝛample 10 ℞ —— 75 ℥
I woꝛke thus. Halfe. 10. is. 5. and his Square is. 25.
that dooe I adde to. 75. and there riſeth. 100. whoſe
roote is. 10. to whiche roote I add. 5. and there com=
                                                    meth

of Cossike nombers.

meth. 15. that is the roote whiche I would haue.

And that I maie proue by triall in this sorte. 10. rootes giue. 150. vnto whiche if I adde. 75. there will amounte. 225. whiche is a Square nomber : and hath 15. for his roote.

The fourthe example is. 105.$\cancel{9}$ —|— 8.$\approx$ .where *The fourthe* I take firste the halfe of. 8. that is. 4. and it in Square *example.* giueth. 16. whiche I adde to. 105. and there amoun= teth. 121. beyng a Square nomber, and the roote of it 11. vnto whiche I shall adde. 4. for halfe the nomber of rootes : and so there riseth. 15. as the roote that I seke for. And to approue it I take. 8. times. 15. whiche *The proofe.* is. 120. and adde it vnto. 105. and so commeth. 225. For the square, and the roote of it is. 15.

Master. The like order of worke shall you vse, in *Other for=* all nombers *Cossike* compounde, whē any. 2. nombers *mes in like* with immediate denominatiōs *Cossike*, are equalle to *sorte.* one of the nexte denomination, in order aboue them.

As. 1.$c\!\!\!/e$. is equalle to. 3.ζ. —|— 10.$\approx$.

And again. 1.$\int\zeta$. equalle to. 6.$\zeta\zeta$ —|— 40.$c\!\!\!/e$.

Likewaies. 1.ζ $c\!\!\!/e$. equalle to. 3.$\int\zeta$ —|— 28.$\zeta\zeta$ But in al these the roote shal beare name of the grea= ter quantie.

Scholar. By the former order of worke, I shall in *The firste* the firste of these. 3. examples, take halfe. 3. (bicause it *example.* is the nomber of the middell quantite). And that is $\frac{3}{2}$. and that shall I multiplie squarely, and so will there rise $\frac{9}{4}$, vnto whiche I shall adde 10 or $\frac{40}{4}$. And that ma= keth $\frac{49}{4}$. whiche is a square nomber, and his roote is $\frac{7}{2}$. vnto whiche I must put the firste halfe, that is $\frac{3}{2}$, and then will it be $\frac{10}{2}$, or els 5. whiche is the *Cubike* roote of that nomber. 3.ζ —|— 10.$\approx$. beyng equalle to 1 $c\!\!\!/e$

For proofe whereof, I multiplie. 5. *Cubikely,* and it *The proofe.* maketh. 125. Then doe I multiplie it squarely, and it will be. 25. Now. 3.ζ. is. 75. and. 10.$\approx$. maketh. 50 whiche bothe added together, giue. 125.

In

The Arte

In the seconde example, where. $1.\sqrt{3}$. is equalle to 6.33 —— $.40.c$. I shall take halfe. 6. (whiche is the nomber of the middell quantitie) and that is. 3. and the square of it is. 9. whiche I must adde vnto 40 and thereof commeth. 49. whiche is a square nomber and hath. 7. for his roote, vnto whiche I adde 3. and so haue I 10 for the *Surfolide* roote, of $6\,3\,3$ —— $40\,c$

And for proofe I saie, if. 10. bee the roote, then is 100. the square, & 1000. the *Cube*, the $3\,3$ is 10000. And the *Surfolide*. 100000. Wherfore. $6.3\,3$ make 60000. and. $40.c$. yelde. 40000. And bothe thei together doe make. 100000. whiche is the quanti= tie of the *Surfolide*.

In the thirde example. $1.\,3\,c$. is equalle to. $3.\sqrt{3}$. —— . $28.3\,3$. whose *zenzicubike* roote, I seke in this sorte.

Firste I take halfe. 3 (as the nomber of the middell quantitie) that is $\frac{3}{2}$, & that maketh in square $\frac{9}{4}$. whiche I adde vnto 28 (that maketh $\frac{112}{4}$) & it yeldeth $\frac{121}{4}$ whiche is a square nomber, and his roote is $\frac{11}{2}$. vnto whiche I adde $\frac{3}{2}$, and it will be $\frac{14}{2}$, or. 7. whiche is the *zenzicubike* roote vnto the foresaid nomber. $3.\sqrt{3}$ —— $.28.3\,3.$

For proofe whereof I multiplie. 7. zenzicubikely, and it maketh 117649. Then must the $\sqrt{3}$. be 16807 and. $3.\sqrt{3}$. 50421. Again the $3\,3$. is. 2401. and so $28.3\,3$. shall bee. 67228. And those bothe together yelde. 117649.

Master. Yet one other forme is there, that in all thinges, saue in one poincte onely : followeth thesame rule. And that is whē the 3 denominations doe not go immediatly together, but yet are equally distante. As $3\,3$. 3. and. 9. where the distaunce is one onely quantitie. Likewaies. $3\,c$. c. and. 9. whiche dif= fer by. 2. quantities. And in like sorte. $c\,c$. $\sqrt{3}$ and z. are distante by. 3. quantities. And so of other, how many so euer bee omitted, so that the difference bee
equalle

of Cofsike nombers.

equalle. In all whiche you shall worke, as you did in
the former rule, till you haue eanded all that worke.
But then haue you here, one thing more to bee confi=
red. For the laste nomber, whiche you haue founde, is
not the roote, but a rooted quantitie. And his roote is
the roote that you seke for.

Scholar. Doe you meane the square roote of that
quantitie, or some other?

Master. It maie be any kinde of roote, in diuerse
nombers, but not in one nomber. Wherfore for your
certaintie marke this rule.

If the denominations of your nombers, differ one=
ly by one, then is it a square nöber, that you doe finde
by the practise of the laste rule. And therfore shall you
take his square roote, for the roote of your nomber.

But if the denomination differ by. 2. quantities,
then shall you extracte a *Cubike* roote, out of your laste
nomber. And if the distaunce bee. 3. quantities, the
roote must bee a *zenzizenzike* roote : and for. 4. quan=
tities distante, a *Surfolide* roote, and so forthe.

As for example. 1.ᶾᶾ. is equalle to. 80.ᶾ ——— . *An example.*
2000.℈. Now for to finde the roote of. 80.ᶾ ——— .
2000.℈. I worke thus. Firste I take the halfe of 80.
(bicause it is the number of the middle quantitie) and
that halfe is. 40. whiche I multiplie Square, and it
maketh. 1600. to it I adde. 2000. and it will bee
3600. whiche is a square number, ℞. 60. is his roote :
to that. 60. I shall adde the forefaied. 40. and then
will it bee. 100. whiche nomber in the firste rule, had
been the true roote. But here considering the distäce
is of one quantitie, I muste extracte his square roote,
whiche is. 10. And that is the *zenzizenzike* roote,
that my nomber containeth.

An other example. 1.ᶾℭ. is equalle to. 400. ℭ . *The seconde*
——— 57344.℈. I take 200. for the halfe of the mid= *example.*
dell quantities nomber, and multiplipng it square, I

The Arte

finde. 40000. whiche J put to. 57344. and then J haue. 97344 : whiche is a Square nomber, and his roote is 312 vnto whiche J shall adde the halfe of 400 and so will it bee. 512. But now must J take the $Cu=$ *bike* roote of this nomber (that is. 8) for my roote, that J desire : Bicause the denominations in the nomber, differ by. 2. quantities.

Scholar. J see very well the order of this worke : And the proofe is in like sorte, whiche J maie practise by my self at any tyme. Wherfore J praie you, pro= cede forthe to other rules.

The seconde sort of equal nombers.

Master. This is sufficiente for the firste sorte. Now for the seconde sorte, in nombers *diminute or re= sidualle* where. $\mathfrak{z}$. is equalle to. $\wp$ ——— $\mathfrak{Z}\!e$. the forme of worke is like vnto the other, in all poinctes saue in one. For in stede of the laste addition, you shall vse in these nombers, Subtraction. As here for example,

Example.

when J saie. 1.$\mathfrak{z}$. is equalle to. 60.$\wp$. ——— .4.$\mathfrak{Z}\!e$. to finde the roote, firste J take the halfe of. 4. (bicause it is the nomber of the middell signe) and that halfe beyng. 2. doeth make in square. 4. whiche J put to 60 and so is it. 64. a square nomber, and hath. 8. for his roote. From whiche roote (by the order of this rule) J must abate. 2. that is the halfe of the firste nomber of rootes. And then will there remaine. 6. for the verie roote of. 60.$\wp$. ——— 4.$\mathfrak{Z}\!e$. beyng equalle to. 1.$\mathfrak{z}$.

The proofe.

Scholar. That is sone proued. For. 6. beeyng the roote, then. 4.$\mathfrak{Z}\!e$. maketh. 24. whiche beyng abated out of. 60. leaueth 36 and that is the iuste square vn= to. 6. as the equation saieth.

The seconde example.

Master. An other example is this. 1.$\sqrt{\mathfrak{z}}$. is equall to. 162.$c\!\!\!/\!\!\!e$. ——— .9.$\mathfrak{z}\mathfrak{z}$.

Scholar. That can J woorke, thus : Firste J take the halfe of. 9. (bicause it is the nomber of the middell signe) and it is $\frac{9}{2}$, whiche J multiplie squarely, and it will be $\frac{81}{4}$ that must bee added to. 162. or $\frac{648}{4}$, and then will

of Cossike nombers.

will there amounte $\frac{729}{4}$, whiche is a Square nomber, and hath for his roote $\frac{27}{2}$ out of whiche, by this rule, I must abate $\frac{9}{2}$, and then riseth $\frac{18}{2}$, that is. 9. whiche is the very roote to. 162. ℞ —— 9.℞℞. beyng equall to. 1.℞.

And for the proofe, I multiplie. 9. *Cubikely*, and it *The proofe.* giueth. 729. so that. 162. ℞. doe make. 118098. out of whiche I must abate. 9.℞℞. that is, 59049. (by thesame roote, sith. 1.℞℞. is 6561). And then will there remaine. 59049. whiche is the iuste quantitie of. 1.℞.

Master. Yet one example more shall you haue of *The thirde* a thirde sorte. *example.*

When. 1 ℞ ℞ is equalle to. 275456.℞ —— 26 ℞ I demaunde of you, what is the valewe of. 1.℞ ?

Scholar. I searche it thus. The nomber of the middell signe is. 26. whose halfe I must take, and first multiplie it squarely, and there will rise 169. whiche I adde to. 275456. and it will bee. 275625. whiche is a square nomber, and hath for his roote. 525. from whiche nomber I must abate halfe the nomber, of the middell signe, that is, 13. and so there will remaine 512. whose *Cubike* roote I must extracte, bicause the denominations differ by. 2. quantities, and that roote will be. 8. whiche is the *Cubike* roote to. 512. but to the nomber propounded, it is the *zenzicubike* roote.

Master. This is inoughe for the worke of the seconde sorte. Now for the thirde sorte of equation, *The thirde* where. ℞. is equalle to ℞.———.℞. I will giue you *sorte of equal* a briefe admonition onely, though it differ from bothe *nombers.* the other. 2. rules, in forme of woorke. For as the equalitie maie be in diuerse sortes, so some tymes you maie vse the woorke of the firste sorte, by Addition of halfe the nomber of the middle signe : and some times you shall worke by subtraction. Wherein this is the difference, from the seconde rule. That there you doe

        Ee.ii. subtracte

The Arte

subtracte halfe the nomber of the middell signe, from the roote whiche you fonde. And in this thirde rule, you shall subtracte the roote from the halfe, and not the halfe from the roote. Forbicause that that roote, is euer lesser then that halfe.

And in this rule, this is specially to bee obserued : that the Square of halfe the nomber, of the middell signe, will euer moæ bee greater, then the nomber of the lesser signe : And therfoæ shall the nomber of the lesser signe, bee abated out of that square. And the re= mainer will bee a Square nomber, with whiche you shall woozke, as I haue taught you befoæ.

And farther in this rule, it is commonly seen, that euery soche equalle nomber, hath. 2. valuations foæ his roote. I meane that any of those. 2. nombers, will bee as the roote in this equation. Foæ otherwaies no nomber can haue. 2. rootes of one denomination.

Scholar. I vnderstande you thus. That no nom= ber can haue. 2. square rootes, oæ. 2. *Cubike* rootes, and so foæthe : Els one nomber maie haue. 3. oæ. 4. rootes. As. 64. hath. 8. foæ his Square roote : 4. foæ his *Cubike* roote : and. 2. foæ his *zenzicubike* roote.

Master. You take it well. And farther foæ the easie knowledge of those. 2. nombers, oæ rootes : Thei must bee soche, as beeyng added together, will make the nöber of the middell signe : and beeyng multiplied together, wil make the nomber of the least signe. And so maie you finde theim without farther multiplica= tion, oæ extraction of rootes.

The firste example.

Foæ example, I sette first. 1. ℨ. equalle to. 16. ℞. ——.63.♃. where I maie espie quickely, that. 63. cã haue no moæ partes to his composition, but. 3.7.9.21 And if I take. 3. and. 21. then their addition will bee greater then. 16. but 7. and. 9. maketh iuste 16. by ad= dition, and. 63. by multiplication. And therefoæ thei shall be the. 2. rootes.

Scholar.

of Cofsike nombers.

Scholar. I will proue that by examination, thus. If. 7. be the roote, then is. 49. the square. And. 16. ze make. 112. out of whiche I must abate. 63. and there resteth. 49. equalle with the Square : so is that a true roote. Then for. 9 : his square is. 81. And. 16. ze. doe yelde 144 frō whiche I shal abate 63. And the remai= ner will be. 81. equalle to the square. And so is that al so a true roote.

Master. Now worke it by the other rules, that I taught you.

Scholar. Firste I take. 8. as halfe the nomber of the middell signe, and that multiplied Square, doeth giue 64 from whiche I shall abate 63 and then doeth there remain but. 1. whiche is coumpted as a Square nomber, and his roote to bee. 1. also, whiche if I adde to. 8. it will make. 9. that is one of the rootes : And if I abate it from. 8. it will leaue. 7. whiche is the other roote. And thus I see one worke cōfirmeth the other.

Master. Take this for the seconde exāple. 1 ꝫ cℓ *The seconde* is equalle to. 8.ſꝫ. ——— .12.ꝫꝫ. what is the roote *example.* saie you?

Scholar. To finde it, firste I loke for the partes of 12. And thei be. 2.3.4.6. of whiche. 2. and. 6. doe serue my purpose, for their addition maketh. 8. and so doeth not. 3. and. 4. Wherefore I saie, that. 2. maie bee the roote, and so maie. 6. But for farther trialle of it : I woorke it by the other rule, saiyng halfe. 8. is. 4. and his square is. 16. From whiche I abate. 12. and there remaineth. 4. whose roote is. 2. that I maie adde to. 4 and so haue I. 6. for one roote : or els abatyng it from 4. I shall haue. 2. for the other roote.

The proofe is manifeste for. 6. beeyng a roote, the *The proofe.* zenzicube is. 46656. The Surſolide is. 7776. And the zenzizenzike is 1296. So that 8.ſꝫ. doe make 62208 And. 12.ꝫꝫ. are. 15552. whiche being abated out of 62208 do leaue 46656. the true quantitie of 1 ꝫ cℓ

Ee.iii.   And

The Arte

And so is that worke good, 6. beyng a roote.

Now if. 2. be sette for a roote : then is the. ჳჳ 16. the √ჳ, 32, and the. ჳ cℓ 64. And so are. 8. √ჳ. equall to. 256. And. 12. ჳჳ. yelde. 192. Wherfore abating 192. out of. 256. there resteth. 64. the iuste quantitie of. 1. ჳ cℓ. And so is that worke also good, and. 2. a true roote.

The thirde example.

Master. Now proue this thirde example, where 1 b√ჳ is equalle to. 2000. ჳჳ ——— 470016 ℨℯ.

Scholar. Halfe the nomber of the middell signe is 1000. And the square of it is. 1000000. From whiche I shall abate. 470016. and there will remaine 529984. whose square roote by trialle of extraction, I finde to be 728. whiche I maie other adde to. 1000 and so there riseth. 1728. whiche I finde to bee (as it ought) a *Cubike* nomber. And his roote to be. 12.

But and if I abate 728. from 1000, there will remain. 272. whiche is no *Cubike* nomber.

Master. So that here semeth to be but one roote. And yet these. 2. nombers. 1728. and. 272. kepe soche a rate, that beeyng multiplied together, thei make 470016. whiche is one of the nombers, and beeyng added together, thei make 2000. whiche is the other nomber of the same *Cossike residualle.*

But now proue in other like nōbers, whiche haue some distaunce, betwene their denominations, whe=

The fourthe example.

ther it will so happen still. As namely in this, where 1. b√ჳ. is equalle to. 12. ჳჳ. ——— 32. ℨℯ.

Scholar. Halfe. 12. is. 6. and his Square. 36. from whiche abatyng. 32. there is lefte. 4. whose roote is. 2 And if I adde that 2. to. 6. it maketh. 8. whiche is a *Cu= bike* nomber, and hath. 2. for his roote. But if I abate 2. from. 6. there remaineth. 4. whiche is no *Cubike* nō= ber, and therfore hath no soche roote. And yet these. 2. nombers. 4. and. 8. by addition, make the middell nō= ber, and by multiplication, thei make the laste nōber.

Master.

of Cofsike nombers.

Mafter. Proue yet ones againe in a nomber, *The fifte* where one quantitie onely is omitted. As when 1.√ℨ *example.* is equalle to. 24.℞. ——— 135.ℤ℞.

Scholar. 12. maketh in fquare. 144. from whiche I fhall deducte. 135. and then refteth. 9. whofe fquare roote is. 3. whiche if I adde to. 12. it will bee. 15. and hath no fquare roote, as here is required. But if I a= bate. 3. from. 12. then remaineth 9 whofe fquare roote is. 3. and ferueth to the nomber, as I haue here pro= ued in my tables. And. 9. and. 15. kepe the cuftoma= ble rate. For by addition thei make. 24. And by mul= tiplication, thei yelde. 135.

But in all thefe examples, where the denominati= ons be are a diftaunce, I can finde but one roote, and not. 2. As it was in the other exāples of thefame rule.

And in fome of theim, the greater nomber contai= neth the roote : but in other, the leffer nomber hath the roote.

Mafter. Bicaufe I can not ftaie now, about this varietie, I will remitte it till an other tyme. But this by the waie, I muft admonifhe you, that I doe folowe here, the common forme of writers, in callyng thefe rootes, that rife in equatiō, where as thei are not the rootes of thofe nombers, but are the value of a roote. For of a *Cofsike* nomber, the roote muft neades bee a *Cofsike* nomber alfo. And foche as by multiplication will make the rooted nomber : But fo can not thofe nombers doe.

And here will I make an eande, of the workes
of *Coffike* ombers. And now will I ap=
plie them to practife in the rule
of *equation*, that is com=
monly called *Al=*
gebers rule.

⦿ The

The Arte

The rule of equation, common=
ly called Algebers Rule.

The rule of
equation.

Etherto haue I taughte you, the common formes of worke, in nom= bers *Denominate.* Whiche rules are vsed also in nöbers *Abstracte,* & like= waies in *Surde* nombers. Although the formes of these workes be seue= ralle, in eche kinde of nomber. But now will I teache you that rule, that is the principall in *Cossike* workes : and for whiche all the other dœ serue.

This Rule is called the Rule of *Algeber,* after the name of the inuentoure, as some men thinke : or by a name of singular excellencie, as other iudge. But of his vse it is rightly called, the rule of *equation* : bicause that by *equation* of nombers, it dœth dissolue doubte= full questions : And vnsolde intricate ridles. And this is the order of it.

The somme of the rule of equation:

Hen any question is propoüded, apperteinyng to this rule, you shall imagin a name for the nom= ber, that is to bee soughte, as you remember, that you learned in the rule of false position. And with that nomber shall you procede, accordyng to the question, vntil you finde a Cossike nomber, equalle to that nom= ber, that the question expresseth, whiche you shal

reduce

of Cossike nombers.

reduce euer more to the leaste nombers. And then diuide the nomber of the lesser denomination, by the nomber of the greateste denomination, and the quotient doeth aunswere to the question Ex= cept the greater denominatiō, doe beare the signe of some rooted nōber. For then must you extract the roote of that quotiente, accordyng to that signe of denomination.

Scholar. It semeth that this rule is all one, with the rule of false position : and therefore mighte so bee called : seyng it taketh a false nōber, to worke withal.

Master. This rule doeth farre excell that other. And dooeth not take a false nomber, but a true nom= ber for his position, as it shall bee declared anon. Wherby it maie bee thoughte, to bee a rule of won= derfull inuention, that teacheth a manne at the firste worde, to name a true nomber, before he knoweth re= solutely, what he hath named.

But bicause that name is common to many nom= bers (although not in one question) and therefore the name is obscure, till the worke doe detect it, I thinke this rule might well bee called, the rule of darke posi= tion, or of straunge position : but not of false position.

And for the more easie and apte worke in this arte wee dooe commonly name that darke position. 1. ℞ . And with it doe we worke, as the question intendeth, till we come to the *equation.*

This rule of *equation,* is diuided by some men, into diuerse partes. As namely *Scheubelius* dooeth make. 3. rules of it. And in the seconde rule, he putteth. 3. seue= ralle cannōs. Some other men make a greater nōber of distinctiōs in this rule. But I intende (as I thinke beste for this treatice, whiche maie serue as farre

The partes of the rule.

Ff.i.   as

The Arte

as their wozkes doe extende) to diſtincte it onely into twoo partes. Whereof the firſte is, *when one number is equalle vnto one other.* And the ſeconde is *when one nom= ber is compared as equalle vnto 2. other nombers.*

Alwaies willyng you to remeber, that you reduce your nombers, to their leaſte denominations, and ſmalleſte fozmes, befoze you pzocede any farther.

And again, if your *equation* be ſoche, that the grea= teſte denomination *Coſsike,* be ioined to any parte of a compounde nomber, you ſhall tourne it ſo, that the nomber of the greateſte ſigne alone, maie ſtande as equalle to the reſte.

And this is all that neadeth to be taughte, concer= nyng this woozke.

Howbeit, foz eaſie alteratiõ of *equations.* I will pzo= pounde a fewe exāples, bicauſe the extraction of their rootes, maie the moze aptly bee wzoughte. And to a= uoide the tediouſe repetition of theſe woozdes : is e= qualle to : I will ſette as I doe often in woozke vſe, a paire of paralleles, oz Gemowe lines of one lengthe, thus : ===== bicauſe noe. 2. thynges, can be moare equalle. And now marke theſe nombers.

1. $14.\mathcal{z}$. ---- .15.$\wp$ ===== 71.$\wp$.

2. $20.\mathcal{z}$. ———— .18.$\wp$ ===== .102.$\wp$.

3. $26.\mathcal{z}$ ---- $10\mathcal{z}$ ===== 9.$\mathcal{z}$ ———— $10\mathcal{z}$ ---- 213.$\wp$.

4. $19.\mathcal{z}$ ---- 192.$\wp$ ===== $10\mathcal{z}$ ---- $108\wp$ ———— $19\mathcal{z}$

5. $18.\mathcal{z}$ ---- 24.$\wp$. ===== 8.$\mathcal{z}$. ---- 2.$\mathcal{z}$.

6. $34\mathcal{z}$ ———— $12\mathcal{z}$ ===== $40\mathcal{z}$ ---- $480\wp$ ———— 9.$\mathcal{z}$

1.   In the firſte there appeareth. 2. nombers, that is
14.$\mathcal{z}$.

of Cossike nombers.

14. Z . —⊢— 15. $\mathit{9}$. equalle to one nomber, whiche is
71. $\mathit{9}$. But if you marke them well, you maie see one
denominatiõ, on bothe sides of the *equation*, which neuer ought to stand. Wherfore abating the lesser, that
is. 15. $\mathit{9}$. out of bothe the nombers, there will remain.
14. Z ══ 56. $\mathit{9}$. that is, by reduction, 1 Z ══ 4. $\mathit{9}$.

Scholar. J see, you abate. 15. $\mathit{9}$. from them bothe.
And then are thei equalle still, seyng thei wer equalle
before. Accordyng to the thirde common sentence, in
the patthewaie :

If you abate euen portions, from thynges that bee equalle,
the partes that remain shall be equall also.

Master. You doe well remẽber, the firste groundes of this arte. For all springeth of those principles
Geometricalle. Wherfore call to your minde likewaies
the seconde common sentence, in thesame booke, and
then haue you another reason, whiche will helpe you
not onely, in the other formes of woorke here, but also very often in the practise of this arte.

Scholar. That is this.

If you adde equalle portions, to thynges that bee equalle,
what so amounteth of them shall be equalle.

Master. These twoo sentences doe instructe you
that when you see on bothe the sides of the *equation*, a-
ny one denominatiõ *Cossike*, you shall marke the signe
that is annexed to the lesser of them bothe : and if it be
the signe of addition. —⊢— . then shall you abate that
lesser nomber, from bothe the partes of the *equation*.
As J did in this firste example. But if the signe be of
abatemente ——— , then shall you adde that lesser nõber, to bothe partes. And so shall you doe, till there be
noe one denomination on bothe partes, but diuerse
and distincte.

So the seconde nomber will be. 20. Z ══ 120 $\mathit{9}$ 2.
and in the leaste termes. 1. Z ══ .6. $\mathit{9}$.

Scholar. J see that you adde. 18. $\mathit{9}$. to bothe par-
Ff.ii.   tes

The Arte

3. tes of the equation. But by that reason, I doubte in
the thirde somme, bicause. 10. ℥ . is in bothe partes
of the equation : in the firste parte with ——— , and in
the seconde parte with ——— , whether I shall adde
10. ℥ , oz abate them.

Master. In soche a case, you maie dooe either of
bothe, at your libertie : and all will be to one eande.

Scholar. If I adde. 10. ℥ . then will it be. 26. ȝ .
——— 20. ℥ . ══════ .9 ȝ . ——— 213. 9 .

Master. And doe you not see. ȝ . on bothe sides of
the equation?

Scholar. I did loke but foz one alteration onely.

Master. If there were twentie like denominati=
ons, you should alter them all. Foz that is the princi=
palle and peculiare reduction, that belongeth to equa=
tions.

Scholar. Then must I abate. 9. ȝ . on bothe par=
tes, and so will there remaine. 17. ȝ . ——— .20. ℥ .
══════ 213. 9 .

Master. Now reduce it by abatyng. 10. ℥ .

Scholar. So it will bee. 17. ȝ . ══════ .213. 9 .
——— .20. ℥ .

And now I remeber, that this is the better fozme
of reduction. Bicause the greater denomination, that
is. ȝ . is alone with his nomber on the one side of the
equation, and the .2. lesser denominations, on the o=
ther side.

Master. Now doe you reduce the other equatiõs,
to their smalleste fozmes?

4. Scholar. In the fourthe example, there is noe de=
nominatiõ, befoze the signe of equation, oz in the firste
parte, but the like is in the seconde parte also, after
the signe of equation. Wherfoze firste, bicause I see
19. ℥ . on bothe sides, I will abate it on bothe sides.
And then will it be thus.

192. 9 . ══════ 10. ȝ . ——— 108. 9 . ——— .38. ℥ .
                                                                    But

of Cofsike nombers.

But bicauſe J ſee 9. yet remainyng on bothe partes, J abate the leſſer, that is. 108 9. from bothe partes, and it will be. 84. 9 . ════ .10. ℥ ──── .38. ᴢℯ .

Maſter. This equation would bee better, if the greater denomination, did ſtande as one parte of the equation alone. Whiche thyng you maie eaſily doe, by addyng. 38. ᴢℯ . to bothe partes : bicauſe ſo moche foloweth ──── , on the one parte.

And euermore when occaſion ſerueth, to tranſlate *Tranſlation* nombers compounde, ──── on the one ſide is equalle *of nombers.* to ──┼── on the other ſide.

Scholar. Then it will be thus.

84. 9 . ──┼── 38. ᴢℯ . ════ .10. ℥ .

Maſter. It were better thus.

10. ℥ ════ .38. ᴢℯ . ──┼── .84. 9 .

And in ſmaller termes.

5. ℥ . ════ .19. ᴢℯ . ──┼── .42. 9 .

But now procede with the examples.

Scholar. The fifthe is eaſily reduced, by abatyng   5. 2. ᴢℯ . on bothe ſides : For ſo will it bee.

8. ℥ . ════ .16. ᴢℯ . ──┼── .24. 9 .

The ſixthe equation will be, by addyng. 12. ᴢℯ . on   6. bothe ſides. 34. ℥ ════ 52. ᴢℯ ──┼── 480. 9 ──── 9 ℥ . But yet J muſt reduce it farther, by addyng. 9. ℥ . on bothe ſides. And then it will ſtande thus.

43. ℥ . ════ 52. ᴢℯ . ──┼── 480. 9 .

Maſter. Now will J ſhewe you the varieties of *Varieties of* equatiõs, taught by *Scheubelius*, bicauſe you maie per= *equations.* ceiue, how thei bee conteined in thoſe .2. formes, na= med by me. As for the manyfolde varieties, that ſome other doe teache, J accoumpte it but an idle bablyng, or (to ſpeake moare fauourably of them) an vnneſſary

The Arte

distinction.

  The firſt equatiõ after *Scheubelius*, ҭ after my mea=
nyng alſo, is, when one nomber is equall to an other :
meanyng that thei bothe muſt be ſimple nombers (*Coſ=
ſike*, and vncompounde. As. 6.ze. equalle to. 18.℈:

$$4.\mathfrak{z}. =\!=\!= .12.ze. \qquad 14.c\!\ell. =\!=\!= .70.\mathfrak{z}:$$

$$15.\sqrt{\mathfrak{z}}. =\!=\!= .90.\mathfrak{z}\mathfrak{z}: \quad 20.\mathfrak{z}c\!\ell =\!=\!= 180.\sqrt{\mathfrak{z}}:$$

$$26.\mathfrak{z}\sqrt{\mathfrak{z}}. =\!=\!= .117.c\!\ell c\!\ell.$$

  In all theſe examples, as you ſee but one nomber,
compared to an other : ſo to finde the quantitie of one
roote, you ſhall diuide the nomber of the leſſer Cha=
racter, by the nomber of the greater Character, and ſo
ſhall the *quotiente* bꝛyng foꝛthe the quantitie of. 1. ze.

  Scholar. It ſemeth at the firſte vewe, that it is a=
gainſt reaſon, to diuide the nomber of the leſſer ſigne,
by the nomber of the greater. But when I conſider,
that if I compare a nomber of crounes, oꝛ any like de=
nomination, to a nomber of ſhillynges in equalitie,
the nomber of crounes, oꝛ other ſoche like, muſt nea=
des be leſſer, then the nõber of ſhillinges. And ſo diui=
ding the nõber of the ſhillinges (oꝛ other leſſer name)
by the nomber of crounes (oꝛ other greater name) the
quotiente will ſhewe, how many ſhillynges make a
croune : and generally, how many of the leſſer, dooe
make one of the greater.

  As if. 20. crounes bee equalle to. 100. ſhillynges,
then. 5. ſhillynges dooeth make a croune. So when
6.ze. bee equall to 18.℈. then. 3.℈. doeth make.1. ze.
And. 4.$\mathfrak{z}$.$=\!=\!=$.12. ze. dooeth cauſe that. 3.℈. muſt
be a roote.

  Maſter. As your examplarie pꝛofe is good, ſo re=
duction will be a ſufficiente pꝛoofe in this.

  Scholar. I ſee it manifeſtly. Foꝛ if. 14. cℓ. bee e=
qualle to. 70.$\mathfrak{z}$. then. 1.cℓ. is equalle to. 5.$\mathfrak{z}$. by that
reduction

of Cossike nombers.

reduction in nombers. And again by reduction in si=
gnes. 1. ℨℯ. is equalle to. 5. ℘.

Likewaies. 15. ∫ℨ. beyng equalle to. 90. ℨℨ. re=
duction by signes and nombers also, will make. 1. ℨℯ
— 6℘. So shall. 20. ℨℯ. — 180. ∫ℨ. be reduced
to. 1. ℨℯ. — 9. ℘. And. 26. ℨ ∫ℨ. — .104. ℯℯ.
will make. 1. ℨℯ. — .4. ℘.

Master. And so generally, when there is noe de=
nomination omitted, betwene those. 2. that bee com=
pared in equalitie, still the diuision of the nomber, of
the lesser denomination, by the nomber of the greater
denomination, will bzyng fozthe in the *quotiente*, the
quantitie of. 1. ℨℯ.

But if there bee any denominations omitted, be=   *The seconde*
twene those. 2. whiche be compared together in equa=   *forme of the*
litie : loke how many denominations are omitted, and   *firste equatiõ*
so many in ozder is the rooted quantitie, whose roote
you must extract, foz the aunswere to the questiõ. Foz
in soche a case, euer moze you shall extracte the roote
of your laste nomber.

As foz example, when. 6. ℯ. be equalle to. 24. ℨℯ.
by the fozmer rule, you shall finde. 4. in the *quotiente*.
But here that. 4. is not the quantitie of a roote, but
is a rooted nomber, whose roote J shall extracte. And
seyng betwene. ℯ. and. ℨℯ. there is no quantitie o=
mitted, but one, that is. ℨ. Therefoze J shall ac=
coumpte. 4. the firste quantitie, that is to saie, a *Square*
nomber, and so take his *Square* roote, beyng. 2. foz the
quantitie of a roote.

Again if. 7. ∫ℨ. be equalle to. 576. ℨℯ. the *quotinete*
will be. 81. and declareth a *zenzizenzike* nomber, bi=
cause there are omitted betwene. ∫ℨ. and. ℨℯ. three
nombers : and *zenzizenzike* is the thirde quantitie :
as you did learne in the beginnyng of this treatice, of
nombers denominate.

Scholar. J perceiue it. And therfoze J must take
the

The Arte

the zenzizenzike roote of. 81. whiche is. 3. and that is
the true roote, where. 7.√ℨ. be equalle to. 567.ℨℯ.

Master. And if thoſe. 7.√ℨ. were accōpted equalle
to. 56.ℨ. the *quotiente* will be. 8. And bicauſe there
are omitted. 2. quantities, that is. ℭ. and. ℨℨ. ther=
foze you ſhall accompte that. 8. to be 1ℭ. oz a ſeconde
quantitie. And his roote *Cubike* is. 2. whiche ſtandeth
as the valewe of a roote, in the fozmer equation.

And it is not poſſible that any other nomber, maie
be placed as a roote, in that equation : oz in any other
fozme of this firſte kinde. Howbeit in one ſozte of e=
quation, of the ſeconde kinde, there maie be. 2. diuerſe
rootes, when one nomber hath. 2. rootes in valewe.
As I taught you befoze in the extraction of rootes.

*The ſeconde
kinde of
equation.*
The ſecōd kinde of equatiō, after *Scheubelius* minde
and myne alſo, is, when one ſimple nomber *Coſsike*, is
compared as equalle to. 2. other ſimple nōbers *Coſsike*,
of ſeueralle denominations, and like diſtaunce.

And in ſoche equation, beyng reduced as is taught
befoze, the roote of thoſe. 2. nombers compounded, as
in one (oz rather the valewe thereof) ſhal be extracted :
As I haue befoze taughte alſo. And that roote doeth
aunſwere to the queſtion.

*The ſeconde
forme of the
ſecond kinde.*
Howbeeit, here is the like obſeruation, as was in
the ſeconde fozme of the firſte kinde. Foz if thoſe. 3.
denominations be not immediate, but doe omit ſome
other betwene them, then ſhall you extracte the roote
of that laſte nomber, in all poinctes, as you did in the
firſte equation.

Examples of the firſte ſozte.

$$4.ℨ. = 6.ℨℯ + 4.ℱ.$$

which beyng reduced, will bee :

$$1.ℨ. = \tfrac{3}{2}.ℨℯ. + 1.ℱ.$$ And the roote wil be. 2

And. $6.√ℨ. = 12.ℨℨ. + 18.ℭ.$

That

of Cossike nombers.

That is by reduction.

1.√ℨ. ＝＝ 2.ℨℨ. ＋ 3.℞. or

1.ℨ. ＝＝ 2.℥ . ＋ 3.℈. And the roote. 3.

5√ℨ ＝＝ 25.ℨℨ ── 30.℞. Or by reduction.

1√ℨ ＝＝ 5.ℨℨ. ── 6.℞. Or.

1.ℨ. ＝＝ .5.℥ ── .6.℈. whose roote is. 3. or. 2.

Likewaies. 2.ℨ ＝＝ 120.℈. ── 8.℥.

Or by reductiō. 1.ℨ ＝＝ 60.℈ ── 4.℥. whose
roote is. 6.

Examples of the seconde sorte.

5.ℨℨ ＝＝＝ 60.ℨ ＋ 320.℈.

That maketh by reduction.

1.ℨℨ. ＝＝ 12.ℨ. ＋ .64.℈.

And the square roote. 4.

Likewaies. 8.ℨ℞. ＝＝＝ .40.℞. ＋ .30208.℈.

Or by the orderly reduction.

1.ℨ℞ ＝＝ 5.℞. ＋ 3776.℈. whose Cubike
roote is. 4.

Again in residualles.

8.ℨ℞ ＝＝ 864.ℨ. ── 24.ℨℨ.

That maketh by reduction.

1.ℨ℞ ＝＝ 108.ℨ. ── 3.ℨℨ. Or els.

1.ℨℨ ＝＝ 108.℈ ── 3ℨ. whose roote is. 3.

So. 9.b√ℨ ＝＝ 90.ℨℨ. ── 144.℥.

Or by reduction. 1.ℨ℞ ＝＝ 10.℞. ── 16.℈.
whose roote is. 8. or. 2.

But now bicause *Scheubelius* dooeth make. 2. seue=
ralle equations of these. 2. formes : And giueth. 3. di=
uerse rules, or canons for eche of them, J will declare
his. 6. canons to be all contained in this seconde kind
of equation.

                                   Gg.i.   He

The Arte

He maketh his diuision thus. When. 1. nomber is compared as equalle to. 2. other, other that one nö=ber is of the smalleste denomination. And then is it of the firste Canon. As. 1.ʒ —— 8.℞ ════ 65.℈. oʒ els that one nomber, is of the greateste denominatiö : As. 3.℞. —— 4.℈ ════ 1.ʒ. And then is it of the seconde Canon : Or els thirdely, the alone nomber is of the middle denominatiö : and then is it of the thirde Canon. As. 1.ʒ —— .12.℈. ════ 8.℞.

The like foʒme be vseth, foʒ the nombers of deno= minations distaunte.

Wherby you maie perceiue, that in my rule there is noe foʒme of nombers, like thē of the firste Canon, nother yet of the thirde : but onely of the seconde. But then again in my rule, there are. 2. soʒtes of examples whiche he hath not. And if you compare them well to gether, you shall perceiue, that thei bee agreable to= gether.

As foʒ exäple. In his firste canon, this is the foʒme 1.ʒ —— 6.℞. ════ .27.℈. whiche equation in my rule, by tradation, is expʒessed thus.

1.ʒ ════ 27.℈ —— 6.℞. bicause I doe still set the greateste denomination alone.

Again in his thirde Canon, this is an example.

1.ʒ —— 15.℈ ════ .8℞ and that nomber doe I translate into this foʒme 1ʒ.════8.℞ —— 15℈.

Now where as he giueth seueralle rules, foʒ euery Cannon, I saie foʒ them all : extracte the roote of that compounde nomber. Foʒ all his rules doe teache no= thyng els.

Scholar. I doe vnderstande the diuersitie, and a= gremente of your rules and his. But foʒ my exercise, I dooe couette some apte questions, appertainyng to these equations.

A question
of ages.
Master. Take this foʒ the firste question.

Alexander beyng asked how olde he was, I am. 2.

                                                          yeres

of Cossike nombers.

peres elder (quod he) then Epheſtio. Pea, ſaied Ephe=
ſtio. And my father was as olde as we bothe, and, 4,
peres moare. And my father hauyng all thoſe peres,
ſaied Alexander, was, 96, peres of age. J demaunde
now of you, how olde was eche of them.

Scholar. J praie you aunſwere the queſtion your
ſelf, to teache me the forme.

Maſter. J will begin with the youngeſte mannes
age, and that will J call, 1, ℥ , whiche is the common
ſuppoſition in all ſoche queſtions. Then is Alexan=
ders age, 2, peres moare, that is, 1, ℥ —+— 2, ℈, And
thoſe bothe together dooe make, 2, ℥ , —+— , 2, ℈,
whereunto if you put, 4, moze, then haue you the age
of Epheſtio his father, that will be, 2, ℥ , —+— , 6, ℈,
And all theſe put together, that is, 4, ℥ —+— , 8, ℈,
will make 96 whiche is the equation that ſhall open
the queſtion.

1, ℥ . is the
common ſup
poſition.

Wherfore J ſet doune the equation thus,
4, ℥ , —+— 8 , ℈, ==== 96, ℈, And bicauſe J ſee on
bothe ſides, one denomination of, ℈ , J doe abate, 8 , ℈,
frō bothe ſides : & then there remaineth, 4 ℥ ==== 88 ℈
And by reduction oz diuiſion, 1, ℥ , ==== 22, ℈,

Scholar. Then maie J eaſily ſaie, that Epheſtio
was, 22, peres olde, ſeyng you did putte, 1, ℥ , foz his
age : and now, 1℥ , is founde to be, 22, And therby all
the other peres be manifeſte. For Alexander beyng, 2
peres elder, muſt be, 24, And Epheſtio his fa=
ther had in age, 22, and, 24 : and, 4, moze, that
is, 50, peres. All whiche make, 96, So is that
queſtion fully aunſwered.

The proofe.

| 22 |
| 24 |
| 50 |
| 96 |

Maſter. An other queſtion is this. J had
a ſomme of money owing vnto me : whereof J did re=
ceiue at one tyme $\frac{1}{4}$ and afterward J receiued $\frac{2}{3}$ of that
reſidue, whiche remained vnpaied. And ſo remained
the reſte of the debte 27, l̷. J would knowe what was
the firſt debte, & what wer the, 2, ſeuerall paiementes

A queſtion
of debte.

The Arte

Scholar. This muste I obserue still, to name the firste doubtfull thyng. 1. ze. wherefore I saie that the firste debte was 1. ze. whereof I receiued $\frac{1}{4}$. And so did there remain. $\frac{3}{4}$. ze. of whiche reste, againe I receiued $\frac{2}{5}$, that is $\frac{6}{20}$. of the whole somme, or $\frac{6}{20}$ ze. And that beyng abated also, then did there remaine $\frac{9}{20}$ ze. whiche you named to be. 27. K. Then if $\frac{9}{20}$. ze. bee equalle to 27. K, diuide. 27. K. by $\frac{9}{20}$, and the *quotiente* will bee $\frac{540}{9}$, that is. 60. whiche was the whole debte : And then is it plaine, that $\frac{1}{4}$. of it is. 15. and $\frac{2}{5}$. of the residue is. 18. whiche maketh. 33. and then remaineth. 27.

Master. There is nothyng better then exercise, in attainyng any kynde of knowlege : And therfore I will proue you with diuerse questions, to make you the moare experte in this rule. And this is one.

A question of pauyng.

There is a flooze Paued with Square Bricke, the lengthe of that flooze beyng longer then the breadth, by $\frac{1}{7}$, and the whole Pauemente containeth. 3584. brickes : I require to knowe the bredthe and lengthe.

Scholar. The lesser quätitie, whiche is the bredth I doe name. 1. ze. And then the lengthe will bee, by your proportion. $1\frac{1}{7}$. ze. Now must I multiplie the bredthe by the lengthe (for that is the orderly worke in all flatte formes, to finde out the whole platte) that is here. 1. ze. by. $\frac{8}{7}$. ze. and there will amounte the whole platte. $\frac{8}{7}$. z. whiche by your supposition is equalle to. 3584.

Wherfore accordyng to your rule, I diuide. 3584. by $\frac{8}{7}$, and the *quotiente* will be. 3136. whiche is a *Square* nomber, bicause there is one denomination omitted in this equatiö. For betwene z and ꝗ. there is omitted. ze. And therfore must I extracte the square roote of. 3136. and it will bee the quantitie of. 1. ze. that I woorke in my tables, and finde it. 56. whiche must be the bredthe : for that I named. 1. ze. Then the length must be moare by $\frac{1}{7}$. of it : and so shall it be. 64.

Now

of Cossike nombers.

Now for to confirme my woorke, I multiplie. 56. by. 64 and it will make. 3584. whiche is the nomber that you did name.

Master. That question is well aunswered : And if you had put. 1. ℞ . for the lengthe, as you might do, then the bredthe will be $\frac{7}{8}$. ℞ . and the square $\frac{7}{8}$. ʒ . and so. 1. ℞ . would bee. 64. as you maie proue at leiser : but in the meane time, what saie you to this questiõ? *An other woorke of that question*

There is a capitain, whiche hath a greate armie, & would gladly Marshall thẽ, into a square battaile, as large as mighte bee. Wherefore in his firste proofe of square forme, he had remainyng. 284. to many. And prouyng again by puttyng. 1. moare in the fronte, he founde wante of. 25. men. How many souldiars had he, as you gesse? *A question of an armie.*

Scholar. I call the firste fronte. 1. ℞ . and then multiplyng it Squarely : I shall haue for the whole battaile. 1. ʒ . and so by your saiyng, there was lefte 284. men, wherefore the whole nomber of men, was. 1. ʒ . —— .284.9.

Now for the seconde proofe, when the fronte was increased by. 1. man : I shall set for former fronte, and 1. manne moare, that is

1. ℞ —— 1.9. And mul=
tiplyinge that nomber,
squarely : there will arise
for the whole armie.

1. ʒ —— 2 ℞ —— 19.
out of whiche I muste a=
bate 25 that, you saie, did

$$1. ℞ \;+\; .1.9.$$
$$1. ℞ \;+\; .1.9.$$
$$\overline{}$$
$$1. ʒ \;+\; .1. ℞ .$$
$$\qquad\qquad 1. ℞ \;+\; 1.9$$
$$\overline{}$$
$$1. ʒ \;+\; 2. ℞ . \;+\; 1.9.$$

wante, to make vp that square battaile. And then it will be. 1. ʒ . —— 2. ℞ . —— 24.9.

Now haue I one nomber of menne, expressed by. 2 *Cossike* nombers : Of necessitie therefore must these. 2. nombers be equalle : seyng thei represente one armie.

Wherfore I set them thus.

The Arte

1.℥. —— 284.℈. ===== 1.℥. —— 2.℞ ———— 24.℈.

And findyng, 1. ℥. on bothe partes of the equation, J doe abate it, & then standeth, 284℈ ===== 2℞ ———— 24℈.
Yet again J fee.℈. on bothe sides of the equation, and therfore, feing the leffer of them hath the figne of fub=traction, J doe adde. 24. to bothe nombers, and then will there be. 308. ===== 2.℞. that is. 154 ===== 1.℞

So that feing 1℞ was fet for the first fronte : the fame front muft be. 154. whofe Square is. 23716. vnto whiche J mufte adde the. 284. that did abounde, and then will the whole nomber be. 24000.

For farther trialle whereof, J take the feconde fronte to be. 155. that is. 1. moare then the firfte : and his Square will bee 24025. And fo is there. 25. moare then the iufte nomber of the armie, as the que=ftion fuppofed.

154.
154.
616.
770
154
23716.
284.
24000.

An other
woorke of
that queftiõ.

Mafter. That queftion maie be wrought alfo, by namyng the feconde fronte. 1.℞. and then will his fquare bee. 1.℥. but feyng there wanteth. 25. menne, to make that Square battaile, the nomber shall bee
1.℥. ———— 25.℈.

Then for the firfte front, you muft fet. 1. man leffe, as the queftion importeth, & that will be. 1.℞———1.℈ whofe fquare will be 1.℥. —— + 1.℈. ———— 2.℞.

1.℞. ———— .1.℈.
1.℞. ———— .1.℈.
———————————————
1.℥. ———— .1.℞.
———— .1.℞. —— + .1.℈.
———————————————
1.℥. —— + .1.℈. ———— .2.℞.

vnto whiche J muft adde the. 284. menne that did a=bounde, whē that battaile was framed, and then will
                                                                    the

of Cossike nombers.

the nomber be. 1. ℨ. ———— .285. ᵱ. ———— .2. ℨℯ. And
it muſt bee equalle to. 1. ℨ. ———— .25. ᵱ. Wherfore to
reduce that equation, firſte I adde on bothe ſides 25. ᵱ
& then reſteth. 1. ℨ. equalle to. 1. ℨ ——— 310 ——— 2. ℨℯ
Then I adde. 2. ℨℯ. bicauſe I will haue noe ——— in
the equation : and it will be,

1. ℨ ——— 2. ℨℯ. ════ .1 ℨ. ——— 310. ᵱ. Thirdely I
abate. 1. ℨ. on bothe ſides of the equation : and then
remaineth. 2. ℨℯ ════ .310. ᵱ. that is. 1. ℨℯ. ════ 155. ᵱ.
whereby it appeareth that the ſeconde fronte was. 155
and the firſte fronte. 154. & ſo forthe, as you wrought
it before.

An other queſtion is this.

There is a kyng with a greate armie : And his ad= *A queſtion*
uerſarie corrupteth one of his heraultes with giftes, *of an armie.*
and maketh hym ſwere, that he will tell hym, how
many Dukes, Erles and other ſouldiars there are in
that armie. The Heraulte lothe to leaſe thoſe giftes,
and as lothe to bee vntrue to his Prince, diuiſeth his
aunſwere, which was true, but yet not ſo plain, that
the aduerſarie could therby vnderſtande that, whiche
he deſired. And that aunſwere was this.

Looke how many Dukes there are, and for eche of
them, there are twiſe ſo many Erles. And vnder eue=
ry Erle, there are fower tymes ſo many ſoldiars, as
there be Dukes in the fielde. And when the muſter of
the ſoldiars was taken, the. 200. parte of them, was
9. tymes ſo many as the nomber of the Dukes.

This is a true declaratiō of eche nomber, quod the
Heraute : and I haue diſcharged my othe. Now geſſe
you how many of eche ſorte there was.

Scholar. Although the queſtion ſeme harde, I ſee
many tymes, that diligence maketh harde thynges
eaſie, and therfore I will attempte the worke of it.

And firſte for the nomber of Dukes, I ſette. 1. ℨℯ.
then will the nomber of Erles bee. 2. ℨ. that is. 1. ℨℯ
by

The Arte

by. 1. ℞ multiplied twise : And the number of soldiars are. 8. cℰ. that is. 2. ₴. multiplied by. 1. ℞. fower tymes, but bicause the. 200. parte of the soldiars is. 9. tymes so moche as the number of the Dukes, therfore must the. 200. parte of. 8. cℰ be equalle to. 9. ℞. And so consequently. 8. cℰ ═══ 1800. ℞ and 1. cℰ ═══ 225. ℞. oʒ. 1 ₴ ═══ 225. ℘.

For if J set $\frac{8}{200}$ and. 9. as equalle together, & would by the arte of fractions, bʒynge the same pʒopoʒtion in whole nombers, J shall haue foʒ. 9. this fraction $\frac{1800}{200}$. And seyng the denominatoʒs, be all one in $\frac{8}{200}$ and in $\frac{1800}{200}$ the pʒopoʒtion consisteth betwene the numeratoʒs.

Then to pʒocede, if. 225. be equalle to. 1. ₴. J shall take the square roote of. 225. foʒ. 1. ℞. and that is. 15 whiche must be the number of Dukes.

And so haue J the firste nomber, wherefoʒe the seconde nomber, that is the number of Erles, must bee 15. tymes. 15. twise : that is. 450. And the number of soldiars shall be. 4. tymes. 15. multiplied by. 450. that is. 27000. And foʒ iuste trialle of this wooʒke, J take the. 200. parte of the soldiars that is. 1350. and J finde it to bee. 9. tymes. 15. that is. 9. tymes so moche as the number of the Dukes. And so is that question solued, and tried.

450.
60
27000

An other question of walles.

Master. This is an other question. There is a grounde inclosed with. 4. walles, beyng like iambes and of one heighthe. The longest. 2. walles are in pʒopoʒtion to the shoʒteste, as. 5. to. 3. And vnto the height thei bee double Sesquialter. Now multipliyng the longeste by the shoʒteste, and that totalle by the heighte, there will rise. 39930. foote. J demaunde then, what is the lengthe and the heighte of eche walle?

Scholar. The least quãtitie is the heighte, whiche J call. 1. ℞. and vnto it the longeste walle is double Sesquialter.

of Cossike nombers.

Sesquialter: that is, $2\frac{1}{2}$. $\mathcal{Z}$. Now that same longeste is
in proportion *Superbipartiente quintas*, to the shorteste
walle. So must the shorter wall be, $1\frac{1}{2}$ $\mathcal{Z}$. Then must
I multiplie all those, 3. nōbers together, that is, 1. $\mathcal{Z}$,
by, $1\frac{1}{2}$ $\mathcal{Z}$, whereof doeth come, $\frac{3}{2}$ $\mathcal{Z}$. then shall I mul=
tiplie that totalle, by $\frac{5}{2}$ $\mathcal{Z}$, and it will be $\frac{15}{4}$ $c\mathcal{Q}$, or $3\frac{3}{4}$ $c\mathcal{Q}$
whiche must be equalle, by the woordes of the questi=
on, to. 39930.

So by reducyng them to one denomination. $\frac{15}{4}$ $c\mathcal{Q}$.
shall be equalle to $\frac{159720}{4}$ that is. 15. $c\mathcal{Q}$ ══════ 159720 $\mathcal{9}$.
and, 1. $c\mathcal{Q}$. ══════ .10648. wherfore I shall extracte the
Cubike roote out of. 10648. and that is the quantitie
of. 1. $\mathcal{Z}$. or the heighte of the walle.

In my Tables I woorke that extraction of *Cubike*
roote, and finde it to be. 22. And therfore must the lō=
geste walle bee double *Sesquialter* to it, that is. 55. And
the shorteste walle will be. 33.

For proofe whereof I dooe multiplie. 22. with. 55.   *The proofe.*
and it maketh. 1210. whiche nomber I shall multi=
plie by. 33. and it will be. 39930. accordyng to the sup=
position of the question.

Master. You doe chose still the leaste nomber, to
be equalle to. 1. $\mathcal{Z}$. as the easieste forme. Howbeit you
maie put. 1. $\mathcal{Z}$. for the lengthe of any of the walles.

And if you sette it for the longeste walle, then the   *An other*
shorteste walle will be $\frac{3}{5}$ $\mathcal{Z}$. and the heighte $\frac{2}{5}$. $\mathcal{Z}$ and   *forme of*
all those. 3. nombers will make, by multiplication to=   *woorke.*
gether $\frac{6}{25}$ $c\mathcal{Q}$. equalle to. 39930. And so will. 6. $c\mathcal{Q}$. be
equalle to. 998250. $\mathcal{9}$. and, 1. $c\mathcal{Q}$. ══════ .166375. $\mathcal{9}$.
whereof the *Cubike* roote is 55. and aunswereth to the
quantitie of. 1 $\mathcal{Z}$.

But if. 1. $\mathcal{Z}$. be set for the measure of the shorteste   *The thirde*
walle, then the longeste walle will bee $\frac{5}{3}$. $\mathcal{Z}$. And the   *forme of*
heighte. $\frac{2}{3}$. $\mathcal{Z}$. And so all. 3. nombers multiplied toge=   *woorke.*
ther will make $\frac{10}{9}$ $c\mathcal{Q}$. ══════ .39930. So shall .10 $c\mathcal{Q}$.
be equall to. 359370. And. 1. $c\mathcal{Q}$. ══════ .35937. where=
Hh.i.   of

The Arte

of the *Cubike* roote is. 33. and is the value of. 1. ℨℯ. in this position.

Scholar. This varietie of woozke, is not onely pleasaunte, but it maketh the reason of the woozke to appeare moare plainly. So that I could neuer be werie to heare soche questions.

Master. Then will I propounde one oz 2. moare befoze we passe from this kinde of equation. Whereof one shall be somewhat like that last. And this it is.

A question of Bricke.

A Bzickeleiar had a pile of Bzicke, whiche he sold by the yarde. The lengthe of it was $\frac{7}{2}$ to the bzedthe, that is *Tripla sesquialtera*. And the heighte was fiue tymes so moche as the lëgthe. This pile the owner sold foz. 980. crounes. By soche rate that he had foz euery yarde so many Crounes, as the Pile had yardes in bzedthe. Now is the question, what was the lengthe, bzedthe, and heighte of this pile?

Scholar. I suppose the bzedthe to bee. 1. ℨℯ. then was the length $3\frac{1}{2}$ ℨℯ. and the heighte. $17\frac{1}{2}$ ℨℯ. These 3. sommes dooe I multiplie together, and thei make $\frac{245}{4}$ ℭℯ. whiche standeth as equalle to all the yardes in the whole pile. But yet what that is, I knowe not.

Wherfoze to procede farther, I consider that euery yarde coste as many crounes, as the bzedthe contained yardes. Now the bzedthe being 1. ℨℯ. I must saie, that euery yarde did coste. 1. ℨℯ. of crounes. And then by the Golden rule : if. 1. yarde coste. 1. ℨℯ. of Crounes, what shall $\frac{245}{4}$ ℭℯ. coste?

Woozkyng by the rule, I

1.		1. ℨℯ.
$\frac{245}{4}$ ℭℯ		$\frac{245}{4}$ ℨℨ.

finde that it shall cost $\frac{245}{4}$ ℨℨ. And the question doeth suppose that it coste. 980. crounes. Wherfoze I must saie, that. 980. crounes, are equalle to $\frac{245}{4}$ ℨℨ. And consequently. 245. ℨℨ. $=\!=\!=$ 3920.℈. wherfoze diuidynge the nomber of the lesser name, by the other, the *quotiente* will be 16. whose *zenzizenzike* roote is 2

And

of Cofsike nombers.

And that therfoze muſt be the value of a roote, and the bzedthe of the pile. So ſhall the lengthe be. 7. yardes, and the heighte. 35. yardes.

Foz trialle of it, I multiplie the lengthe, by the bzedthe, and that totalle by the heighte, and ſo haue I 490. foz all the yardes of Bzicke. Then conſideryng that euery yarde coſte. 2. crounes, bicauſe. 2. yardes is the bzedthe of the pile : the nomber of crounes muſt be twiſe. 490. that is. 980. And ſo is the woozke good. *The proofe.*

Maſter. Now woozke that queſtion, by ſettynge 1. ℥. foz the lengthe. *An other forme of woorke.*

Scholar. If the lengthe be. 1. ℥. the bzedthe muſt bee ⅔. ℥. that is *Subtripla ſeſquialtera* to. 1. ℥. And the heighte muſt bee. 5. ℥. All whiche ſommes make by multiplication $\frac{10}{7}$ ℥.

Then farther, if 1. yarde coſte $\frac{2}{7}$ ℥. $\frac{10}{7}$ ℥. ſhall coſte $\frac{20}{49}$. ℥℥. whiche is equalle to 980. And ſo is. 20. ℥℥. equal to. 4 8 0 2 0. and by diuiſion 1. ℥℥. ═══ .2401. whoſe *zenzizenzike* roote is. 7. And that is the lengthe of the walle, and is the value of. 1. ℥.

$$1. \diagup \frac{2}{7}\,℥.$$
$$\frac{10}{7}\,℥. \diagup \frac{20}{49}\,℥℥.$$

The reſte of this woozke, is like as befoze.

Maſter. Yet pzoue the thirde waie. *A thirde forme of woorke.*

Scholar. The heighte beeyng. 1. ℥. the lengthe muſt be the firſt part of it, that is ⅕ ℥. And the bzedth $\frac{2}{35}$ ℥. All theſe make by multiplication $\frac{2}{175}$ ℥. Then foz the pzice, if. 1. yard coſte $\frac{2}{35}$ ℥. what ſhall $\frac{2}{175}$ ℥. coſte? By the Golden Rule there is founde, $\frac{4}{6125}$. ℥℥. whiche is equalle to 980. And ſo ſhall. 4. ℥℥. be equalle to. 6002500. And. 1. ℥℥. ═══ 1500625. whoſe *zenzizenzike* roote is. 35. And that is the value of. 1. ℥. and the heighte of the Pile.

$$1. \diagup \frac{2}{35}\,℥.$$
$$\frac{2}{175}\,℥. \diagup \frac{4}{6125}\,℥℥.$$

Maſter. One queſtion moare will I pzopounde,

The Arte

and so eande with this equation.

A question of a Testament.

A poore man died, whiche had fower children, and all his goodes came to. 72. crounes : whiche he would haue parted so, that the seconde & thirde childe should haue. 7. times so moche as the firste. And that the portions of the thirde and fourthe childe should bee. 5. tymes so moche as the secondes parte : And that the first and the fourthe, should haue twise as moche as the thirde. If you worke the solution wel, you maie seme worthy to be master of those wardes.

Scholar. I trust to obtaine moare benefite by the question, then by that office. Wherefore I will giue good hede vnto it. And for the firste nöber, I set. 1. ℈ then muste the seconde and thirde portions make together. 72. ℈ . And the fourthe must bee all the reste of the. 72. that is. 72 ——— 8 ℈ . Now the thirde must be halfe the firste & the fourthe, that is. 36 ——— $3\frac{1}{2}$ ℈ . And the third & fourthe, is. 5. tymes the second. wherfore the seconde shall be the. 5. part of. 108 ——— $11\frac{1}{2}$ ℈ that is. $21\frac{3}{5}$ ——— $2\frac{3}{10}$ ℈ , whiche nomber I shall set in order with Letters, as here I haue dooen for my owne ease, and aide of memory. And then shal I adde them all together. Whereof there commeth.

$129\frac{3}{5}$ ——— $12\frac{4}{5}$ ℈ , whiche is equalle to 72. First therfore I do adde all that foloweth ——— to bothe partes of the equatiö. And so haue I $129\frac{3}{5}$ === $12\frac{4}{5}$ ℈ ——— 72. But bicause there are nombers Absolute on both sides, I shall abate the lesser somme, that is. 72. from bothe partes, and then will there bee lefte, $57. \frac{3}{5}$ === . $12\frac{4}{5}$. ℈ . that is. 288. === 64. ℈. And by diuision $4\frac{1}{2}$. === . 1. ℈ .

A	1. ℈
B	$21\frac{3}{5}$ ——— $2\frac{3}{10}$ ℈
C	36. ——— . $3\frac{1}{2}$ ℈
D	27. ——— . 8. ℈
$129. \frac{3}{5}$	——— $12\frac{4}{5}$ ℈

The proofe.

So shall the firste mannes portion bee $4\frac{1}{2}$. And the seconde and thirde mannes portion. 7. times so moche
that

of Cossike nombers.

that is. 31½. Whereby it followeth, that the fourthe manne, shall haue the reste of 72. that is. 36.

$$\left.\begin{array}{ll} A & 4\frac{1}{2}. \\ B & 11\frac{1}{4} \\ C & 20\frac{1}{4} \end{array}\right\} 31\frac{1}{2}.$$

Then seeyng the thirde manne, hath halfe so moche as the first and the fourthe, his portio shall be 20¼.

$$\begin{array}{l} D \quad 36. \\ \hline \quad 72. \end{array}$$

And then by diuerse reasons, the seconde mänes part shall bee 11¼. And all these partes added together, doe make iuste 72. Wherfore the woorke is good.

Master. You haue wroughte it well. And yet maie you woorke it thus. Firste sette doune. 1.℥. for the firste mannes parte. And then for the seconde and thirde ioyntly. 7.℥. so shall the fourthe manne haue 72.♀.⸺.8.℥. And bicause the seconde mannes parte is ⅕. of the thirde and fourthe mannes portion, if you ioyne all their. 3. partes together, the seconde mannes portion will be ⅙ of that totalle. Put therfore 7.℥, whiche is the partes of the second and the third vnto. 72 ⸺ 8℥, whiche is the fourthe mannes parte, and the totalle will be. 72.♀ ⸺ 1.℥. whose firste parte is 12.♀.⸺ ⅙℥, for the seconde mannes share. Whiche somme if you abate out of. 7.℥. there wil remain for the thirde mannes parte 7⅙℥⸺ 12.♀.

An other forme of woorke.

And so haue you euery mannes portiõ allotted to hym duely. As I haue here set it forthe for you. And all thei added together, doe make. 72.

$$\begin{array}{l} A \quad 1.℥. \\ B \quad 12.♀.\ ⸺\ \frac{1}{6}℥ \\ C \quad 7\frac{1}{6}℥\ ⸺\ 12.♀ \\ D \quad 72.\ ⸺\ .8.℥. \\ \hline \quad\quad 72. \end{array}$$

Scholar. But here is noe equatiõ yet, though the partes be diuided iustly.

Master. Now shall you see it.

The question saieth, that the thirde mannes portion is halfe the portions of the firste and fourthe man. wherefore seyng the firste and fourthe mannes portions doe make. 72⸺ 7.℥. the thirde mannes por=

tion

The Arte

tion beeyng doubled, ſhall make as moche. But the double of the thirde mãnes parte, is $14\frac{1}{3}$ ℥ —— 24ℛ. and therfoꝛe J ſaie, that thoſe. 2. nombers be equalle, that is. 72.ℛ. —— 7.℥. ═══ $14\frac{1}{3}$℥. —— 24.ℛ.

Firſte adde. 7. ℥. to eche parte, and it will bee 72.ℛ. ═══ $21\frac{1}{3}$℥. —— 24.ℛ. Then adde. 24.ℛ. on bothe ſides, and there will be. 96.ℛ. ═══ $21\frac{1}{3}$℥. that is by reduction. 288. ═══ .64.℥. as you made it. And then all agreeth.

Likewaies foꝛ the equation, you maie ſet the third mannes poꝛtion, with the halfe of the firſte & fourthe mennes partes. And ſo will. $7\frac{1}{6}$.℥. —— .12.ℛ. be equalle to. 36.ℛ. —— $3\frac{1}{2}$℥. And by reduction, $10\frac{2}{3}$℥. ═══ 48.ℛ. That is in other termes of whole nomber. 32℥. ═══ .144. And by diuiſion it will bee 1. ℥. ═══ .4$\frac{1}{2}$. And thus will we eande the examples of the firſte equation, foꝛ this tyme. And will ſhewe you ſome queſtions of the ſeconde equation.

Examples of the ſeconde equation,
by queſtions propounded.

A queſtion
of ſilkes.

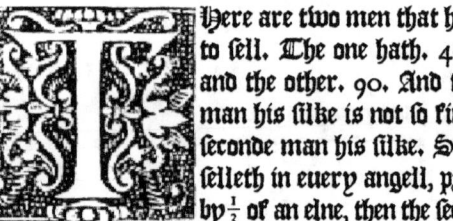 Here are two men that haue ſilke to ſell. The one hath. 40. elnes, and the other. 90. And the firſte man his ſilke is not ſo fine as the ſeconde man his ſilke. So that he ſelleth in euery angell, pꝛice moꝛe by $\frac{1}{3}$ of an elne, then the ſeconde mã doeth. And at the eande, bothe their moneis made but 42. angelles. Now J demaunde of you, how moche eche man ſolde foꝛ an angell?

Scholar. J will folowe my olde foꝛme, in putting 3.℥. foꝛ the leaſte quantitie, whiche is the ſeconde mannes ſomme, and then ſhall the firſte mannes ſomme be. $1\frac{1}{3}$ ℥.

Maſter. You are deceiued all readie. Foꝛ you ſet
1.℥.

of Cofsike nombers.

1.℥. foz an elne. Seyng you name $\frac{1}{3}$ of an elne, to be $\frac{1}{3}$.℥. And so were the pofition neadeleffe, and like=waies all the woozke.

Scholar. J fee my faulte : but J knowe not how to amende it. Foz that. 1.℥. maie bee a parte oz partes of an elne : and fo maie it be moare then. 1. oz. 2. elnes fo that J ought not to haue fet $\frac{1}{3}$ (whiche is certainly referred, in this queftion, to an elne) as the parte of a doubtfull quantitie, but rather as the parte of a quan=titie certaine. Whereas. 1. ℥ .is euer put foz a nom=ber vnknowen.

Mafter. To helpe you herein, J will fet the firfte nombers, as you began theim. The feconde man his nombers of elnes, fhall bee. 1.℥ . as you did name it, and then fhall the firfte man=nes poztion be as moche, and $\frac{1}{3}$ of an elne moare : whiche $\frac{1}{3}$ J

A 1. ℥ ──────── $\frac{1}{3}$ ℈.
B   1.℥.

maie befte call $\frac{1}{3}$ ℈. And fo fhall it bee diftaunte from 1.℥. clerely in all woozke *Arithmeticall.*

But now to pzoceade, J fhall diuide eche mannes nomber of elnes, whiche he had, by the nomber of el=nes, whiche he folde foz an angelle, and the *quotiente* will declare how many angelles eche man had recei=ued. So that the firfte mannes nomber of elnes, bee=yng. 40. fhall bee the numeratoz, and the fomme of meafure, whiche he folde foz an Angelle, that is 1℥ ──────── $\frac{1}{3}$℈. fhall bee the denominatoz. And fo is the diuifion eanded. And that frac=tion is the *quotiente.*

Scholar. Now J perceiue the woozke. And by like reafon : the fe=conde mannes fomme of elnes bee=yng. 90. fhall bee the numeratoz, and. 1.℥. beyng the fomme of meafure,

$$\frac{40.}{1.℥ ── \frac{1}{3}℈.} \quad \frac{90.}{1.℥.}$$

folde foz one Angelle, fhall be the denominatoz, that is in one frac=tion $\frac{90}{120}$: acozdingly as J haue fette bothe nomber s here

The Arte

here diſtinctly.

Maſter. It were moare eaſe for you in workyng,
if you did tourne that fractiō of $\frac{1}{3}$ into an intere vnitie.

Scholar. That will eaſily be doen, by multipliyng
euery nomber, of that whole fraction by .3. And then
will it be $\dfrac{120.}{3.\text{æ}.\ \underline{\quad}\ .1.\text{ɋ}}$, whiche is all one in value with

$\dfrac{40.}{1.\text{æ}\ \underline{\quad}\ \frac{1}{3}\text{ɋ}.}$ And this I conſider farther, that as
theſe .2. fractions, ſeuerally dooe ex=
preſſe the ſommes of angelles, that eche of them recei=
ued, ſo ioynctly bothe together, dooe declare the full
ſomme, of all their angelles. Wherefore I ſhall adde
theim both together. And thei will make.

$\dfrac{390\text{æ}\ \underline{\quad}\ 90\text{ɋ}}{3\text{ȝ}\ \underline{\quad}\ 1\text{æ}}$ As here in woorke I haue expreſſed.

$$390.\text{æ}.\ \underline{\quad}\ .90.\text{ɋ}.$$

120.	90.
$3.\text{æ}.\ \underline{\quad}\ .1.\text{ɋ}.$	$1.\text{æ}.$

$$3.\text{ȝ}.\ \underline{\quad}\ .1.\text{æ}.$$

And by your ſuppoſition, their bothe ſommes of An=
gelles made. 42. So that thoſe. 2. ſommes are e=
qualle : and therefore am I come to an equation. In
whiche I ſee a nomber abſolute, equalle to a fraction
Coſſike compounde.

Maſter. When ſo euer that, or the like dooeth
chaunce, you ſhall reduce the whole nōber, to the like
denomination : and then their numeratoꝛs will bee
equalle.

Scholar. Then ſhall I multiplie. 42. by the deno=
minatoꝛ $3\text{ȝ}\ \underline{\quad}\ 1\text{æ}$ & it wil be $126\text{ȝ}\ \underline{\quad}\ 42\text{æ}$
whiche muſte bee equalle to. $390.\text{æ}.\ \underline{\quad}\ 90.\text{ɋ}.$
That is in leſſer termes.

$$21\text{ȝ}\ \underline{\quad}\ 7\text{æ}.\ \equiv\ .65.\text{æ}.\ \underline{\quad}\ .15.\text{ɋ}.$$
Where firſte I dooe abate. $7.\text{æ}.$ on bothe ſides : and
there remaineth then. $21\text{ȝ}.\ \equiv\ .58.\text{æ}.\ \underline{\quad}\ .15.\text{ɋ}.$

But

of Cossike nombers.

But now J remēber your admonitiō, that bicause the nomber annexed to the greateste signe, is moare then. 1. J shall diuide all the nombers by it, and sette their *quotientes* in their stede, with their signes. And so will the nomber of the greateste signe, euermoze be 1. And this equation will be 1. $\mathfrak{z}$. $= \frac{58}{21} \mathcal{Z}$. $+ \frac{15}{21} \mathcal{P}$. Where J must extracte the square roote of the later part, accozding to your doctrine, and it will be. 3. As it appereth in this wozke folowing, whiche J did frame in my tables.

$\frac{29}{21}$. in square doeth make $\frac{841}{441}$, vnto whiche J muste adde $\frac{315}{441}$, whiche is all one with $\frac{15}{21}$, by reduction to one denominatiō. So is the full additiō $\frac{1156}{441}$. whose square roote is $\frac{34}{21}$. vnto whiche J shall adde $\frac{29}{21}$, and it will bee $\frac{63}{21}$, that is. 3.

Master. This is well doen. Now wozke the same questiō, as it was propioned, and you shall easily finde all the other nombers to bee true, and agreable to the question.

Scholar. Seyng the seconde manne solde. 3. elnes foz an angell, the firste manne did sell. 3. elnes and $\frac{1}{3}$. So. 40 (whiche is the somme of elnes of the first man his silke) diuided by. $3\frac{1}{3}$. doeth yelde. 12. and sheweth how many angelles that man receiued. *The proofe.*

Again foz the seconde man, whiche had. 90. elnes, diuide that. 90. by. 3. and so shall you finde. 30. foz the nomber of his Angelles. And that. 30. and. 12. dooe make. 42. it neadeth not to be prooued.

Master. Now againe foz your exercise, suppose the firste mannes somme to be. 1. $\mathcal{Z}$. *An other forme of woorke.*

Scholar. Then muste the seconde manne sell foz an angelle. 1. $\mathcal{Z}$ $— \frac{1}{3} . \mathcal{P}$. And their nombers of elnes, diuided by those nombers will make. $\frac{40}{1\mathcal{Z}}$. and $\frac{90}{1\mathcal{Z} — 3\mathcal{P}}$, whiche bothe added together, will bee $\frac{390 \mathcal{Z} — 40\mathcal{P}}{3\mathcal{Z} — 1\mathcal{Z}}$ equalle to. 42. $\mathcal{P}$. That is by reduction.

390. $\mathcal{Z}$. $—$ .40 $\mathcal{P}$. $==$ .126. $\mathfrak{z}$. $—$ .42. $\mathcal{Z}$.

And

The Arte

And by addition of. 42. z. on bothe partes.

432. z ——— 40. θ. ═══ .126. ζ. And by diuifion it will be. $\frac{24}{7}$ z ——— $\frac{20}{63}$ θ. ═══ 1. ζ.

So that now J muft extracte the roote of that com= pounde *Cofsike* fraction, thus. $\frac{12}{7}$ fquarely, dooe make $\frac{144}{49}$ out of whiche J fhall abate $\frac{20}{63}$. And therfoze, firfte of all J doe reduce thē to one denomination, ⁊ thei make $\frac{9072}{3087}$, and $\frac{980}{3087}$. wherefoze if J abate the leffer out of the greater : there will remaine $\frac{8092}{3087}$. that is in leffer ter= mes $\frac{1156}{441}$. and is a fquare nomber, whofe roote is. $\frac{34}{21}$ vn= to whiche if J adde $\frac{12}{7}$ that is $\frac{36}{21}$. it will make $\frac{70}{21}$, oz $\frac{10}{3}$. that is the valewe of. 1. z. And is the firfte mannes nomber of elnes, agreably as J tried it befoze. And fo doe bothe wozkes agree.

But now commeth to my remembzaūce, that this nomber, whofe roote J did extracte, in this lafte wozke is of that fozte, where. z. ——— θ. is equalle to. ζ. And therfoze hath in it. 2. rootes : thone by addition, as this, whiche J now founde : And the other by fub= traction, whiche in this example, by abatyng $\frac{34}{21}$ out of $\frac{36}{21}$, will bee $\frac{2}{21}$. But how J maie frame that roote, to a= gree to this queftion, J doe not fee.

Mafter. That varietie of rootes dooeth declare, that one equation in nomber, maie ferue foz. 2. feue= ralle queftions. But the fozme of the queftion, maie eafily inftruct you, whiche of thofe. 2. rootes, you fhall take foz your purpofe. Howbeit fometymes you fhall take bothe. As foz example again, marke this que= ftion.

A queftion of money.

A gentilman, willyng to pzoue the cunnyng, of a bzaggyng *Arithmetician*, faied thus : J haue in bothe my handes. 8. crounes : But and if J accoumpte the fomme of eche hande by it felf feuerally, and put ther= to the fquares and the *Cubes* of thē bothe, it will make in nomber. 194. Now tell me (quod he) what is in eche hande : and J will giue you all foz your laboure.

Scholar.

of *Cofsike* nombers.

Scholar. Soche incoragementes, would make me ſtudie harde, and trauell very willyngly in learned exerciſes : though learnyng bee moſte to be loued, foꝛ knowledges ſake. But foꝛ to finde the true aunſwere thus J doe pꝛoceade.

Firſte J ſuppoſe the one nomber in one hand, to be 1.℥. And then muſt the other nedes be 8.℈ —— .1.℥ Then doe J make theim bothe Squares. And foꝛ the firſte J haue. 1.₴. and foꝛ the ſeconde. 1.₴. —+— 64℈ —— 16.℥. Thirdely J multiplie them bothe *Cubi= kely* : and ſo haue J foꝛ the firſte. 1.℀. and foꝛ the other 24.₴. —+— .512.℈. —— 1.℀. —— 192.℥. Then muſt J adde bothe the nöbers, with their ſqua= res, and their *Cubes,* into one ſomme. As here in woꝛk

1.℥. —+—	.1.₴. —+—	.1.℀.
8.℈.	——	.1.℥.
1.₴. —+—	.64.℈.	.16.℥.
24.₴ —+—	.512℈ — 1℀	192.℥.
26.₴ —+—	584.℈. ——	.208.℥.

it is ſet foꝛthe. Where foꝛ eaſe J haue ſet.1.℥, 1.₴. and. 1.℀. (whiche is the Roote, the Square and the Cube of one nomber) all in one line : and the other Roote, Square, and Cube, J haue ſet ſeuerally. And ſo all thei doe make. 26₴ —+— 584℈ —— 208℥ whiche is equalle to. 194. by the intente of the que= ſtion. Wherefoꝛe J adde firſte. 208.℥. to bothe par= tes, and there remaineth.

26.₴. —+— 584.℈ === 208.℥ —+— 194.℈.

Then J abate. 194. from bothe ſides, and ſo reſtethe the equatiö thus. 26.₴. —+— 390.℈ === 208.℥ That is by diuiſion. 1.₴. —+— .15.℈. === 8.℥. And by tranſlation of. 15.℈. toſette. 1.₴. alone, it wil be. 1.₴ === 8℥ —— 15.℈. And now haue J the exacte and complete equation, where J muſt ſeke foꝛ

The Arte

the value of. 1. z_e . by extractyng the roote. Therfore
firste I take halfe of. 8. and that is. 4. whose square
is. 16. out of whiche I abate. 15. and the remainer is
1. whiche I maie either adde to. 4. and so haue I. 5. o=
ther, I maie abate it from 4 and so haue I 3. Whiche
nombers also accordyng to the same rule, beyng added
together dooe make. 8. that is the nomber of the mid=
dell denomination. And beyng multiplied together,
thei dooe make. 15. that is the other parte of the same
compounde *Cossike* nomber.

Master. And if you had marked that firste, you
might easily haue found bothe those nombers, by the
partes of. 15. whiche can be none other, but. 5. and. 3.

And farther, seyng thei. 2. doe make. 8. and. 8. is the
nomber (named in the questiõ) that thei should make,
therfore you shall take them bothe. And name whiche
of them you liste to be. 1. z_e . And the other shall be of
necessitte, the reste of. 8.

The proofe.

Scholar. To examine theim, by the order of the
question, I doe proceade thus. 3. with his Square. 9.
and his *Cube*. 27. dooeth make. 39. And. 5. with his
square 25 and his Cube. 125. doe yelde 155. And bothe
thei together doe bryng forthe. 194. accordyng to the
saiyng of the question : and therfore it is certain, that
the woorke is good.

An other woorke for equations.

Master. Before you passe any farther, I will ad=
monishe you of one waie, whiche I ofte vse in reduc=
tion of soche equations, as this is, when there is noe
denomination on the one side, but the like is on the o=
ther side, with a greater nomber annexed to it. Then
maie you abate all the lesser nõbers, out of their grea=
ters, and the reste shall bee accoumpted equalle to no=
thyng. Whiche chaunce can neuer happen : excepte
there bee some nombers on the greater side, with the
signe of abatement. ——— .

As here you had.

of Coſsike nombers.

26ʓ. ——— 584♈. ——— 208⅊. ═══ 194.♈.
Bicauſe on the one ſide, there is noe nōber but 194 ♈
and on the other ſide, there is. 584.♈. beeyng a grea=
ter nomber, and of theſame denomination : therefoꝛe
maie you abate. 194. from bothe ſides, and then re=
maineth. 26ʓ ——— 390♈ ——— 208⅊ ═══ 0
Wherfoꝛe you maie well conſider, that the nombers
whiche be ioined with ——— , are equalle to the nom=
bers that bee ſet with ——— . And therfoꝛe the one a=
batyng the other iuſtly, dooe remaine together as e=
qualle to nothyng.

Wherefoꝛe it is reaſonable, that ſeeyng the nom=
bers with ——— bee equalle to the nombers with
——— that J maie tranſlate the nombers with ———
from that ſide of the equation, and ſet them on the cō=
trary ſide, with the ſigne of ——— . And ſo in this exā=
ple it will bee. 26.ʓ. ——— .390♈. ═══ 208.⅊.
And this foꝛme ſhall eaſe you moche, in reducynge of
equations.

Scholar. J thanke you moche. And J will not foꝛ=
get to vſe it, as occaſiō ſhall happen. But J pꝛaie you
pꝛopounde yet ſome moare queſtions, that J maie ſee
their diuerſe varieties.

Maſter. There were twoo ſeueralle men, which
had certaine ſommes of angelles, in ſoche rate, that
the ſeconde manne his ſomme, was *triple ſeſquiquarta*
to the firſte : and if their. 2. ſommes were multiplied
together, and to that totall the 2 firſte ſommes added,
there would amounte. 142½ . J demaunde of you,
what was eche of their ſommes in angelles?

Scholar. The firſte mannes ſomme J call. 1.⅊ .
And the ſeconde mannes ſome ſhall be. 3¼⅊ . which
2. ſommes beeyng multiplied together, dooe make
3¼ʓ. vnto whiche J muſt adde bothe the firſte nom=
bers, that is 4¼⅊ . And it will be 3¼ʓ ——— 4¼⅊
equalle to. 142½. All whiche nombers, J ſhal bꝛing
              Ji.iii.   into

A queſtion of money.

The Arte

into whole nombers, if I multiplie theim by. 4. And
so will it be. 13. γ. ┼ 17. ζ ═══ 570. And by
reducyng the greateste denomination *Cossike*, to an v=
nitie. 1. γ. ┼ $\frac{17}{13}\zeta$. ═══ 43$\frac{11}{13}$. And laste of all, by
translatyng the nomber of. ζ. to set. 1. γ. alone, on
one side of the equation, it will be. 1. γ ═══ 43$\frac{11}{13}$ ℈
─── $\frac{17}{13}$. ζ. where I must extracte the value of the
roote thus. $\frac{17}{26}$. squarely dooe make $\frac{289}{676}$. vnto whiche I
shall adde the. 43$\frac{11}{13}$ (it beeyng firste multiplied by. 52.
to bryng it to the denomination of. 676. And so ma=
kyng $\frac{29640}{676}$) And it will be $\frac{29929}{676}$ whiche is a square nom=
ber (as I haue proued in my Tables) and his roote is
$\frac{173}{26}$. from whiche roote I must abate $\frac{17}{26}$, and then wil
there remain $\frac{156}{26}$, that is. 6.

And that. 6. is the value of. 1. ζ, and standeth for
the firste mannes nomber. So that the seconde man=
nes nöber must be as. $\frac{13}{4}$ to it : that is *tripla sesquiquarta*.
And so shall it be. 19$\frac{1}{2}$.

The proofe.   Master. Now proue those nombers, accordyng to
the question.

Scholar. 19$\frac{1}{2}$ multiplied by. 6. doeth make. 117.
vnto whiche I shall adde. 25$\frac{1}{2}$. amountyng of their. 2
addiriös, and all will be. 142$\frac{1}{2}$, accordyng to the pur=
porte of the question.

Master. So is your woorke good. Yet woorke it
*An other
worke of the
same questiö.* again, by chaungyng the position.

Scholar. I maie put. 1. ζ. to betoken the seconde
manne his somme. And then shall the firste mannes
somme bee $\frac{4}{13}$. ζ. whiche bothe multiplied together
doe make $\frac{4}{13}$ γ. And then addyng the. 2. firste sommes
to it, it wil bee $\frac{4}{13}$ γ ┼ 1$\frac{4}{13}$ ζ. And that is equalle
to. 142$\frac{1}{2}$. All whiche nombers will bee reduced to
whole nombers, by multiplication conueniente. And
so will it be. 8. γ. ┼ 34. ζ. equalle to. 3705 : that
is by reduction. 1. γ. ┼ .4$\frac{1}{4}$ ζ ═══ 463$\frac{1}{8}$ ℈.
and by translation of the termes.

1. γ.

of Cossike nombers.

1 ȝ. ═══ 463 $\frac{1}{8}$.℈. ─── 4 $\frac{1}{4}$.ꝛ. out of whiche nomber I shall extract the value of the roote, in this sorte.

Firste I saie $\frac{17}{8}$ multiplied Square, doeth make $\frac{289}{64}$, vnto whiche nomber I must adde. 463 $\frac{1}{8}$, reduced as it ought, and it will bee in all $\frac{29929}{64}$. whiche is a square nomber, and hath for his roote $\frac{173}{8}$. from whiche I must abate $\frac{17}{8}$. And then will there remain $\frac{156}{8}$, that is 19 $\frac{1}{2}$, for the value of. 1.ꝛ. And so consequently for the second mannes nōber: whiche was named in this position, 1.ꝛ. And this maie bee proued as the other was.

Master. What saie you then to this question? There is a straunge iorneye appoincted to a manne. The firste daie he must goe 1 $\frac{1}{2}$ mile, and euery daie after the firste, he must still augemente his iorney, by $\frac{1}{6}$ of a mile. So that his iorney shall procede by an *Arithmeticalle* progression. And he hath to trauell for his whole iorney. 2955. miles. I demaunde in what nōber of daies, shall he eande his iorney?

A question of iorneyng.

Scholar. I knowe not how to proceade in this question.

Master. Doe you not heare me name it, an *Arithmeticalle* progression? Wherby you might be adsured, that it doeth appertaine to that rule. And accordyng to the canons of that rule, must you woorke this question. But for your better instruction, I will helpe you in this woorke.

Firste aunswere to the question, by the common position: and saie that the tyme of his iorney is. 1.ꝛ. of daies. And then shall all the *excesses* (whiche maie also be called the *nomber of the spaces*) be. 1.ꝛ ─── 1℈ The *common excesse* was supposed to bee. $\frac{1}{6}$. of a mile. And therefore the *somme of all the excesses* muste bee (1ꝛ ─── 1℈)/6 that is to saie, the nomber of all the *excesses* multiplied by $\frac{1}{6}$, that is here, the sixte part of the nomber

The Arte

nomber of the *exceſſes.*

And bicauſe that *the firſte nomber is.* $1\frac{1}{2}$. J muſt adde it vnto the *ſomme of the exceſſes,* and ſo haue J the *laſte nomber* of that progreſſion. Wherefore addyng. $1\frac{1}{2}$. (whiche is $\frac{3}{2}$, oʒ in like denomination with the other, $\frac{9}{6}$) with $\dfrac{1\,\mathcal{Z}\!\!\!=\!\!\!-\!\!\!-\!\!\!-1\,\vartheta}{6}$ it will make $\dfrac{1\,\mathcal{Z}\!\!\!=\!\!\!+\!\!\!-\!\!\!-8\,\vartheta}{6}$. And that is the *laſte nomber* of the progreſſion.

Now you remember, that in progreſſion *Arithme-* *ticall,* if you adde *the firſte* nomber to *the laſte :* and mul= tiplie that totalle, by the *number of halfe the places,* there doeth amounte the ſomme totalle of that progreſſion.

And therfoʒe in this exaple, if you adde. $1\frac{1}{2}$ (whiche is the firſte nöber in the progreſſion) vnto $\dfrac{1\,\mathcal{Z}\!\!\!=\!\!\!+\!\!\!-\!\!\!-8\,\vartheta}{6}$ (that is the laſte nomber of the progreſſion) there wil amounte $\dfrac{1\,\mathcal{Z}\!\!\!=\!\!\!+\!\!\!-\!\!\!-17\,\vartheta}{6}$, whiche beeyng multiplied by the nomber of halfe the places, that is $\frac{1}{2}\mathcal{Z}$. (Foʒ all the number of places is. $1.\mathcal{Z}$) there will riſe, $\dfrac{1\,\mathcal{Z}\!\!\!=\!\!\!+\!\!\!-\!\!\!-17\,\vartheta}{12}$, whiche is the totalle ſomme of all the miles : and therfoʒe is equalle to. 2955.

Scholar. All the reſte, and this againe can J dooe now. Seyng $\dfrac{1\,\mathcal{Z}\!\!\!=\!\!\!+\!\!\!-\!\!\!-17\,\vartheta}{12}$. is equalle to. 2955. J will firſte bʒyng the whole nomber to the like denomina= tion, with the fraction, and it will bee. $\dfrac{1\,\mathcal{Z}\!\!\!=\!\!\!+\!\!\!-\!\!\!-17\,\vartheta}{12}$ $=\dfrac{35460}{12}$. And then omittyng the like denomina= tions. $1.\mathcal{Z}\!\!\!\!\!\mathcal{Z}\!\!\!-\!\!\!+\!\!\!-17.\mathcal{Z}\!\!\!=$. $====35460.\vartheta$. That is by tranſlation $1.\mathcal{Z}\!\!\!\!\!\mathcal{Z}====35460.\vartheta.\!\!\!-\!\!\!-17\,\mathcal{Z}\!\!\!=$. whoſe roote in value J ſhall find out thus. J mul= tiplie $\frac{17}{2}$ ſquarely, and it will be $\frac{289}{4}$, vnto whiche J ſhal adde. 35460. & it will make $\frac{142129}{4}$, whiche is a ſquare nomber, and hath foʒ his roote $\frac{377}{2}$, frö whiche J ſhall abate $\frac{17}{2}$, and then remaineth $\frac{360}{2}$, that is. 180. whiche is the value of. $1.\mathcal{Z}\!\!\!=$. And expʒeſſeth the nomber of da= yes, whiche the queſtion requireth.

The proofe.   Maſter. The pʒoofe in this, and the like queſti= ons, is, to ſet fooʒthe the progreſſion with all his ter= mes.

of Cossike nombers.

mes. Excepte you will for shortnesse, sette doune the
firste terme, whiche in this example is. $1\frac{1}{2}$: and then
by the nomber of the *excesses*, oz distaunces (whiche is
euer one lesse then the nöber of places) multiplie the
quantitie of one *excesse* : and put to it the firste terme :
and so haue you the laste terme. Then hauyng the
firste terme and the laste, with the nomber of *excesses*
you knowe how to finde the totalle.

As in this example, the nomber of *excesses* beeyng
179. And the quantitie of one *excesse* beyng $\frac{1}{6}$. their
multiplication giueth $\frac{179}{6}$. vnto whiche if you adde the
firste nomber, that is $1\frac{1}{2}$, it will be $\frac{188}{6}$. And that is the
laste nomber of that progression. Then to trie the to=
talle somme of the miles, adde the firste nomber. $1\frac{1}{2}$ to
the laste, and thei will make $\frac{197}{6}$, that you shall multi=
plie by halfe the nomber of the places, whiche in our
example are. 90 (sith the whole nomber is. 180) and
there will amounte. 2955. accordyng as the question
saieth.

Scholar. This is sufficient for this question. And
at some idle time, I will not sticke to trie it out, by set
tyng the progression foorthe at large. In the meane
tyme I praie you for better exercise, giue me some
moare questions.

Master. There is a nomber, whiche I haue for=   *An other*
gotten : and it is diuided into. 2. partes, whereof the *question.*
one I haue forgotten also, but the other was. 4. And
yet this I remember, that if the parte, whiche I haue
forgotten, be multiplied by it self, and then also with
4. those. 2. sommes will make. 117. Now would I
knowe what was the whole nomber, and also what
is the parte, whiche I haue forgotten.

Scholar. I suppose the whole nomber to be. $1\mathcal{x}$.
And bicause. 4. is his one parte, the other parte must
neades bee. $1\mathcal{x}$. ——— 4. Then doe I accordyng to
the question, multiplie. $1\mathcal{x}$. ——— 4. firste by it self,
                        Kk.i.   and

The Arte

and it will make. 1 ℨ —|— 16.♃ ——— 8.℥. Secon=
darily, J doe multiplie it (that is. 1.℥. ——— 4) by .4
And it giueth. 4.℥. ——— 16.

Then adde J bothe theſe nombers together, and it
will be. 1.ℨ ——— 4.℥. whiche by the queſtion ſhall
be equalle to. 117.

$$1.ℨ. \; —\!\!|\!\!— \; .16.♃. \; —— \; .8.℥.$$
$$4.℥ —— .16.♃.$$
___
$$1.ℨ. \; —— \; .4.℥.$$

But then muſt J vſe the accuſtomed tranſlation,
to bꝛyng the greateſte quantitie in denomination, to
ſtande alone. And ſo will it bee.
$$1.ℨ. \; === \; .4.℥. \; —\!\!|\!\!— \; .117.♃.$$
Where J muſt ſerche foꝛ the value of a roote. And
therfoꝛe J multiplie. 2. by it ſelf ſquarely, and ſo haue
J. 4. vnto whiche J adde. 117. and it maketh. 121.
whoſe roote is. 11. vnto whiche J muſte adde. 2. and
there commeth. 13. as the value of. 1.℥ and the quan=
titie of the whole nomber.

The proofe.

Foꝛ pꝛoofe of this woꝛke, J abate. 4. out of. 13. and
there reſteth. 9. as the other parte. Then doe J multi=
plie. 9. by it ſelf, and therof riſeth. 81. Alſo J doe mul=
tiplie. 9. by. 4. and it maketh. 36. whiche bothe toge=
ther, doe make. 117. as the queſtion would.

An other
woorke.

Maſter. Set. 1.℥. foꝛ the vnknowen parte, and
then wooꝛke it, to ſee the diuerſitie of the wooꝛkes.

Scholar. Jf. 4. bee one parte, and. 1.℥. the other
parte, then will the whole nomber be. 1.℥ —|— 4 ♃
Wherefoꝛe firſte J multiplie. 1.℥. by it ſelf, and it
yeldeth. 1.ℨ. Then dooe J multiplie. 1.℥. by. 4. and
it giueth. 4.℥. whiche bothe ſommes together, dooe
make. 1.ℨ. —|— .4.℥. whiche is equalle to. 117.
And by tranſlatiõ. 1.ℨ. === .117.♃ ——— 4.℥.

Wherefoꝛe J doe multiplie. 2. ſquarely, and it gi=
veth

of Coſsike nombers.

veth. 4, whiche added to. 117. maketh. 121. and the roote of that is. 11. from whiche J ſhall abate. 2. and there will reſte. 9. as the other parte of the nomber. This is verie plain, & the proſe of it as it was before.

Maſter. Then aunſwere to this queſtion.

There are 3 nōbers in proportion *Geometricall.* And one of the extremes is. $20\frac{1}{4}$, the other extreme, with the double of the middell terme, doeth make 22. Now would J knowe of you, what thoſe. 2. nombers bee?   *A queſtion of proportion.*

Scholar. For trialle, J name the other extreme, 1.℥. And bicauſe it, with the double of the middle terme dooeth make. 22. the middell terme ſhall bee 11. ———— . $\frac{1}{2}$. ℥. for his double is. 22. ———— 1 ℥. whiche with. 1. ℥. doeth make. 22.

Then to proceade, J knowe the propertie of thoſe nombers in proportion *Geometricall* to bee ſoche, that *the multiplication of bothe the extremes is equalle to the ſquare of the middell terme,* wherefore J multiplie the. 2. extremes together, and there will riſe. $\frac{81}{4}$ ℥. Then dooe J multiplie. 11 ———— $\frac{1}{2}$ ℥. by it ſelf in Square, and it will bee. 121. 9. ——|—— . $\frac{1}{4}$ ℨ. ———— .11 ℥, whiche muſt bee equalle to $\frac{81}{4}$ ℥. or. $20\frac{1}{4}$ ℥. Then to reduce it, J adde. 11. ℥. on bothe ſides, and it will be. $31\frac{1}{4}$ ℥. ═══ $\frac{1}{4}$ ℨ. ——|—— .121 9. and by tranſ= ſlation. $\frac{1}{4}$ ℨ ═══ $31\frac{1}{4}$ ℥. ———— .121 9. That is 1. ℨ. ═══ 125. ℥. ———— 484. 9.

Now reſteth nothyng, but to ſearche the value of 1. ℥. Wherfore J take $\frac{125}{2}$ and multiplie it Square, and ſo haue J $\frac{15625}{4}$. from whiche J muſt abate. 484. that is $\frac{1936}{4}$. And there will remain $\frac{13689}{4}$ whoſe roote is $\frac{117}{2}$, whiche J ſhall abate from $\frac{125}{2}$, and there will re= main $\frac{8}{2}$, that is. 4. for the other extreme.

Then for the middell terme, thus ſhall J doe. Mul=   *The proofe.* tiplie. 4. and. $20\frac{1}{4}$ together, and there will riſe. 81. whoſe roote is. 9. and is the middell nomber. That 9 doubled will make. 18. and 4. ioined therto, giueth 22

Bk.ii.   So

The Arte

So are thofe. 3. termes in progreſſion *Geometricall*, ac=
coꝛdyng to the conditions limited in the queſtion.

Maſter. Proue the woꝛke now, how it wil frame
if. 1.℥. be ſet foꝛ the middell nomber. Foꝛ it wer fol=
lie, to trie whether this queſtion, would admitte ad=
dition of the. 2. laſte nombꝛes. Although the rule doe
declare that in ſoche ſoꝛte of equations, there is dou=
ble valuation to eche roote.

Scholar. Yet I beſeke you, let me examine it a li=
tle, to ſee the cauſe, why I maie not adde them, and ſo
take the roote.

Maſter. I muſt bere with you ſo moche. By ad=
dition you ſee, there will riſe $\frac{242}{2}$, that is 121. And then
the middell nomber will be. 49. $\frac{1}{2}$. And ſo the pꝛopoꝛ=
tion is $\frac{22}{9}$. that is *Dupla ſuperquadripartiens nonas*. Where
as in the other. 3. nombers. 4. 9. 20 $\frac{1}{4}$. the pꝛopoꝛtion
is *Dupla ſeſquiquarta*.

But in the queſtion is one cōdition, that ſecludeth
the roote, that riſeth by additiō. Foꝛ the double of the
middell terme, with the other vnknowen extreme,
ſhould make. 22. As in. 4. and. 9. it doeth. But in 49 $\frac{1}{2}$
and 121. it would be 220. that is 10. tymes ſo moche.

Scholar. And if you had ſaied in the queſtion, that
the double of the middell nomber, with the other ex=
treme, would haue made. 220. then I ſhould haue ta=
ken this later roote by additiō, and not the firſte roote
by ſubtraction.

And ſo I perceiue the varietie of conditions in the
queſtion dooeth limite, whiche of the. 2. rootes I ſhall
of neceſſitie take, and leaue the other.

An other
woorke.
But now to varie that woꝛke, I will ſet. 1.℥. foꝛ
the middell terme. And then the double of it, with
the other terme, will make. 22. The double of. 1.℥.
is. 2.℥. So muſt the other terme be 22 ℈——— 2.℥.

Then to ſeke out an equation, I multiplie the. 2.
extremes together, that is. 22 ℈——— 2 ℥. by 20 $\frac{1}{4}$.
And

of Cossike nombers.

And there riseth. $445\frac{1}{2}$. ——— $40\frac{1}{2}$. $\mathbb{Z}\!e$. And the square of. 1. $\mathbb{Z}\!e$. beyng the middell terme, is sone perceiued to be. 1. $\mathbb{Z}$. And so the firste equation is,

$$1\,\mathbb{Z} \;=\!=\!=\; 445\tfrac{1}{2}\,\mathbb{9}. \;\;\text{———}\; .40\tfrac{1}{2}\,\mathbb{Z}\!e.$$

Wherefore I take halfe. $40\frac{1}{2}$, that is, $\frac{81}{4}$, whose Square is $\frac{6561}{16}$. And vnto it I putte $445\frac{1}{2}$. whereby there commeth $\frac{13689}{16}$. whose roote is $\frac{117}{4}$. from whiche roote I must abate $\frac{81}{4}$, and so remaineth $\frac{36}{4}$. that is. 9. As the value of. 1. $\mathbb{Z}\!e$. And for the middle nomber.

Then for the proofe : if. 9. bee the middell nomber, the square of it, which is. 81, shall bee equalle to the multiplications of the extremes. Wherefore if I diuide. 81. by $20\frac{1}{4}$, the *quotiente* beyng. 4. declareth the other extreme. *The proofe.*

Master. You seme experte inough in this forme of woorke. Therfore I will procede to other questiōs, that differ somewhat from these.

There are. 2. menne talkyng together of their monies, and nother of theim willyng to expresse plainly his somme, but in this sorte. The nomber of angelles in my purse, saieth the firste manne, maie bee parted into soche 2. nombers, whiche beyng multiplied together, will make. 24. And their *Cubes* beeyng added together, will make. 280. Then, quod the other man. And the like maie I saie of my money, saue that the *Cubes* of the. 2. partes, will make. 539. Now I desire to knowe, what monie eche of them had. *A double question.*

Scholar. The firste mannes some, I set to be $1\,\mathbb{Z}\!e$ whiche I must parte into twoo soche partes, that thei bothe multiplied together, maie make. 24.

Master. You erre verie moche. For it is not possible, that the partes of any *Cossike* nomber multiplied together, can make an absolute nomber. Wherefore in soche cases, where you perceiue that there is required, after the firste position, any multiplication to make an absolute nomber, you shall call the firste nō=

BB.iii.   bers

The Arte

bers, by some other name of pleasure. As here you maie call the firste mannes somme. A. And the second mannes somme. B. and then in their partition, vse the name of. 1. z.

And as thei are twoo questions in one, so shall you make seueralle woozkes foz them.

Scholar. Then shal I saie, that the firste mannes somme is. A. and it is diuided as he declared. Wherefoze foz one nomber of that diuision, I set. 1. z. And then the other shall be $\frac{24\;9}{1z}$. foz as the one nomber multiplied by the other, doeth make. 24. So. 24. 9. diuided by the one of them, must neades bzyng fozthe the other.

Master. That is well remembzed of you. Foz as 4. and. 5. by multiplication, doe make. 20. So. 20. diuided by. 5. bzingeth foz the 4. and diuided by. 4. it yeldeth. 5.

Scholar. So $\frac{20}{5}$ is but. 4. and $\frac{20}{4}$ is. 5.

Master. Go foz th then with the rest of the wozke.

Scholar. The *Cube* of. 1. z. is. 1. c. and the *Cube* of $\frac{24\;9}{1z}$ is $\frac{13824\;9}{1c}$ whiche. 2. nombers I maie not adde together, vntill I haue reduced theim vnto one denomination : whiche thyng I shall doe, by settyng. 1. c. as a fraction thus $\frac{1c}{19}$. And then woozkyng after the rate of fractiõs, in the firste reduction thei will stande thus. $\frac{18\,c}{1c} + \frac{13824\;9}{1c}$. And by farther addition thus. $\frac{18\,c + 13824\;9}{1c}$

And hetherto the woozke of bothe these. 2. mennes sommes, are indifferente and agreynge. So that this one woozke serueth foz theim bothe. But now thei will differ. Foz in the firste mannes woozdes, and so in the wozke foz him $\frac{18\,c + 13824\;9}{1c}$ is equalle to 280 : but in the seconde mannes woozke, it must be accompted equalle to. 539.

But firste to goe foz ward with the firste man. Seyng $\frac{18\,c + 13824\;9}{1c}$ is equalle to. 280. Therefoze by reduction

of Cossike nombers.

reduction to one denomination, $\frac{13\mathcal{R}\ +\ 13824\,\dot{g}}{1\mathcal{R}}$ is e-
qualle to $\frac{280\mathcal{R}}{1\mathcal{R}}$. And remouyng the common denominator, the numerators shal kepe the same proportion:
and therfore, $13\mathcal{R}\ +\ 13824.\dot{g}.$ shall be equalle
to. $280\mathcal{R}$. And by translation, to leaue the greateste
denomination alone, $13\mathcal{R} = 280\mathcal{R}\ -\ 13824\,\dot{g}$
Where I shall seke the value of. 1.ℨ. whiche shall
not be here accoumpted the square roote, but the *zenzicubike* roote, or the *Cubike* roote of the square roote,
accordyng to the greateste denomination.

Wherfore. 140. in square, maketh 19600. from
whiche I must abate 13824. And there doeth remain
5776 whose square roote is. 76. whiche beyng added
vnto. 140. dooeth giue. 216. and beyng abated from
it, it leaueth. 64. of whiche bothe I must extracte the
Cubike roote, bicause in the equation there are. 2. quātities omitted. So that of. 216. the *Cubike* roote is 6.
And of. 64. the *Cubike* roote is. 4. Here I see bothe
rootes serue so my purpose, that I shall take thē both.

Master, And good reason. For as in setting 1ℨ
for your position, you could not tell whether it were
the greater parte, or the lesser, so maie you not now
applie it to either of theim bothe, but take bothe rootes for the. 2. partes of your nomber.

Scholar. So doeth the firste mannes nomber ap- *The proofe.*
peare to be. 10. seyng the partes bee. 4. and. 6. whiche
I maie examine thus. That thei make. 24. by multiplication, it is easily seen. And that their *Cubes* added
together, doe make. 280. is sone perceiued: seyng the
Cube of. 4. is. 64: and the *Cube* of. 6. is. 216. whiche. 2.
nombers by addition, doe make. 280.

Master. Now proue the seconde mannes woorke. *The worke*
Scholar. In his woorke $\frac{13\mathcal{R}\ +\ 13824\,\dot{g}}{1\mathcal{R}}$ is equalle *of the second*
to 539. And by reduction to one denomination, it is *parte.*
equalle to $\frac{539\mathcal{R}}{1\mathcal{R}}$. So that. $13\mathcal{R}\ +\ 13284.\dot{g}.$ is
equalle to. $539.\mathcal{R}$. and by translation.

$1.3\mathcal{R}.$

The Arte

1.ℨℭ. ⸺⸺⸺ .539.ℭ. ⸺⸺⸺ .13824.ℱ. whofe *zenzicubike* roote J feke, thus : $\frac{539}{2}$ doth make in fquare $\frac{290521}{4}$, from whiche J muft abate $\frac{55296}{4}$ and then remai= neth $\frac{235225}{4}$ whofe roote is $\frac{485}{2}$ vnto whiche J maie adde $\frac{539}{2}$, and then will it bee $\frac{1024}{2}$ that is. 512. whofe *Cubike* roote is. 8. And is one parte of the feconde mannes nomber. And foz the other parte, J fhall abate $\frac{485}{2}$ out of $\frac{539}{2}$ and there remaineth $\frac{54}{2}$, that is, 27. whofe *Cubike* roote is. 3. And is the other parte of the feconde man= *The proofe.* nes nomber. As it maie fone be tried thus. Foz. 3. ty= mes. 8. maketh. 24. and. 27. whiche is the *Cube* to. 3. added with. 512. whiche is the *Cube* to. 8. dooeth make 539, as the queftion intendeth.

A queftion of an armie. Mafter. One other queftion J will pzopounde, of. 2. armies beyng bothe fquare, and of like nomber. And if you abate. 4. from the one armie, and adde. 10. to the other armie, and then multiplie them bothe to= gether, there will amounte. 9853272. J demaunde of you, what is the fronte of thofe fquare battailes.

Scholar. J call the fronte 1ℨℯ. And then muft the battaile bee. 1.ℨ. Now abatyng. 4. from the one, it will bee. 1.ℨ. ⸺⸺ .4.ℱ. Then addyng. 10. to the o= ther, it wil make. 1.ℨ. ⟍⟋ .10.ℱ. And if you mul= tiplie thofe.2. nombers together, there will amounte by it. 1.ℨℨ ⟍⟋ 6.ℨ. ⸺⸺ 40.ℱ. whiche fomme muft be equalle to. 9853272.

1.ℨ. ⟍⟋ .10.ℱ.

1.ℨ. ⸺⸺ .4.ℱ.

1.ℨℨ. ⟍⟋ .10.ℨ. ⸺⸺ .4.ℨ.

⸺⸺ 4.ℨ. ⸺⸺ .40.ℱ.

1.ℨℨ. ⟍⟋ .6.ℨ. ⸺⸺ .40.ℱ.

And if you adde. 40.ℱ. to bothe partes of the equa= tion, it will be. 1.ℨℨ ⟍⟋ 6 ℨ. equalle to. 9853312

And

of Cossike nombers.

And by tranſlation. 1.ƶ.ƶ. ═════ 9853312.────── 6.ƶ.
out of whiche laſte equation, I ſhall ſearche for the
value of. 1℥. by multiplipng firſt. 3. ſquarely, where=
of commeth. 9. and then addyng it to. 9853312. And
ſo commeth. 9853321. whoſe roote is. 3139. from
whiche I muſt abate. 3. And then remaineth. 3136.
whiche is the full nomber and Square of the one ar=
mie. And hath for this roote. 56.

For as here is one onely quantitie omitted, ſo the
firſte nomber, whiche in other queſtiōs of immediate
equations, was the verie roote, in theſe interrupte e=
quations is a rooted nomber, and is here a ſquare nō=
ber : whoſe roote therfore, I haue drawen accordyng=
ly. And for triall of this woorke. 56. in ſquare maketh   *The proofe.*
3136. from whiche if you abate. 4. there will reſte
3132. Again if you adde. 10. there will riſe. 3146.
And thoſe. 2. nombers multiplied together, doe make
9853272, as the queſtion intendeth.

Maſter. This you ſee, what vſe is in theſe equa=
tions, yet are there many other equatiōs, whiche here
be not ſpoken of : but here after you ſhall haue moare
largely declared, if you ſhewe your ſelf diligente in
this parte.

And one queſtion I will propounde, & aſſoyle with   *A queſtion
of ſtraunge
equation.*
out woorke for brefeneſſe, that you maie ſee there is
moare behinde. There is a nomber whoſe Square
abated by. 16. and the firſte nomber augemented by
8. and then bothe thei multiplied together, will bryng
forthe. 2560.

Scholar. I will proue the woorke of it. And there=
fore ſuppoſe the firſte nomber to be. 1.℥. Then is his
ſquare 1ƶ whiche abated by 16. leueth. 1.ƶ ── 16.9.
and the nōber augemēted by. 8. yeldeth 1℥ ──┼── 8.9.
Theſe. 2. nombers multiplied together, will make
1₵.──┼── .8.ƶ. ────── 16.℥. ────── 128.9. beyng
equalle to. 2560.

                                          Ll.i.   1.ƶ.

The Arte

1.ʒ. —————— .16.℞.

1.℥. ——+—— .8.℞.

1.℀. —————— .16.℥.

8.ʒ. —————— .128.℞.

1.℀. ——+—— .8.ʒ. —————— .16.℥. —————— .128.℞.

And addyng 128.℞. on bothe ſides of the equation,
it will be. 1℀ ——+—— 8ʒ ———— 16℥ ======= 2688℞
Againe addyng. 16.℥. on bothe ſides, it will bee

1. ℀. ——+—— 8ʒ ======= 16℥ ——+—— 2688℞.

Maſter. VVhere at ſtaie you now?

Scholar. J ſee no ſhifte, but other to leaue it, as it
is, 2. nombers equalle to. 2 : other els to make. 1. nom=
ber equalle to. 3. And all that is aboue my cunnyng.
For hetherto J haue learned noe rule for any of them
bothe. So that I can not geſſe, what the firſte nomber
might bee.

Maſter. The nomber is. 12. And his Square is
144. from whiche if you abate. 16. it will bee. 128.
And if you adde. 8. to. 12. it will yelde. 20. Then mul=
tiplyng. 128. by. 20. the ſomme will be. 2560. as the
queſtion declared.

Of other
equations.

But to put you out of doubte, this equation is but
a trifle, to other that bee vntouched. And yet J will
tourne this equation a litle, to giue you ſome light in
it, and other ſoche. As here.

1. ℀. ======= .16.℥. ——+—— .2688.℞. ———— 8ʒ.

where you ſee. 1. ℀. equalle to. 3. other nombers. And
is it not certaine to you, that this equation is true?

Scholar. Yes, J am aſſured thereof.

Maſter. And yet to auoide doubtfulnes the more
trie it by reſolution, accoumptyng. 12. for. 1.℥.

Scholar. VVhere. 12. is. 1.℥ , there. 1.ʒ. is. 144.
and. 1.℀. is 1728. whiche. 1728. muſt bee equalle to
16.℥.

of Cossike nombers.

16. ℞ (that is, 192) and to, 2688, ſaue that you muſt abate, 8. ℨ, that is 1152. Now if I adde 192, to 2688 it will make, 2880, out of whiche abatynge, 1152, there will remaine, 1728, wherby I ſee the equation is iuſte.

Maſter. Then you ſee that the equation is true. And can you doubte, that any nomber, whiche is e= qualle to a *Cubike* nomber, hath in it a *Cubike* roote?

Scholar. It muſt neades be a *Cubike* nomber, that is equalle to a *Cubike* nomber : and therefoze muſte neades haue a *Cubike* roote : although I knowe not how to extracte that roote.

Maſter. Likewaies, when I ſaie :
8. ℨ ℭ. ═══ .12. √ℨ. ──┼── .128. ℱ. it is certaine, not onely that. 12. √ℨ. ──┼── .128. ℱ. containeth in it as moche as. 8. ℨ ℭ. but that the. 8. parte of it is a ℨ ℭ. nomber, and hath a *zenzicubike* roote.

Here the roote is 2.

And farther it is manifeſte, that as euery. ℨ ℭ. nomber, dooeth containe in it certaine. √ℨ. nombers exactly, ſo if any nomber be annexed with thoſe *Surſo= lides* (as here in this example are ſet 128) it is of neceſ= ſitie, that that. 128. muſt containe in it certaine *Surſo= lides* exactly.

So if. 8. ℨ ℭ. bee equalle to
10. √ℨ ──┼── 20. ℨ ℨ ──┼── 400. ℭ ──┼── 31250. ℱ. it muſt neades be that the. 8. parte of this compounde nomber ſhall bee a. ℨ ℭ. nomber. And alſo that the ℨ ℨ. with the other nombers folowyng dooeth con= taine a certain nomber of. √ℨ. nombers. And the. ℭ. in like ſozte includeth a nomber of. ℨ ℨ. nombers. And laſte of all. 31250. ℱ. doeth compzehende certain *Cubike* nombers exactly.

The roote is 5

In like ſozte, when we ſaie, that. 1. √ℨ. is equalle to 6. ℭ. ──┼── .8. ℨ. ──┼── .9. ℱ. All this compounde nomber is a *Surſolide*, and hath a. √ℨ. roote. And 8. ℨ. ──┼── .9. ℱ. includeth certaine *Cubes*. And ſo

The roote here is 3

                               Ll.ii.   doeth

The Arte

doeth. 9.ß. containe exactly. 1.ʒ. oʒ moare.

But of thefe and many other verie excellente and wonderfulle woozkes of equation, at an other tyme J will inſtructe you farther, if J fee your diligence ap= plied well in this, that J haue taughte you.

And therefoze here will J make an

eande of *Coſsike* nombers,

foʒ this tyme.

Of Surde nombers, in diuverſe ſortes
And firſte of Surde nombers
vncompounde.

Ow that you haue ſome=
what learned the arte of *Coſ=
ſike* nombers, with the rule of
equation, it ſemeth good time
and apte place, to teache you
the arte of *Surde* nõbers, whi=
che are diuerſe in name, accoz=
dyng as there are diuerſe na=
tures of rootes, whiche maie
giue theim name.

Foz generally, a *Surde* nomber is nothyng elſe, but   *A Surde*
ſoche a nomber ſet foz a roote, as can not be expzeſſed *nomber.*
by any other nomber abſolute.

As the *Square roote* of. 10, oz of. 8. oz of any nomber,
that is not ſquare. Likewaies the *Cubike roote* of. 4, oz
of. 5. oz of any nomber that is not *Cubike.* So the *zen=
zizenzike roote* of. 8, .12. oz. 20, oz of any nomber that
is no *zenzizenzike,* is called a *Surde* nomber. And in
like maner, any other roote of any nomber, that hath
noe ſoche roote, doeth cauſe that nomber to be a *Surde*
nomber.

Foz if you ſee thoſe ſignes annexed with nombers,
that hath ſoche rootes, thoſe nombers are not *Surde*
nombers pzoperly, but ſette like *Surdes.* As the *Square
roote* of. 4. oz of. 9. oz. 25. ꝛc. The *Cubike roote* of. 8. 27.
oz. 125. ꝛc. whiche ſometymes is vſed foz apte wozke,
as you ſhall ſee here after.

Of Numeration.

He numeration of thẽ doeth conſiſte, in know=
lege of their figures, whiche partly be declared
befoze. But their common and peculiare ſignes
are theſe. √. ᴟ . ᴠ . Although there maie be moare
                                 Ll.iii.   varieties

The Arte

varieties : Pet these for this tyme maie suffice.

The firste, that is, $\sqrt{}$. is customably set, to signifie a *Square roote.* As this, $\sqrt{}$.5. betokeneth the *Square roote* of, 5. And, $\sqrt{}$.12. is the *Square roote* of, 12. Howbeit many tymes it hath with it, for the moare certeintie the *Cossike* signe, ʒ. And is written thus, $\sqrt{}$ ʒ.20. the *Square roote* of, 20. And, $\sqrt{}$ ʒ.56. the *Square roote* of, 56.

The seconde signe is annexed with *Surde Cubes,* to expresse their rootes. As this, $\backslash\!\backslash\!\backslash$.16. whiche signi-fieth the *Cubike roote* of, 16. And, $\backslash\!\backslash\!\backslash$.20. betokeneth the *Cubike roote* of, 20. And so forthe. But many ty-mes it hath the *Cossike* signe with it also : as $\backslash\!\backslash\!\backslash$. ℞ 25 the *Cubike roote* of, 25. And, $\backslash\!\backslash\!\backslash$. ℞.32. the *Cubike roote* of, 32.

The thirde figure doeth represente a *zenzizenzike roote.* As, $\backslash\!\backslash$.12. is the *zenzizenzike roote* of, 12. And $\backslash\!\backslash$.35. is the *zenzizenzike roote* of, 35. And likewaies if it haue with it the *Cossike* signe, ʒ ʒ. As $\backslash\!\backslash$ ʒ ʒ 24 the *zenzizenzike roote* of, 24. and so of other.

Scholar. It were againste reason, to seke reason for those signes, whiche be set voluntarily to signifie any thyng : although some tymes there bee a certaine apte conformitie in soche thynges. And in these figu-res, the nomber of their minomes, seameth disagrea-ble to their order.

Master. In that there is some reason to bee she-wed :for as, $\sqrt{}$. declareth the multiplication of a nom-ber, ones by it self : So, $\backslash\!\backslash\!\backslash$. representeth that multi-plication *Cubike,* in whiche the roote is represented thrise. And, $\backslash\!\backslash$. standeth for, $\sqrt{}$, $\sqrt{}$. that is, 2, figures of *Square* multiplication : and is not expressed with, 4. minomes. For so should it seme to expresse moare then, 2. *Square* multiplications. But of voluntarie si-gnes, it is inoughe to knowe that this thei doe signi-fie. And if any manne can diuise other, moare easie or apter in vse, thei maie well be receiued.

But

of Surde nombers.

But concerning the numeration of *Surde* nombers this shal you marke : that when any compounde signe is putte before a nomber, whiche hath any roote, that maie bee expressed by parte of that signe, that nomber is not absolutely so to bee expressed, onlesse it bee for ease or aptnesse in worke. As. $\sqrt{ zz }$.36. whiche betokeneth the *zenzizenzike roote* of. 36. Seyng it is well knowen, that. 36. hath. 6. for his *Square roote*, it were moare apte expressynge that nomber thus. $\sqrt{z}$ 6. that is the square roote of. 6.

Otherwaies, if the nöber that foloweth the signe, haue a roote agreable to that signe : it is noe *Surde* nomber. As. $\sqrt{z}$. 16. is. 4. and is noe *Surde* nomber. So. $\sqrt[w]{}$.27. is. 3. and neadeth not to bee written in *Surde* forme, excepte it bee for aptnesse of woorke. And by this maie you iudge of all other, as thei come in vse

Scholar. If this bee all that is requisite to numeration, I praie you procede to addition. For that is nexte in order.

Master. That is the common order. Howbeit in vulgare fractions, you remember that multiplication and diuision, are set before addition and subtraction : bicause of the easier formes of woorke, in multiplication and diuisiö. And so in these *Surde nombers*, bicause the woorkes of multiplication, and of diuision, be not onely moare easie, then the woorkes of addition, and of subtraction, but also be requisite to them, therefore will I begin with them, and so come to the other.

Of Multiplication.

Ultiplicatiö in *Surde nombers* vncöpounde hath noe difficultie, if thei be of one denomination : els must thei be reduced to one denomination : and that by multiplication, accordyng to their signes.

But where noe reduction neadeth, you shall multiplie

The Arte

tiplie the nombers together, and sette their common
signe before the number, that resulteth of that multi=
plication.

Examples of square Surdes.

F you wil multiplie.$\sqrt{3}$.15. by.$\sqrt{3}$.
26. it wil make.$\sqrt{3}$.390.
So.$\sqrt{3}$.32. multiplied by.$\sqrt{3}$.48.
dooeth make.$\sqrt{3}$.1536.
And.$\sqrt{3}$.56.multiplied by. $\sqrt{3}$ 21.
doeth yelde.$\sqrt{3}$.1176.

Howbeit some tymes it happeneth, that the nom=
ber, whiche is made by that multiplication, is a nom=
ber absolute, and not a *Surde nomber*.

Examples of soche as make
nombers Absolute.

$\sqrt{}$.12.	$\sqrt{}$.48.
$\sqrt{}$. 3.	$\sqrt{}$. 3.
$\sqrt{}$.36. that is. 6.	$\sqrt{}$.144. that is. 12.
$\sqrt{}$.12.$\frac{1}{2}$.	$\sqrt{}$.28$\frac{4}{5}$.
$\sqrt{}$.4.$\frac{1}{2}$.	$\sqrt{}$.7.$\frac{1}{5}$.
$\sqrt{}$.56.$\frac{1}{4}$ that is. 7$\frac{1}{2}$.	$\sqrt{}$.207.$\frac{9}{25}$ that is 14$\frac{2}{5}$.
$\sqrt{}$. 240.	$\sqrt{}$.325.
$\sqrt{}$. 15.	$\sqrt{}$. 13.
$\sqrt{}$.3600. that is. 60.	$\sqrt{}$.4225. that is. 65.

And generally when any nomber is multiplied by
an other, if the proportion betwene those. 2. nombers
bee represented by a Square nomber, as by. 4. 9. 16.
25. &c. then dooe thei make a square nomber by their
multiplication.

Examples

of Surde nombers.
Examples of Cubike rootes.

$\sqrt{cc} \cdot 91 \cdot \qquad \sqrt{} \cdot 7 \cdot \frac{2}{3} \cdot \qquad \sqrt{} \cdot 256 \cdot$

$\sqrt{cc} \cdot 12 \cdot \qquad \sqrt{} \cdot \frac{3}{4} \cdot \qquad \sqrt{} \cdot \frac{10}{13} \cdot$

————————————————

$\sqrt{cc} \cdot 1092 \cdot \qquad \sqrt{} \cdot 5 \cdot \frac{3}{4} \cdot \qquad \sqrt{} \cdot 196 \frac{12}{13} \cdot$

Examples of foche as make
Abfolute nombers.

$\sqrt{} \cdot 54 \cdot \qquad\qquad \sqrt{} \cdot 686 \cdot$

$\sqrt{} \cdot 32 \cdot \qquad\qquad \sqrt{} \cdot 4 \cdot$

————————————————

$\sqrt{} \cdot 1728 \cdot$ that is $\cdot 12 \cdot \qquad \sqrt{} \cdot 2744 \cdot$ that is $\cdot 14$

$\sqrt{} \cdot 486 \cdot$

$\sqrt{} \cdot 96 \cdot$

————————————

$\sqrt{} \cdot 46656 \cdot$ that is $\cdot 36 \cdot$

Examples of zenzizenzike rootes.

$\sqrt{} \cdot 15 \cdot \qquad \sqrt{} \cdot 204 \cdot \qquad \sqrt{} \cdot 162 \cdot$

$\sqrt{} \cdot 7 \cdot \qquad \sqrt{} \cdot 26 \cdot \qquad \sqrt{} \cdot 32 \cdot$

————————————————————

$\sqrt{} \cdot 105 \cdot \qquad \sqrt{} \cdot 5304 \cdot \qquad \sqrt{} \cdot 5184 \cdot$ that is $\cdot \sqrt{} \cdot 72 \cdot$

$\sqrt{} \cdot 7\frac{1}{5} \cdot \qquad\qquad \sqrt{} \cdot 27 \cdot$

$\sqrt{} \cdot \frac{4}{5} \cdot \qquad\qquad \sqrt{} \cdot 12 \cdot$

————————————————————

$\sqrt{} \cdot 5\frac{19}{25} \cdot$ that is $\cdot \sqrt{} \cdot 2\frac{2}{5} \cdot \qquad \sqrt{} \cdot 324 \cdot$ that is $\cdot \sqrt{} \cdot 18 \cdot$

Examples of zenzizenzike rootes
that make abfolute nombers.

$\sqrt{} \cdot 32 \cdot \qquad\qquad \sqrt{} \cdot 128 \cdot$

$\sqrt{} \cdot 8 \cdot \qquad\qquad \sqrt{} \cdot 32 \cdot$

————————————————————

$\sqrt{} \cdot 256 \cdot$ that is $\cdot 16 \cdot \qquad \sqrt{} \cdot 4096 \cdot$ that is $\cdot 64 \cdot$

$\mathfrak{Pm.i.} \qquad \sqrt{} \cdot 288$

The Arte

√. 288.

√. 72.
_____

√.20736. that is. 12.

But here is to bee noted, that if you would multi=
plie any *Surde* nomber, by an absolute nomber, oz any
Surde nomber of one denominatiō, by a *Surde* nomber
of an other denomination : you must firste reduce that
Absolute nomber to the like denomination. And so
must you reduce the. 2. *Surde* nombers to one denomi=
nation.

And bicause that this woozke doeth serue often in
this arte, and that in diuerse woozkes, J will set here
the arte of reduction.

Of reduction in Surdes.

Eduction in *Surdes*, is the bzingyng
of sundrie denominatiōs vnto one.
Whiche in absolute nōbers is this
doen. You shall multiplie the abso=
lute nomber, accozding to the signe
of the *Surde*, and then sette befoze it
the like signe. So that if you would
double. √.℥.8. that is to saie, if you would multiplie
it by. 2. you must firste multiplie that. 2. squarely, and
then multiplie those nombers together. That is to
saie, you shall multiplie.√.℥.8. by.√.℥.4. and so is
it doubled.

Likewaies, to *triple* any Square *Surde*, is to multi=
plie it by. 9. And so to *quadriple* any square *Surde*, is to
multiplie it by. 16. And so fozthe.

But if you double any *Cubike* nomber, you shall
multiplie it by. 8. that is the *Cube* of. 2. And so if you
would triple a *Cubike* roote, you muste multiplie it by
27. And if you would *quadriple* it, you shall multiplie

it

Of Surde nombers.

it by. 64. And so of other like woozkes.

Again, if you will double any *zenzizenzike* roote, you must multiplie it by. 16. And if you will *triple* it, you shall multiplie it by. 81. And so if you will *quadri=ple* it, you must multiplie it by 256. And in like maner euer moare, foz the nomber absolute, you shall set his *zenzizenzike* nomber. Like as in Squares, foz any nomber absolute, you shall set his square. And in *Cubes* you shall take his *Cube.*

Scholar. This is plaine inoughe : yet J pzaie you put an example oz twoo, of eche kinde.

Master. Take these examples foz square rootes.

√. 38.	√.ʒ. 128.	√. 3264.
2.	6.	12.
√. 152.	√.ʒ.4608.	√.469976.

Examples in Cubike rootes.

₥√. 52.	₥√. 163.	₥√. 4806.
2.	5.	8.
₥√. 416.	₥√. 20375.	₥√. 2460672.

Examples in zenzizenzike nombers.

₩. 69.	₩. 251.	₩. 1385.
2.	4.	5.
₩. 1104.	₩.64256.	₩.2250625.

Scholar. This J perceiue well. But now in *Surde* nombers of diuerse denominations, what the ozder of reductiõ is, J pzaie you to set fozth with some exãples

Master. These examples with their declaration, maie sufficiently serue foz a shewe, if J would multi=plie. ₥√.12. by. √.5. J must firste multiplie the nom=ber of one signe, accozdynge to the signe of the other

Mm.ii.   nomber,

The Arte

nomber, and so alter them bothe. Whiche woorke is like the reduction of fractions, to one common deno= mination. As here I muste multiplie. 5. *Cubikely*, and 12. must be multiplied squarely, and then shall I adde bothe signes in one, for their common signe. So shall I haue for them the. ℨ·ℭ. roote of. 144. to be multi= plied by the *zenzicubike* roote of 125. And so will there come of that multiplication, the *zenzicu= bike* roote of. 18000. As here by example doeth appeare.

$$\sqrt{.\text{ℨ}\,\text{ℭ}.} \quad 144.$$
$$\sqrt{.\text{ℨ}\,\text{ℭ}.} \quad 125.$$
$$\overline{\sqrt{.\text{ℨ}\,\text{ℭ}.} \quad 18000.}$$

Likewaies if I would multiplie. √. ℨ·ℨ. 250. by ⩔. 34. I shall firste multiplie. 250. *Cubikely*, and it will bee. 15625000. And 34. must I multiplie *zen= zizenzikely*, and it will yelde. 1336336. Wherefore multipliyng theim together, and addyng thereto the common denomination, it will bee the. ℨ·ℨ·ℭ. roote of. 2088025000000.

This woorke is aptly represented in figure, after this sorte. And then shall you multiplie crosse waies the nomber of the one, by the signe of the other. And

so maie you dooe in all other like nombers, of diuerse denominations.

This reduction doeth serue for any other woorke, as well as for multiplication.

Of Diuision.

Diuision is as easie as multiplication. For in it there is noe regard had to the signes. But the one nomber diuided by the other as if thei were nöbers absolute. And then the firste signe added to the *quotiente*. For the more lighte and certaintie, I haue set here, exam= ples of eche sorte.

And

of Surde nombers.

And firſt examples of ſquare rootes.

$\sqrt{}$. 72.   ($\sqrt{}$.9. that is. 3. $\sqrt{}$. 128.
$\sqrt{}$. 8. $\sqrt{}$. 4.   ($\sqrt{}$.32.

$\sqrt{}$. 457$\frac{1}{3}$.   ($\sqrt{}$.21$\frac{7}{9}$.
$\sqrt{}$.   21.

Examples of Cubike rootes.

$\sqrt{}$. 96.   ($\sqrt{}$.24. $\sqrt{}$. 1664.   ($\sqrt{}$.52.
$\sqrt{}$. 4. $\sqrt{}$.   32.

$\sqrt{}$. 5624.   ($\sqrt{}$.74$\frac{5}{19}$.
$\sqrt{}$.   76.

Examples of zenzizenzike rootes.

$\sqrt{}$. 54.   ($\sqrt{}$.9. that is. $\sqrt{}$.3.
$\sqrt{}$. 6.

$\sqrt{}$. 286.   ($\sqrt{}$.6$\frac{17}{21}$. $\sqrt{}$. 5892.   ($\sqrt{}$.109$\frac{1}{9}$.
$\sqrt{}$. 42. $\sqrt{}$. 54.

And this maie ſuffice foz Diuiſion. The pzofe of it
is by the contrary kinde. Foz Multiplication pzoueth
Diuiſion : and Diuiſion trieth Multiplication.
    Scholar. All this is eaſie inoughe to remember.

Of *Addition.*

Maſter.

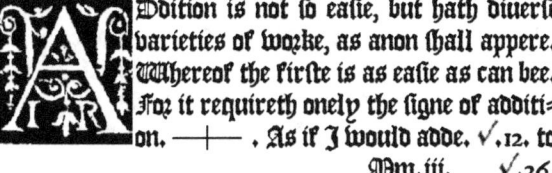 Ddition is not ſo eaſie, but hath diuerſe *The firſte*
varieties of wozke, as anon ſhall appere. *forme of*
Whereof the firſte is as eaſie as can bee. *Addition.*
Foz it requireth onely the ſigne of additi=
on. —+— . As if J would adde. $\sqrt{}$.12. to
                          Mm.iii. $\sqrt{}$.26.

The Arte

√.26. J ſhall ſet it thus. √.26. ——┼—— √.12. And ſo
√.20. put vnto. √.54. maketh. √.54. ——┼—— .√.20.
This foꝛme ſerueth chiefly foꝛ rootes of diuerſe na=
mes. As. w.ȝȝ.20. ——┼—— . w.cℰ. 30. where
√.cℰ.30. is added to w.ȝȝ. 20. And ſo of al other.

    The ſeconde foꝛme is not ſo eaſie : and yet many ti=
mes it is moare certaine. And this is the oꝛder of it.

    You ſhall ſette doune your. 2. nombers, that you
would adde together, foꝛſeyng that thei be of one de=
nomination. Then ſhall you adde in plaine foꝛme,
their nombers together, puttyng thereto the ſigne of
the roote. And kepe that as a parte of the addition.
Again you ſhall multiplie the. 2. firſte nombers toge=
ther. And their totalle you ſhall multiplie by. 4. And
befoꝛe that ſhall you ſette the ſigne of the roote. And
it ſhall ſtande as the ſeconde parte of that addition.
So that thoſe. 2. partes, ſhall be added with the ſigne
——┼—— . And then is the woꝛke eanded. Example
hereof. J would adde the. 2. firſte ſommes, that is,
√.12. to. √.26. wherfoꝛe J
ſet them thus. And then doe
J adde thē bothe plainly to=
gether, and thei make. √.38
whiche J ſet by, as one part
of the addition. Then doe J
multiplie √.26. by. √.12. and
there riſeth. √.312. whiche
J muſt double, oꝛ multiplie
by. 2. And therfoꝛe ſeyng the

√.26 ——┼—— √ 12	
	√.26.
√.38.	√.12.
	√. 312.
	√. 4.
√.1248.	
√.38. ——┼—— √.1248.	

woꝛke is in ſquare rootes, J ſet the ſquare of 2. with
the ſigne of.√. foꝛ. 2. and then multipliyng theim to=
gether, J haue. √.1248. whiche is the ſeconde parte
of the roote. Wherfoꝛe addyng thoſe. 2. partes toge=
ther, with the ſigne. ——┼—— . there commeth. √.38.
——┼—— .√.1248. as the totalle of that addition.

    Scholar. As me thinketh, the firſte foꝛme of addi=
tion,

of Surde nombers.

tion ſerueth better foʒ theſe nombers, then this ſe=
conde foʒme. Foʒ it is moare eaſie to vſe, in any kinde
of wooʒke, and moare ſpedily doen : and it ſemeth that
this laſte nomber, is moare obſcure then the firſte.

Maſter. Yet is this wooʒke good, and very neceſ=
ſarie. Foʒ in theſe nombers, and ſoche other like, it
ſerueth onely (as appereth) to alter the ſtate of the nō=
bers, whereby thei maie bee commenſurable, with o=
ther, then thei were befoʒe that alteration. But in
ſome nombers, and that very many, it reduceth them
to one ſimple foʒme of roote. As by the examples folo=
wyng you ſhall perceiue.

| An example. | | | Theſame example other waies wroughte. | *A thirde forme of ad= dition.* |

Where firſte I haue ſet foʒthe. 2. examples of one
addition, that you maie ſee the agremente of thē both

And firſte I would adde. √.28. with. √.7. where=
foʒe I dooe ioyne. 28. and. 7. in one ſomme, whiche I
ſet a parte, as the firſte poʒtion of the addition. Then
I doe multiplie. 28. by. 7. And thereof commeth. 196.
whiche is a ſquare nōber, and hath. 14. foʒ his roote.
So that I maie vſe now. 2. woʒkes. Foʒ other I maie
continue my wooʒke, as I haue doen (agreable to the
firſte example) in multipliyng that. √.196. by. √.4.
(whiche is but doubling) and ſo there cometh. √.784
                                                      whiche

The Arte

whiche is a nomber abſolute : bicauſe it hath a roote, accordyng to his ſigne, whiche roote is. 28. and maie be ſet for. √.784.

Now in the ſeconde woorke, bicauſe the firſte mul= tiplication of. 28. by. 7. doeth make a ſquare nomber, I doe take the roote of that nomber for it : ſeyng it is all one thyng to ſaie. √.196, and. 14. for. 14. is the roote of. 196. And then hauyng the roote, I muſte double it, accordyng to the rule, or multiplie it by. 2. and there of commeth. 28. whiche I ſhall adde with 35. And ſo haue I. 63. whoſe roote containeth the ad= dition of. √.28. and. √.7.

Scholar. This woorke ſemeth ſtraunge : and far= theſte from common reaſon, of all other woorkes in this arte.

Maſter. I mighte eaſily by demonſtration make you, to perceiue as moche reaſon in this woorke, as cã be in any : for it dependeth of the. 38. Theoreme of the patthewaie. But haſte of other buſineſſe, maketh me to omit the demonſtration at this tyme. whiche ſhor= tly you ſhall haue, for all the equations, and other woorkes likewaies.

But for this preſente tyme, it ſhall be ſufficiente to woorke an example in *rationall* nombers, as if thei wer *Surde* nombers : that therby you maie perceiue the or= der, and the truthe of the woorke.

Wherfore I take theſe twoo nombers. √.36. and √.49. to bee added together. Where I doe firſte adde the twoo nombers plainely together : And thei make 85. for the firſte parte of the addition. Then dooe I multiplie. 49 by. 36. and there riſeth. 1764. whiche is a ſquare nomber. And therefore maie I vſe. 2. wor= kes, as you ſee. In the firſte I multiplie that Square nomber by. 2. or by. √.4. whiche is all one : and there doeth amounte. 7056. a Square nomber alſo, whoſe roote is. 84.

The

of Surde nombers.

The firſte forme.		The ſeconde forme.	
√.36. —— . √.49.		√.36 —— √.49	
√. 49		√. 49.	
√. 36	49.	√. 36.	
294	36.	294.	
147	85.	147.	
√. 1764		√. 1764.	
√. 4		That is. 42.	
√. 7056		2.	
√.85. —— .√.7056.		84.	
D₂.√.85. —— .84		√.85 —— 84.	

That is. √.169.

D₂   13.

In the ſeconde woozke I take the roote of. 1764.
whiche is 42 and doublyng it, I haue 84. agreable to
the other woozke. Then doe I ſette thoſe. 2. nombers
doune with —— , and putte to them the ſigne. √. in
token that I muſte take the roote of that compounde
nomber : and not of any one parte of it.

Scholar. That haue I marked well : Foz 85. hath
no roote, nother 84. hath any roote. But 85 —— 84
that is. 169. hath. 13. foz his roote.

And ſo I ſee, that the roote of. 36. whiche is. 6. And
the roote of. 47. that is. 7, beeyng bothe added toge=
ther will make. 13. that is the roote of. 169.

Maſter. Yet one other forme of eaſie woozke, I
will ſhewe you, whiche is bothe pleaſaunte and pzo=
fitable : But is not generalle, foz it ſerueth onely foz
nombers commenſurable, I meane ſoche nombers, as by
one common diuiſoz, maie bee bzought into Square
nombers. With whiche nombers, you ſhall woozke
thus.

Of nombers commenſura ble, a fourthe forme.

Pn.i.   Firſte

The Arte

Firſte diuide theim by the common diuiſoʒ : and ſet foʒ them their rootes. Then adde thoſe. 2. rootes together, and multiplie it ſquarely. And that ſquare beyng multiplied by the common diuiſoʒ, will bʒynge foʒthe the Square of bothe the rootes. As here followeth in example.

Where I would adde √.384 vnto √.150 which nombers I doe examin, til I maie finde their commō, and leaſte diuiſoʒ, whiche here is. 6. Then diuidyng them by that. 6. I haue foʒ 384. a ſquare nomber. 64. And foʒ. 150. I haue an other ſquare, that is. 25. Of

√.384	+	√.150.
6.)	64	25.
	8	5.
	13.	
	13.	
	169.	
	6.	
	1014.	

whiche bothe ſquares I ſet doune the rootes : and the common diuiſoʒ alſo. Then doe I adde bothe rootes together, and thereof commeth. 13. whoſe Square is 169. that I doe multiplie by. 6. whiche is the commō diuiſoʒ, and it will bee. 1014. whoſe roote doeth contain bothe the rootes befoʒe named. As you ſhall ſee is pʒoued anon by Subtraction.

Scholar. In the meane ſeaſon I conſider, that one of theſe foʒmes, maie confirme the other. And therefoʒe if I wooʒke this laſte example, by one of the other foʒmes, and finde theſame totall, it muſt neades be that the wooʒke is good. Whiche I pʒoue thus.

Firſte ſettyng doune the nombers, in foʒme of the eaſieſte Addition. And then addyng theim together, I finde. 534. whiche I ſette a ſide, as one parte of the nomber, that I doe ſeke foʒ.

Then dooe I multiplie the. 2. nombers together, and thei make. 57600. whiche I dooe multiplie again by 4. And there riſeth. 230400. being a ſquare nomber, and hath. 480. foʒ his roote. Wherefoʒe I

ſet

of Surde nombers.

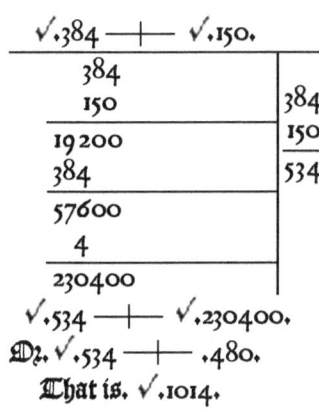

√.384 ——+—— √.150.

384	
150	384
	150
19200	534
384	
57600	
4	
230400	

√.534 ——+—— √.230400.

Dʒ. √.534 ——+—— .480.

That is. √.1014.

set. 534. and. 480. to=
gether, with the figne
of Addition, thus.

534 ——+—— 480. And
the roote of that nom=
ber, is equalle to bothe
the firſte rootes. But
conſidering that bothe
thoſe nombers, which
bee ioyned laſte of all
with ——+——, are nom=
bers rationall and ab=
ſolute, I maie adde thē
in one, ⁊ ſo thei make

1014. agreably to the other woozke. Wherefoze I
iudge them bothe to be good.

Maſter. You might haue wzought this woozke
otherwaies, bicauſe the firſte nomber, that riſeth of
the multiplication is a ſquare nomber.

Scholar. Then I perceiue, I mighte haue taken
the roote of it, whiche is. 240. and doublyng it, I
ſhould haue. 480. As I had in the other woozke. And
ſo all doe agree in one.

But my chief doubte now is, how to knowe thoſe
nombers that bee *commenſurable* : Foz if I ſhall ſtande
long in ſearchyng foz that, I might ſoner woozke the
other fozme of woozke, then to make that trialle of *com=
menſurableneſſe.*

Maſter. The eaſieſte waie is, to diuide the grea=
ter nomber, by the leſſer, as if thei were bothe nom=
bers abſolute : ⁊ the *quotiente* will declare their *Squares.*

As if you doubte, whether. 384. and. 150. bee *com=
menſurable,* diuide. 384. by. 150. and the *quotiente* will
be $2\frac{14}{25}$, that is $\frac{64}{25}$. Then diuide whiche of the. 2. firſte
nombers you liſte, by his like nomber in the *quotiente* :
And the common diuiſoz will amounte. So if you di=
uide

Pn. ii.

*The findyng
of nombers
commenſu=
rable.*

The Arte

viðe the greater nomber. 384.by the greater nomber
in the *quotiente*, whiche is. 64, you ſhall finðe the new
quotiente 6. whiche. 6. is the common nomber. Or if
you ðiuiðe. 150. by 25. the common nomber. 6. will be
the *quotiente*.

But and if the *quotiente* be a whole nomber, and no
fraction, and be a Square nomber, then is it the leſſer
ſquare. Wherefore if you ðiuiðe the leſſer nomber of
the. 2. by the *quotiente*, the common nomber will ap=
peare in the ſeconðe *quotiente*. And then if you ðiuiðe
the greater of the. 2. nombers, by that common nom=
ber, his *quotiente* will ſhewe you the other *Square*.

And if ſo happen, that the *quotiente* of the firſte ðiui=
ſion be not a ſquare nomber, then are thoſe nombers
incommenſurable.

So. √.32. anð. √.128. bee *commenſurable* : anð the
quotiente of their ðiuiſion is. 4. whiche is the leſſer
ſquare. And. 8. appeareth to be the common nomber.
And the greater ſquare is. 16.

Howbeit by this nomber it maie eaſily bee eſpieð,
that ſome nombers maie be reſolueð, into moꝛe ſqua=
res then one. As theſe. 2. nombers, beyng ðiuiðeð by. 2
ðooe giue. 16. anð. 64. And beeyng ðiuiðeð by. 8, thei
bꝛyng foꝛthe. 4. anð. 16.

But foꝛ their aððition, what Squares ſo euer you
take, that reðounðe by one common ðiuiſoꝛ, the triall
will be like, and the roote one.

Scholar. J pꝛaie you let me pꝛoue that varietie.

Maſter. Then pꝛoue it in ſoche nombers, where
you maie finðe moare varietie. As theſe bee. √.288.
anð. √.1152.

Scholar. If J ðiuiðe. 1152. by. 288. the *quotiente*
will bee. 4. whiche J muſt take foꝛ the leaſte Square.
Then by it J ðiuiðe. 288. and the *quotiente* will be. 72.
as the common ðiuiſoꝛ. By whiche if J ðiuiðe. 1152.
there will riſe. 16. as the ſeconðe ſquare. Then ſet J
                                                              the

of Surde nombers.

the nōbers in oꝛder thus.
And vnder. 1152. I set the
one Square. 16. And vn=
der. 288. I putte the other
ſquare. 4. And vnder eche
of theim his roote. Then
adde I the Rootes toge=
ther, whiche maketh. 6.
whoſe ſquare is. 36. And
that beyng multiplied by
72. the common nomber,
doeth yelde. 2592. whoſe

$$\sqrt{.1152} \quad\text{———}\quad \sqrt{.288}.$$

	16	4.
72)	4	2.
	6.	
	6.	
	36.	
	72.	
	72.	
	252.	
$\sqrt{.}$	2592.	

roote doeth containe bothe the other. 2. rootes by ad=
dition.

But now how I ſhall finde any other Squares in
thoſe nōbers, to make any farther trial, I knowe not.

Maſter. Diuide alwaies one of the nombers, by
ſome ſquare nōber, that will parte it exactly, without
any remainer. And marke the *quotiente*. Foꝛ by it ſhal
you diuide the other nōber, and if the *quotiente* in that
laſt diuiſion, be a ſquare nomber, then haue you your
purpoſe. Els muſte you pꝛoue with an other Square
nomber.

Scholar. I vnderſtande you. And therfoꝛe in theſe
nombers, I will make trialle with. 9. by whiche I di=
uide. 288. And finde the *quotient*. 32. Then by theſame
32. I diuide 1152. and the *quotiente* is. 36. So haue I 9
and. 36. foꝛ the. 2. ſquares, and. 32. foꝛ the cōmon diui=
ſoꝛ. Wherfoꝛe I ſet the nōbers in oꝛder as thei ought.
And vnder them I place the. 2. ſquare nombers with
their rootes. Then addynge the rootes together, I
finde. 9. whiche I multiplie ſquare, and it yeldeth. 81.
that. 81. I doe multiplie by the common nomber. 32.
and there amounteth. 2592. As it did befoꝛe in the o=
ther woꝛke. Whereby I perceiue that theſe woꝛkes
doe confirme one an other.

Pn.iii.   And

The Arte

$\sqrt{} .1152$ ———┼——— $\sqrt{} .288.$

$$
\begin{array}{rr}
36 & 9 \\
32)\quad 6 & 3 \\
9 & \\
9 & \\
\hline
81 & \\
32 & \\
\hline
162 & \\
243 & \\
\hline
\sqrt{}\quad 2592 & \\
\end{array}
$$

And therefoze I will pzoue, how many varie= ties of this wozke, I may finde in these nombers. And foz my purpose, I will diuide the lesser of the. 2. nombers, by as many Squares as I can, foz that seameth to be the readieste waie. And firste I pzoue with. 16. And so the *quotient* is. 18. by whi= che. 18. I diuide. 1152. and the *quotiente* is. 64. whiche is a square nöber. So that I haue that varietie moze.

Then again I pzoue with. 25. But I see, that will not frame. Wherfoze I assaie with. 36. And finde the *quotiente* 8. by whiche I diuide the greater square, and the *quotiente* is. 144. a square nomber also. And ther= foze I note that foz an other varietie.

Thirdly, I pzoue with. 49. but that wil not agree. Then attempte I with. 64. And that serueth as euil. Nexte that I assaie. 81. 100. and. 121.but none of them will diuide. 288. wherefoze I passe vnto. 144. whiche is twise contained in 288. by that. 2. I diuide 1152. and finde the *quotiente*. 576. whiche is a Square nomber also. And so haue I. 3. other varieties beside the. 2. fozmer woozkes : whiche. 3. varieties, foz my re= membzaunce I set doune, thus.

$\sqrt{} .1152.$

of Surde nombers.

$\sqrt{\cdot}$.1152 ──┼── $\sqrt{\cdot}$.288 | $\sqrt{}$ 1152 ──┼── $\sqrt{\cdot}$.288

	64		16
18)	8		4
		12	
		12	
		144	
		18	
		1152	
		144	
	$\sqrt{\cdot}$.	2592 .	

	144		36
8)	12		6
		18	
		18	
		324	
		8	
	$\sqrt{\cdot}$.	2592	

$\sqrt{\cdot}$.1152 ──┼── $\sqrt{}$ 288

	576		144
2)	24		12
		36	
		36	
		1296	
		2	
	$\sqrt{\cdot}$.	2592	

Master. Then for to gratifie you, I will sette doune. 2. other nombers with 6 varieties. Whiche maie seame to suffice for this worke, without more exaples. And bicause you know the order to trie the I will sette theim doune without any explication, other declaration. As here you see.

$\sqrt{\cdot}$.28800 ──┼── $\sqrt{\cdot}$.7200 | $\sqrt{\cdot}$.28800 ──┼── $\sqrt{\cdot}$.7200

	14400		3600
2)	120		60
		180	
		180	
		32400	
		2	
	$\sqrt{\cdot}$.	64800	

	3600		900
3)	60		30
		90	
		90	
		8100	
		8	
	$\sqrt{\cdot}$.	64800	

$\sqrt{\cdot}$.28800

The Arte

√.28800 —— √.7200		√.28800 —— √.7200	
1600	400	900	225
18) 40	20	32) 30	15
60		45	
60		45	
3600		2025	
18		32	
28800		4050	
36		6075	
√. 64800		√. 64800.	

√.28800 —— √.7200		√.28800 —— √.7200	
576	144	400	100
50) 24	12	72) 20	10
36		30	
36		30	
1296		900	
50		72	
√. 64800		√. 64800	

Scholar. This varietie is pleaſante.

Maſter. I will ſatiſſie your delite better at moꝛe leiſare. But yet one thyng moare will I ſaie, befoꝛe we eande this ſoꝛte of Additiõ : that if you would adde any roote to it ſelf. As. √.6. to. √.6. oꝛ. √.10. to. √.10. ꝛc. you ſhall onely *quadriple* the nomber : and ſo haue you doen.

Scholar. I ſee good reaſon in that : Foꝛ addition of any nomber to itſelf, is but doublyng that nomber oꝛ multiplication by. 2. And that muſt be doen by that *quadriplation,* as you taught befoꝛe.

Addition of cubike rootes

Maſter. Now will I ſet foꝛthe ſome examples of addition in *Cubike* rootes. Foꝛ the woꝛke is like vnto this laſte foꝛme in Square rootes, ſaue that the mul= tiplications,

Of Surde nombers.

tiplications, whiche were Square in that woorke,
muſt be *Cubike* in this worke. And that onely in nom=
bers *commenſurable*. For nombers *incommenſurable* bee
added with the ſigne. ——┼——. without moare worke.

J call ſoche *Cubike* rootes *commenſurable*, whiche be=
yng diuided by any common nomber, will make *Cu*=
bike nöbers in their *quotiente*. As. √√.24. and. √√.81.
whiche diuided by. 3. doe make. 8. and. 27. bothe beyng
Cubike nombers. So. √√.320. and. √√.135. beyng di=
uided by. 5. doe make. 27. and. 64. bothe *Cubike* nom=
bers. Likewaies. √√.2744. and. √√.1000. be *com*=
menſurable, bicauſe thei make. 343. and. 125. bothe *Cu*=
bike nombers : Jf thei be diuided by. 8.

Cubike roo=
tes commen=
ſurable.

Scho. J praie you make your exäples with theſe.

Maſter. There nedeth noe wordes in this worke
it is ſo like the Addition of ſquare rootes. And there=
fore marke theſe examples well.

√√. 81. ——┼—— √√. 24.	√√.320. ——┼—— √√. 135
27 8	64 27
3.) 3 2	5.) 4 3
5	7
5	7
125	343
3	5
√√. 375	√√. 1715

√√.2744 ——┼—— √√.1000

     343   125.
8.)   7   5.

     12
     12
   1728
     8
√√.   13824

Do.i.   Scholar.

The Arte

Scholar. Here is noe diuerſitie, from the foʒmer woʒkes, but in ſettyng the *Cubike* roote, foʒ the ſquare roote. And in multipliyng the addition of the. 2. roo= tes *Cubikely.*

Maſter. That is all. And therefoʒe will J ſtande noe longer aboute it : But will pʒoceade to an other foʒme of addition, whiche ſerueth alſo foʒ *Cubike roo= tes commenſurable.* The rule is this. Set doune the *Cubike* rootes, with thier common diuiſoʒ, and the *Cu= bes* that riſe therby, and their rootes alſo. All this you did in this foʒmer woʒke. But now peculiarly in this rule, you ſhall ſet doune. 3. other nombers oʒderly, vn= der thoſe. 3. foʒmer nombers, The firſte is the ſquare of that laſte *Cubike* roote : the ſecōde is the *triple* of that Square : and the thirde is a nomber reſultyng of the multiplication of that triple by the other roote.

Then take the. 4 extreme nombers, that is thoſe 2 laſte nombers, and the. 2. *Cubes,* and adde theim toge= ther into one ſomme. And that ſomme beyng multi= plied by the common diuiſoʒ, will make a *Cubike* nom= ber, whoſe *Cubike* roote ſhall containe bothe the firſte rootes, whiche you intended to adde. Now marke theſe examples : and cōferre theim well with the woʒ= des of the rule.

∿ .384 ———	∿ .48
64	8
6) 4	2
16	4
48	12
48	96
216	
6	
∿ . 1296	

∿ .15972 ———	∿ .2592.
1331	216
12) 11	6
121	36
363	108
1188	2178

4913
12
9826
4913
∿ . 58956

∿ .52488.

of Surde nombers.

$$\sqrt{.52488.} \quad \text{———}\,+\,\text{———} \quad .\sqrt{.24696.}$$

	5832.		2744.
9)	18		14.
	324		196.
	972		588.
	10584		13608

$$
\begin{array}{r}
32768. \\
9. \\
\hline
\end{array}
$$

$$\sqrt{.} \quad 294912.$$

Scholar. In these examples J see, the woozdes of your rule obſerued. Foz vnder eche *Surde Cubike* roote, there is ſet a true *Cubike* nomber, whiche is founde by the common diuiſoz : then foloweth the roote of that true *Cube* : and beſide it ſtandeth the common diuiſoz. Then in the fourthe roome is the Square of the true *Cubike* roote. And vnder it his nomber tripled (as. 48 vnder. 16, and. 12. vnder. 4) whiche triple bee= yng multiplied by the roote of the other ſide, dooeth make the loweſte nomber in that rowe. So. 48. multiplied by. 2. maketh. 96. whiche is ſet vnder the roote. 2. And. 12. multiplied by. 4. yeldeth. 48. whiche is placed vnder that. 4.

Then thoſe. 4. extreme nombers. 64. and. 48, 8. ¶ 96. doe make by addition 216. whiche ſomme is mul= tipled by. 6, that is the common diuiſoz, and ſo riſeth 1296. whoſe *Cubike* roote compzehendeth bothe the firſte rootes.

Maſter. The like maie you iudge of the other. 2. examples. But bicauſe you maie vnderſtande the certaintie of this woozke the better, J haue here ſette fozthe. 2. examples of true *Cubike* rootes, fozmed like *Surde* nombers.

The Arte

$$\sqrt{\ .4096.} \quad \text{---} \quad .\sqrt{\ .1728.}$$

	512.	216.
8)	8.	6.
	64.	36.
	192.	108.
	864.	1152.

2744
8

$$\sqrt{\ .} \quad 21952$$

$$\sqrt{\ .19683.} \quad \text{---} \quad .\ \sqrt{\ .3375.}$$

	729	125
27)	9	5
	81	25
	243	75
	675	1215

2744
27

19208
5488

$$\sqrt{\ .} \quad 74088$$

Scholar. J perceiue by examinatiõ of woozke in my Tables here, that 4096. is a *Cubike* nomber, and hath 16 foz his roote. So 1728 is a *Cubike* nõber alſo, ɛ his roote is. 12. thoſe bothe rootes added together, doe make. 28. And that. 28. is the *Cubike* roote to. 21952. as the firſte example would.

And foz the ſeconde example, J ſee likewaies that 19683. hath. 27. foz his *Cubike* roote. And. 3375. hath 15. foz his roote. And thei bothe make. 41, whiche is the *Cubike* roote to. 74088. accozdyng to the woozke of the ſeconde example.

Addition of zenzizenzike rootes.

Maſter. Seyng you are conueniently inſtructed, in theſe nombers, wee will goe in hande with zenzizenzike rootes, and their addîtiõ : wherein is no diffeærēce of woazke, but onely foz the multiplicatiõ, whiche muſt be agreable to the nature of the nombers, zenzizenzikely. And the reduction by the common diuîſoz,

of Surde nombers.

foz, in like fozme, into zenzizenzike nombers, whē the firſte nombers bee *commenſurable.* But if thei be *incom=menſurable,* then muſt the addition be wzoughte by the ſigne. —+— , without any other buſineſſe.

Examples of zenzizenzikes
beeyng commenſurable.

√.648 —+— √.5000 | √.1280 —+— √.6480
81   625 | 256   1296
8)   3   5 | 5) 4   6
         8 | 10
         8 | 10
      4096 | 10000
         8 | 5
√.   32768 | √.   50000

√.38416 —+— √.65536

2401   4096
16)   7   8
          15
          15
       50625
          16
      303750
       50625
√. 810000

In the firſte and ſe=conde examples the nō=bers are *Surdes,* but in the thirde example thei are rationall nombers, framed like vnto *Surdes* to the intente that you mighte the better per=ceiue the fozme of the wozke. Foz 38416. is a *zenzizēzike* nomber, & hath. 14. foz his roote

So. 65536. is a *zenzizenzike* nomber, and hath. 16. foz his roote. And theſe. 2. rootes do make. 30. whiche is the *zenzizenzike* roote vnto. 810000. And there=foze maie it bee truely ſaied, that. √.810000 doeth containe the twoo firſte rootes.

Scholar. J pzaie you pzocede to Subtraction. Foz all this J doe well perceiue.

                              Do.iii.   Of

The Arte
Of Subtraction.

Master.

Ubtraction doeth differ from addition, in litle moare then the signe ——— , whiche signe serueth generally, for all nombers *incommensurable*. And consideryng there is litle difficultie in Subtraction: If you re= member well the arte of Addition, J wil lightly passe it ouer in thesame examples, that J haue wrought in Addition, bicause it maie bee a proofe of that woorke : and that woorke also a confirmation of this.

Onely this shall you obserue in this rule peculiar= ly : that as in the seconde forme of Addition, you must adde the rootes together, before you multiplie theim. So here you shall Subtracte the lesser roote, from the greater, before you doe multiplie theim.

Example of Subtraction, with ——— .

√.12. abated out of √.26. maketh. √.26 ——— √.12. and so of other.

Examples of the seconde
forme of Subtraction

√.63. ——— . √.28.		The seconde forme
63		of that woorke.
28	63	√.63. ——— .√.28.
—————	28	63
504	91	28
126		—————
—————		1764
√.1764		whose roote is. 42.
√. 4		42
—————		2
√.7056		—————
√91 ——— √.7056		84
Dr. √.91 ——— 84.		√.91 ——— 84.

That is. √.7.

√.169

of Surde nombers.

√.169 ———— √.36

```
        169
         36
      ─────────
        1014
         507
      ─────────
√.   6084
√.      4
      ─────────
       24336
√.205 ———— √.24336.
or. √.205 ———— √.156.
```

```
169
 36
─────
205
```

An other forme of
that woozke.

√.169. ———— .√.36.

```
        169
         36
      ─────────
√.   6084
whose roote is. 78.
         78
      ─────────
          2
      ─────────
        156
√.205 ———— 156.
```

That is. √.49.

Scholar. J see in all thefe examples, you take the=
fame nombers, that you had befoze in Addition. And
firfte you fet the totalle, out of whiche you abate one
of the nöbers, that befoze were added, & the remainer
bzingeth forthe the other. Foz in the firfte of thefe. 2.
examples. √.28. is abated out of. √.63. and there re=
maineth. √.91. ———— 84. that is. √.7. foz. 84. taken
out of. 91. leaueth. 7. And in the feconde exaple. √.39
abated out of. √.169. doeth leaue remainyng. √.49.

Mafter. The thirde forme of Subtraction, is like
the thirde forme of Addition : faue that we fet ————.
foz ———|——. And here wee mufte abate the leffer roote
frö the greater (as J faid) befoze we doe multiplie that
nomber by it felf. As by this exaple, you maie perceiue
Where J dooe Subtracte. √.105. out √.1014, and
the remainer is. √.384. Now marke the woozke

```
√.1014 ———— √.105.
─────────────────────
      169            25.
6)    13             5.
               8
               8
           ─────────
              64
               6
           ─────────
√.   384
```

Here you fee all thinges
agree, with the forme of
Addition, faue ————. foz
———|——. and when J begin
to gather the nomber, that
ftädeth in the middle, whi=
che J multiplie by it felfe,
and J dooe not make that
nomber,

The Arte

nomber, by addyng bothe rootes together : For so. 13.
and. 5. would make. 18, but J abate. 5. out of 13. and so
there doeth remain. 8. with whiche J procede as J did
in Addition. And then commeth forthe the remainer.
√.384.

Scholar. J vnderstande it very well. And J praie
you that for a proofe, J maie varie the other examples
of addition. Partly for my exercise, and partly for exa=
minatiõ of the former additions, by the cõtrary kind.

Master. VVith good will.

Scholar. Then will J set them, and worke them,
as here foloweth.

But firste J will begin, with the worke of this last
example, after the seconde forme of Subtraction : for a
double confirmation of it.

√.1014 ——— √.150		¶ An other forme of theſame worke.
1104	1014	√.1014 ——— √.150.
150	150	1014
50700	1164	150
1014		50700
√. 152100		1014
√. 4		152100
√.608400		whoſe roote is. 390.
√.1164——— √.608400		390 :
Dr.√.1164——— 780		2
		√.1164——— 780.

That is.√.384.

And now here are the variations of the other ex=
amples.

√.2592.

# of Surde nombers.

√.2592 ——— √.288.
36   4.
72)   6   2.
       4
       4
      16
      72
      32
     112
√.   1152

√.2592 ——— √.288.
81   9.
32)   9   3.
       6
       6
      36
      32
      72
     108
√.   1152

√.2592 ——— √.288.
144   16.
18)   12   4.
       8
       8
      64
      18
     512
      64
√.   1152

√.2592 ——— √.288.
8)   324   36.
     18   6.
      12
      12
     144
       8
√.   1152

√.2592 ——— √.288.
1296   144.
2)   36   12.
      24
      24
     576
       2
√.   1152

√.2592 ——— √.1152.
1296   576.
2)   36   24.
      12
      12
     144
       2
√.   288

Other examples varied, for proofe of the like. 6. examples in Addition.

Pp.i.   √.64800

The Arte

√.64800 ———— √.7200	√.64800 ———— √.7200.
32400 3600	8100 900.
2) 180 60	8) 90 30
120	60
120	60
14400	3600
2	8
√. 28800	√. 28800

√.64800 ———— √.7200.	√.64800 ———— √.7200.
3600 400.	2025 225.
18) 60 20	32) 45 15.
40	30
40	30
1600	900
18	32
12800	√. 28800
16	
√. 28800	

√.64800 ———— √.7200.	√.64800 ———— √.7200.
1296 144.	900 100.
50) 36 12	72) 30 10.
24	20
24	20
576	400
50	72
√. 28800	√. 28800

Subtraction
of Cubike
rootes.

Maſter. Like difference is there in Subtraction
of *Cubike* rootes *commenſurable.* And therfoze I ſet the
examples onely, without any larger declaration.

ᴡᴡ.375.

of Surde nombers.

$\sqrt{}$.375 ——— $\sqrt{}$.81.		$\sqrt{}$.1715 ——— $\sqrt{}$.135.	
125	27.	343	27.
3) 5	3	5(7	3.
2			4
2			4
8			64
3			5
$\sqrt{}$. 24		$\sqrt{}$. 320	

$\sqrt{}$.13824——— $\sqrt{}$.1000	
1728	125
8) 12	5
7	
7	
343	
8	
$\sqrt{}$. 2744	

Another woorke of Subtraction for Surde Cubes.

In the seconde forme of addition of *Surde Cubes*, you remember that you added 4 nombers together. But in Subtraction, you shall adde to eche roote severallie that, that commeth of his owne multiplication, with the other *triple*. And then shall you Subtracte the lesser nomber, out of the greater. And the remainer you shall multiplie by the common divisor. And so shall you haue the roote that remaineth of the Subtration. As in example.

$\sqrt{}$.1296 ——— $\sqrt{}$.48.		$\sqrt{}$58956——— $\sqrt{}$15972	
216	8	4913	1331
6) 6	2	12) 17	11
36	4	289	121
108	12	867	363
72	216	6171	5537
64		216	
6		12	
$\sqrt{}$. 384		$\sqrt{}$. 2592	

Pp.ii. $\sqrt{}$294912

The Arte

√ .294912 ——— √ .24696

$$
\begin{array}{r}
32768 \\
9)\quad 32 \\
1024 \\
3072 \\
\hline
18816
\end{array}
\qquad
\begin{array}{r}
2744 \\
14 \\
196 \\
588 \\
\hline
43008
\end{array}
$$

$$
\begin{array}{r}
5832 \\
9 \\
\hline
√.\quad 52488
\end{array}
$$

Scholar. In all these examples I se the confirmation of the former additiō. And in these laste woorkes, this I see peculiare from ad= ditiō, that the *Cube* is added with the loweste nomber in that rowe (as in the firste example. 216. is added with 72. and maketh. 288 : And. 8. is added with. 216. that yeldeth. 224.) And then is the lesser abated from the greater (as. 224. from 288.) And the remainer (whi= che there is. 64) set in the middle vnder bothe the re= wes of nombers. And then is multiplied by the com= mon nomber, to make the remainer.

So in the firste example, the remainer is. √.384. where. √.48. is abated out of. √.1296. And in the seconde example where. √.15972. is subtracted out of. √.58956. the remainer is √.2592. Like= waies in the thirde example. √.24696. is abated out of. √294912 & leaueth remainyng. √52488

Master. But now in addition there foloweth. 2. other examples, whiche by subtraction maie bee pro= ued thus : as here you see.

√ 21952 ——— √ 4096

$$
\begin{array}{r}
2744 \\
8)\quad 14 \\
196 \\
588 \\
\hline
2688
\end{array}
\qquad
\begin{array}{r}
512 \\
8 \\
64 \\
192 \\
\hline
4704
\end{array}
$$

$$
\begin{array}{r}
216 \\
8 \\
\hline
√.\quad 1728
\end{array}
$$

√ 74088 ——— √ 19683

$$
\begin{array}{r}
2744 \\
27)\quad 14 \\
196 \\
588 \\
\hline
3402
\end{array}
\qquad
\begin{array}{r}
729 \\
9 \\
81 \\
243 \\
\hline
5292
\end{array}
$$

$$
\begin{array}{r}
125 \\
27 \\
\hline
√.\quad 3375
\end{array}
$$

Scholar.

of Surde nombers.

Scholar. J see, in these examples of Subtraction : that the firste nomber is the totalle, oz laste nomber in addition. And the seconde nomber, whiche foloweth ———— . is the nomber to be abated : and then laste and loweste of all, is the remainer, whiche was one of the firste sommes in addition.

And though there remaine. 3. other exāples of *zen=zizenzike* nombers, J see no difficultie in theim, but that J can woozke them : As here J haue set thē foorth.

$$\sqrt{32768} \;\text{———}\; \sqrt{.648}$$

4096	81
8) 8	3
5	
5	
625	
8	
$\sqrt{.}$ 5000	

$$\sqrt{.50000} \;\text{———}\; \sqrt{1280}$$

10000	256
5) 10	4
6	
6	
1296	
5	
$\sqrt{.}$ 6480	

$$\sqrt{810000} \;\text{———}\; \sqrt{65536}$$

50625	4096
16) 15	8
7	
7	
2401	
16	
$\sqrt{.}$ 38416	

Master. Seeyng you are experte inough in the 5. woozkes of these *Surdes* vncōpounde, J wil teache you the like wozkes in cō= pounde *Surdes.*

Scholar. Js there thē noe reduction, nother ex= traction of rootes, to bee *Of reduction and extracti= on of rootes.* taughte in these vncompounde *Surdes?*

Master. As foz reduction, J haue taughte you all readie in multiplication, as moche as is required in these nombers.

And foz extraction of rootes, you maie sone vnder= stande, that here can be none. Foz then were thei not *Surde* nombers. And therfoze J saie vnto you befoze,

The Arte

that. $\sqrt{}.\mathfrak{z}.100.$ is not a *Surde* nomber, although it be written like a *Surde* nomber, bicause it hath a Square roote, accordyng to his signe : and that is. 10. Likewa= yes. $\sqrt{}.256.$ is no *Surde* nomber : for his Square roote is knowen to be. 16.

Scholar. J might haue cosidered as moche, by the definition of *Surde* nombers, that their rootes can not be assigned in nombers absolute. And therfore J see that. $\sqrt{}.125.$ is noe *Surde* nöber, sith his *Cubike* roote is. 5. And. $\sqrt{}.256.$ is a nomber *rationalle*, and no *Surde* nomber : for his *zenzizenzike* roote is. 4.

Master. But. $\sqrt{}.64.$ is a *Surde* nomber, and yet hath. 64. a *Square* roote, and a *Cubike* roote also, but not a *zenzizenzike* roote, accordyng to his signe. And therfore ought better to be written thus. $\sqrt{}.8.$

Scholar. J praie you to procede to *Surde* nombers compounde.

Of Surde nombers compounde.

Master.

 *Vrde* nombers compoude, are made not onely of. 2. or. 3. or moare *Surde* nombers uncompounde, but also of *rationalle* or *Abstracte* nombers ioy= ned with *Surde* nombers. As. $\sqrt{}.10$ ——— $\sqrt{}.12.$ and. 8. ——— $\sqrt{}.6.$ like= waies. $\sqrt{}.20.$ ——— 3. and. $\sqrt{}.40.$ ——— $\sqrt{}.14.$ ——— .3.

Compounde Surdes.

But here shall you marke, that J call compounde nombers, not onely soche as haue the signe of. ——— , but also soche as haue the signe of ——— for although in nature of the number $\sqrt{}.10$ ——— $\sqrt{}.5.$ be not com= pounde, but abated, yet in name he is compoude, and augemented. For. ——— . doeth as well augemente
the

of Surde nombers.

the name, as ———|—— doeth.

Scholar. It femeth reafonable. Foꝛ when I faie, √ .12 . ———— √ .7. the name is compounde, an well as if I had faied. √ .12. ———|—— √ .7. although the quanti= tie bee not fo greate. Foꝛ ———— doeth euer abate the quätitie of the nöber, though it do increafe the name.

Mafter. Yet foꝛ a difference, the nombers that be compounde with ——|—— be called *Bimedialles* : and thofe that be compounde with ————, be named *Refidualles*. And if the *Bimedialles* haue all their nombers and par= tes of one denominations, then bee thei called onely by their generalle name *Bimedialles*. But if their par= tes be of. 2. denominations, then are thei named *Bino= mialles* pꝛoperly. Howbeit, many vfe to all *Binomialles* all compounde nombers that haue ——|—— . And fo wil I let the names paffe.

Euclides definitions doe not very aptly agree to this place, as at an other tyme I will ſhewe you, and ther= foꝛe I doe omitte them foꝛ this tyme.

But touchyng our pꝛincipalle intente, whiche is to declare the pꝛactike wooꝛke of *Binomialles*, and *Reſi= dualles*, there is litle difficultie, if you marke well that whiche is taught befoꝛe. Foꝛ as *Binomialles* and *Reſi= dualles*, bee made of *Surdes*, oꝛ els of *rationalle* nombers with *Surdes*, fo the wooꝛke of the compounde nombers dependeth of the wooꝛke of the fimple nombers, and is all one with them. And concernyng the fignes ——|—— and. ————. here is no moare to bee faied, then was taughte in *Coſike* nombers compounde.

Scholar. Yet of euery kinde, it maie pleafe you to fet foꝛthe fome eramples.

Mafter. I thinke that mete, without many woꝛ= des els. Not foꝛgettyng by the waie, that *vniuerſalle rootes*, are not accompted emongefte thefe compounde *Surdes* : but are referued to their peculiare treatice, as rootes of compounde *Surdes*.

Of

(marginal notes:) *Bimedialles.* *Refidualles.* *Binomialles.*

The Arte
Of Numeration.

Numeration is moare plain, then that I neade to stande in declaryng it, otherwaies then by examples : As here you see.

Examples of Binomialles.

6. ——+—— √.8.   That is 6 moze the *Square* roote of 8.

√.20 ——+—— .3.   Is the *Square* roote of. 20. moare. 3.

ⱳ.30 ——+—— ⱳ.9. Signifieth the *Cubike* roote of. 30. moze the *zenzizenzike* roote of. 9. And so of other.

Examples of Residualles.

24. ———— √96.   That is. 24. abating the roote of 96

√.150. ———— .9.   Is the *Square* roote of 150. abating 9

ⱳ5208—— √35.   The *zenzizenzike* roote of. 5208. saue the *Square* roote of. 35. And so fozthe.

Scholar. So I see any *Surdes* maie bee compounde with other : And any nöbers *rationalle* ioined with thē.

Of Addition.

Master. Addition is as plaine. Foz as the partes bee, so shall the Addition bee, accozdyng as you haue learned befoze.

Examples of Binomialles.

√.50 ——+—— 10	15. ——+—— √.15.	√.1264 ——+—— 8.
√.2. ——+—— .8.	18 ——+—— √.60.	28 ——+—— √316
√.72 ——+—— 18	33. ——+—— √.135.	36 ——+—— √2844

ⱳ.48. ——+—— √.5.	ⱳ.32 ——+—— √10.
ⱳ.243. ——+—— √.45.	ⱳ.4. ——+—— √19.
ⱳ.1875 ——+—— √.80.	ⱳ108 ——+—— √29 ——+—— √760

Examples

Of Surde nombers.
Examples of Residualles.

√.75. ——— .4	14——— √.3.	250——— √.108.
√. 3 ——— 1.	16 ——— √.27	√.44 ——— 76.
√.108 ——— 5.	30 ——— √.12.	174 ——— √.275.

ᴗᴗ.72. ——— ᴗ.96	ᴗ.32. ——— √.5.
ᴗᴗ. 9. ——— ᴗ.6.	ᴗ.32. ——— √.24.
ᴗᴗ243 ——— ᴗ162	ᴗ.512 ——— √.29 —+— √480

Examples of Binomialles with Residualles.

√.80. —+— 6.	30. —+— √.20	.561 ——— √512
√. 5. ——— 2.	12. ——— √.5.	√288 —+—340
√.125 —+— 4.	42 —+— √.5.	901 ——— √1568

√.63. ——— ᴗᴗ160	ᴗᴗ.320 ——— √.56.
√. 7. —+— ᴗᴗ.20.	ᴗᴗ.40 —+— √.24.
√.112 ——— ᴗᴗ684	ᴗᴗ1680 ——— √.80 ——— √5376

Scholar. I see that you make seueralle Adiditons in all these nombers. For you adde still like nombers with their matches. So that here is nothyng diuerse from the woozkes of simple *Surdes.* Although in euery thirde example, there appeare moare difficultie, then there is in deede: When I consider the like transposition in *Cossike* nombers. For the woozke addeth like nombers together.

Of Subtraction.

Master. In Subtraction there is as litle diuersitie. As these examples will sufficiētly declare: whiche be set as trialles of the former Additions.

Qq.i.   Examples

The Arte
Examples of Binomialles.

$$\sqrt{.72} \;\rule[0.5ex]{1em}{0.4pt}\!+\!\rule[0.5ex]{1em}{0.4pt}\; 18$$
$$\sqrt{.\,2} \;\rule[0.5ex]{1em}{0.4pt}\!+\!\rule[0.5ex]{1em}{0.4pt}\; 8.$$
$$\sqrt{.50} \;\rule[0.5ex]{1em}{0.4pt}\!+\!\rule[0.5ex]{1em}{0.4pt}\; 10$$

$$36 \;\rule[0.5ex]{1em}{0.4pt}\!+\!\rule[0.5ex]{1em}{0.4pt}\; \sqrt{2844}$$
$$\sqrt{.1264} \;\rule[0.5ex]{1em}{0.4pt}\!+\!\rule[0.5ex]{1em}{0.4pt}\; 8.$$
$$28 \;\rule[0.5ex]{1em}{0.4pt}\!+\!\rule[0.5ex]{1em}{0.4pt}\; \sqrt{.316.}$$

$$33 \;\rule[0.5ex]{1em}{0.4pt}\!+\!\rule[0.5ex]{1em}{0.4pt}\; \sqrt{.135}$$
$$15 \;\rule[0.5ex]{1em}{0.4pt}\!+\!\rule[0.5ex]{1em}{0.4pt}\; \sqrt{.15.}$$
$$18 \;\rule[0.5ex]{1em}{0.4pt}\!+\!\rule[0.5ex]{1em}{0.4pt}\; \sqrt{.60}$$

$$\mathcal{W}.1875 \;\rule[0.5ex]{1em}{0.4pt}\!+\!\rule[0.5ex]{1em}{0.4pt}\; \sqrt{.80.}$$
$$\mathcal{W}.\,48. \;\rule[0.5ex]{1em}{0.4pt}\!+\!\rule[0.5ex]{1em}{0.4pt}\; \sqrt{.5.}$$
$$\mathcal{W}.243 \;\rule[0.5ex]{1em}{0.4pt}\!+\!\rule[0.5ex]{1em}{0.4pt}\; \sqrt{.45.}$$

$$\mathcal{WW}.108 \;\rule[0.5ex]{1em}{0.4pt}\!+\!\rule[0.5ex]{1em}{0.4pt}\; \sqrt{.29.} \;\rule[0.5ex]{1em}{0.4pt}\!+\!\rule[0.5ex]{1em}{0.4pt}\; \sqrt{.760.}$$
$$\mathcal{WW}.\,4 \;\rule[0.5ex]{1em}{0.4pt}\!+\!\rule[0.5ex]{1em}{0.4pt}\; \sqrt{.19.}$$
$$\mathcal{WW}.\,32 \;\rule[0.5ex]{1em}{0.4pt}\!+\!\rule[0.5ex]{1em}{0.4pt}\; \sqrt{.10.}$$

Examples of Refidualles.

$$\sqrt{.108} \;\rule[0.5ex]{1.5em}{0.4pt}\; 5$$
$$\sqrt{.\,3} \;\rule[0.5ex]{1.5em}{0.4pt}\; 1$$
$$\sqrt{.\,75} \;\rule[0.5ex]{1.5em}{0.4pt}\; 4$$

$$174. \;\rule[0.5ex]{1.5em}{0.4pt}\; \sqrt{.275.}$$
$$\sqrt{.44} \;\rule[0.5ex]{1.5em}{0.4pt}\; 76$$
$$250. \;\rule[0.5ex]{1.5em}{0.4pt}\; \sqrt{.108}$$

$$30. \;\rule[0.5ex]{1.5em}{0.4pt}\; \sqrt{.12.}$$
$$14. \;\rule[0.5ex]{1.5em}{0.4pt}\; \sqrt{.3.}$$
$$16. \;\rule[0.5ex]{1.5em}{0.4pt}\; \sqrt{.27.}$$

$$\mathcal{WW}.243 \;\rule[0.5ex]{1.5em}{0.4pt}\; \mathcal{W}.162.$$
$$\mathcal{WW}.\,9 \;\rule[0.5ex]{1.5em}{0.4pt}\; \mathcal{W}.6.$$
$$\mathcal{WW}.\,72 \;\rule[0.5ex]{1.5em}{0.4pt}\; \mathcal{W}.96.$$

$$\mathcal{W}.512 \;\rule[0.5ex]{1.5em}{0.4pt}\; \sqrt{.29.} \;\rule[0.5ex]{1em}{0.4pt}\!+\!\rule[0.5ex]{1em}{0.4pt}\; .\sqrt{.480.}$$
$$\mathcal{W}.\,32 \;\rule[0.5ex]{1.5em}{0.4pt}\; \sqrt{.5.}$$
$$\mathcal{W}.\,32 \;\rule[0.5ex]{1.5em}{0.4pt}\; \sqrt{.24.}$$

Examples of bothe together.

$$\sqrt{.125} \;\rule[0.5ex]{1em}{0.4pt}\!+\!\rule[0.5ex]{1em}{0.4pt}\; 4$$
$$\sqrt{.\,5} \;\rule[0.5ex]{1.5em}{0.4pt}\; 2$$
$$\sqrt{.80} \;\rule[0.5ex]{1em}{0.4pt}\!+\!\rule[0.5ex]{1em}{0.4pt}\; 6$$

$$901 \;\rule[0.5ex]{1.5em}{0.4pt}\; \sqrt{1568}$$
$$\sqrt{.288} \;\rule[0.5ex]{1em}{0.4pt}\!+\!\rule[0.5ex]{1em}{0.4pt}\; 340$$
$$561 \;\rule[0.5ex]{1.5em}{0.4pt}\; \sqrt{.512.}$$

42.

of Surde nombers.

$$42 \longrightarrow \sqrt{.5.} \qquad \sqrt{.112} \longrightarrow \text{\tiny ᴠᴠᴠ}\sqrt{.648.}$$
$$12 \longrightarrow \sqrt{.5.} \qquad \sqrt{.7} \longrightarrow \text{\tiny ᴠᴠᴠ}\sqrt{.20.}$$
$$30 \longrightarrow \sqrt{.20.} \qquad \sqrt{.63} \longrightarrow \text{\tiny ᴠᴠᴠ}\sqrt{.160.}$$

$$\text{\tiny ᴠᴠᴠ}\sqrt{.1080.} \longrightarrow \sqrt{.80.} \longrightarrow \sqrt{.5376.}$$
$$\text{\tiny ᴠᴠᴠ}\sqrt{.40.} \longrightarrow .\sqrt{.24.}$$

$$\text{\tiny ᴠᴠᴠ}\sqrt{.320.} \longrightarrow .\sqrt{.56.}$$

Scholar. This is as easie as Addition, saue for. 3. examples, whiche I vnderstande not. For although I see the laste example, of eche of the sortes of nombers, to bee agreable with the like examples in Addition, yet I can not so well perceiue, the order of their Subtraction, as I doe knowe the maner of their Additiō. For by the arte of simple *Surdes*, I see that. $\sqrt{.10}$ and. $\sqrt{.19}$. doe make. $\sqrt{.29} \longrightarrow \sqrt{.760.}$ But when $\sqrt{.29.} \longrightarrow .\sqrt{.760.}$ is set as a totalle, and. $\sqrt{.19.}$ to be Subtracted out of it, how I shall woorke that, and leaue. $\sqrt{.10.}$ for the remainer, I see not.

So in the *residualles*, I knowe how. $\sqrt{.5.}$ and $\sqrt{.24.}$ doe make. $\sqrt{.29} \longrightarrow \sqrt{.480.}$ But I knowe not how $\sqrt{.5}$ abated out of. $\sqrt{.29} \longrightarrow \sqrt{.480.}$ doeth make for the remainer. $\sqrt{.24.}$

And the like doubte is in the thirde sorte of *Surdes*, whiche are mixte nombers. For where I see in Addition $\longrightarrow \sqrt{.24.}$ added with $\longrightarrow .\sqrt{.56.}$ And the totalle to bee. $\longrightarrow .\sqrt{.80.} \longrightarrow \sqrt{.5376.}$ I knowe the reason of the woorke, for the signes. $\longrightarrow .$ and $\longrightarrow .$ by that I learned in *Cossike* nombers : And the reaste is manifeste by Addition of simple *Surdes*. For it is wrought by abatyng. $\sqrt{.24.}$ out of. $\sqrt{.56.}$ But then in Subtraction, how $\longrightarrow \sqrt{.24.}$ beyng Subtracted from $\longrightarrow \sqrt{.80} \longrightarrow \sqrt{.5376}$ shall leaue $\longrightarrow \sqrt{.56}$ I can not iudge. And yet by the signes I gesse (as I learned in *Cossike* nombers) that it is doen by Additiō, bicause the signes doe disagree.

Qq.ii.   Master.

The Arte

Master. In that you remember the former rules, to conferre them aptly with these later workes, I can praise you well. But in that you can not vnderstande the reason of that, whiche was not yet taughte you, I can not greatly blame you. Although I can not praise you, for that you thinke your self to be cunnynger then you are. For in those Additions, that you thinke your self to be experte inough, I dare saie, that you bee disceiued, if you take them to bee nombers of any soche, as hetherto hath been taughte vnto you.

Scholar. I take them for compounde *Surdes*.

Master. Thei are not so : Nother is their woorke agreable, with the woorke of compounde *Surdes*. But thei are the rootes of compounde *Surdes* : And therfore are called *vniuersalle rootes* of *Surdes*. And accordyng to their proper nature, thei ought to bee called rootes of *Surdes*, and not *Surde* rootes. As I will tell you anon. When I will also discusse your doubte.

But before I speake any moare of theim, I will eande the woorkes of these compounde *Surdes* : wherof. 2. kindes yet remaine behinde.

Of Multiplication.

 Ultiplication of compounde *Surdes*. is as easie as can bee. And differeth in nothyng, frō the woorke of simple *Surdes*. Onely this must you marke, as reason would, that you muste multiplie euery parte of the one nōber, by euery parte of the other nōber : as you remember the woorke of compounde *Cossike* nombers.

Scholar. I praie you giue me some examples.

Master. That shall you haue. And that maie suffice for this woorke. Marke them well therfore.

Examples

of Surde nombers.

Examples of Binomialles.

23 —+— √.15.
6 —+— √.8.

138 —+— √.120.
—+— √.540. —+— √.4232.

138 —+— √.4232 —+— √.540 —+— √.120.

√. 120 —+— √.12.
√. 12 —+— √.7.

√1440 —+— √.84.
—+— √.840 —+— 12.

12 —+— √.1440 —+— √.840 —+— √.84.

Examples of Residualles.

5. —— √.10.
5. —— √.10.

25 —+— 10.
—— √.250 —— 250.

35 —— √.1000.

√. 24 —— √.20.
√. 30 —— √.24.

√.720 —+— √.480.
—— 24 —— √.600.

√.720 —+— √.480 —— 24 —— √.600.

Examples of bothe together.

32 —+— √.14.
√.124 —— 6.

√.126976 —+— √.1736.
—— . 192. —— √.504.

√.126976 —+— √.1736 —— 192 —— √.504

𝔔u.iii.   √.52.

The Arte

$$\sqrt{.} \quad 52 \quad \text{———} \quad 17.$$
$$17 \quad \text{———} \quad \sqrt{.52.}$$
$$\sqrt{.15028} \quad \text{———} \quad 289.$$
$$\text{———} \quad 52 \quad \text{———} \quad \sqrt{.15028.}$$

37.

Scholar. Multiplication, as J see, is the easieste woozke of all the other. So that J dooe marke the re= duction, in gatheryng the totalle : whiche is easie i= nough to vnderstand, by that J haue learned in *Cossike* nombers. And Diuision be no harder, it maie sone be learned.

Of Diuision.

Master.

Iuision by one simple nomber, is no moare difficulte : as these exam= ples doe declare. Where the diuisoz is a nomber vncompounde.

$\sqrt{.26}$ ——— 15 diuided by. 5. doeth make. $\sqrt{.}1\frac{1}{25}$ ——— 3.

Againe. $\sqrt{.56}$ ——— $\sqrt{.24.}$ diui= ded by. $\sqrt{.6.}$ doeth yelde. $\sqrt{.}9\frac{1}{3}$ ——— 2.

And so $\sqrt{.75}$ ——— $\sqrt{.48.}$ diuided by. $\sqrt{.3.}$ dooeth bzyng forthe. 5. ——— 4. that is. 1.

Likewaies. $\sqrt{.320.}$ ——— $\sqrt{.180.}$ beyng parted by $\sqrt{.5.}$ doeth make the *quotiente.* 14.

Scholar. J see it so. For at the firste it is. $\sqrt{.64.}$ ——— $\sqrt{.36.}$ that is. 8. ——— 6. whiche maketh. 14.

Master. So maie you woozke all like diuisions. But when the diuisoz is a compounde nomber, then must you vse an another meane : that is to reduce that compounde nöber, to a simple nomber : whiche thing you maie easily doe, by multipliyng any *Binomialle,* by his *Residualle,* oz contrary waies, the *Residualle* by his *Binomialle.*

As

of Surde nombers.

As 6 ╾╾ √.10. multiplied by. 6 ──── √.10 doeth make. 26.

And ſo. √.8 ──── √.5, multiplied by. √.8 ╾╾ √.5 doeth yelde. 8. ──── 5. that is .3.

Scholar. I perceiue a brief waie in this multipli= cation : For I neade not in the firſte example, to mul= tiplie 6. by. √.10. ſith it would amounte to nothyng. In ſo moche as at one multiplication, it would bee ╾╾ , and at an other. ──── . And ſo the one would abate the other, and leaue nothyng for them bothe.

Maſter. That is well marked. And it is ſo gene= rally. Wherefore (as you ſee) the diuiſor by this mea= nes, maie lightly be tourned into a ſimple nomber, or a plaine abſolute nomber.

And now to make the diuidede, in the ſame propor= tion, to this newe diuiſor, that it was vnto the old di= uiſor, you ſhall multiplie it by theſame nomber, by whiche the diuiſor was multiplied. For if any nom= bers bee multiplied, by one common nomber, their newe totalles kepe theſame proportion, that was be= twene the firſte nombers.

Scholar. That muſt neades be ſo. For as. 3. is ſeſ= *quialtera* vnto. 2. ſo if you multiplie them by. 5. thei will make. 15. and. 10. whiche be in *ſeſquialtera* proportion and likewaies will their proportion remain, by what ſo euer nomber thei be multiplied. Wherfore it muſt neades be reaſonable, that if the diuidende and the di= uiſor, be multiplied by any one nomber, ſimple or cō= pounde, thei ſhall kepe the ſame proportion, that thei had before.

Maſter. For more certain vnderſtandyng of this rule, take theſe examples. The firſte is, where √.68 ╾╾ √.54. is ſette to bee diuided by. √.6 ╾╾ √.3.

Here firſte I multiplie the diuiſor by his contrarie, that is his *Binomi*

√.6 ╾╾ √.3.

√.6 ──── √.3

────────

6 ──── 3.

That is. 3.

alle

The Arte

alle. $\sqrt{.6}$——3. And there riſeth. 6 ——3. that is, 3 whiche I ſhall kepe for the newe diuiſor.

Then doe I multiplie the diuidede $\sqrt{.68}$ + $\sqrt{.54}$ by theſame *Reſidualle.*

$$\sqrt{.68.} \quad + \quad \sqrt{54.}$$
$$\sqrt{.6.} \quad —— \quad \sqrt{3.}$$
$$\overline{\sqrt{.408}} + \quad \sqrt{.324.}$$
$$—— \sqrt{.204} —— \sqrt{.162.}$$
$$\overline{\sqrt{.408} + \sqrt{.324} —— \sqrt{.204} —— \sqrt{.162.}}$$

And there doth amoũte, as here in worke is expreſſed.

$$\sqrt{.408} + \sqrt{.324} —— \sqrt{.204} —— \sqrt{.162.}$$

whiche nomber ſhall be taken for the newe diuidede : and muſt be diuided by. 3. that is the newe diuiſor. In whoſe ſtede I ſet. $\sqrt{.9}$. for moare redineſſe in worke. Therfore I ſet thẽ doune in order, as here foloweth.

$$\sqrt{.408} + \sqrt{.324} —— \sqrt{.204} —— \sqrt{.162} \ (\sqrt{.45\tfrac{1}{3}} + 6 —— \sqrt{22\tfrac{2}{3}} —— \sqrt{18.}$$
$$\sqrt{.9} \qquad \sqrt{.9.} \qquad \sqrt{.9} \qquad \sqrt{.9}$$

And then doe I ſeke how often. $\sqrt{.9}$. maie bee founde in. $\sqrt{.408}$. whiche maie bee. $45\tfrac{1}{3}$ of tymes. Where= fore I ſet. $\sqrt{.45\tfrac{1}{3}}$ in the *quotiente.* And then doe I rei= terate the diuiſor, and ſette it vnder. $\sqrt{.324}$. where I finde it. 36. tymes : and therefore ſet I. 6. for it, bicauſe the *quotiente* els would bee. $\sqrt{.36}$. whiche is iuſtly .6. Thirdly, I remoue the diuiſor vnder $\sqrt{.204}$. where it maie bee founde. $22\tfrac{2}{3}$ tymes. For whiche I ſette $\sqrt{.22\tfrac{2}{3}}$ in the *quotiente.* And then ſet I the diuiſor laſt of all vnder. 162. where it is founde. 18. tymes : and for that cauſe I ſet $\sqrt{.18}$. in the *quotiente* : And ſo is the whole *quotiente* $\sqrt{.45\tfrac{1}{3}}$ + 6 —— $\sqrt{.22\tfrac{2}{3}}$ —— $\sqrt{.18}$.

Scholar. This diuiſion is ſtraunge to credite, al= though it be not difficulte to worke.

Maſter. If you doubte of it, you maie vſe the ac= cuſtomable trialle by the contrary kinde.

                                       Scholar.

of Surde nombers.

Scholar. So must it folowe, that if I dooe multi=
plie this *quotiente* by the firste diuisor, the firste diui=
dende will resulte thereof.

And for the proofe of that, I dooe multiplie,

$\sqrt{.45\frac{1}{3}.}$ ——|—— $.6.$ ———— $\sqrt{.22\frac{2}{3}}$ ———— $.\sqrt{.18.}$ by
$\sqrt{.6.}$ ——|—— $\sqrt{.3.}$ But for the moare ease, I doe tourne
all the mixte nombers into onely fractions. And then
doe I multiplie them orderly.

$\sqrt{.\frac{136}{3}}$ ——|—— 6 ———— $\sqrt{.\frac{68}{3}}$ ———— $\sqrt{.18.}$
$\sqrt{.6}$ ——|—— $\sqrt{.3.}$

$\sqrt{.\frac{816}{3}}$ ——|—— $\sqrt{.216}$ ———— $\sqrt{.\frac{408}{3}}$ ———— $\sqrt{.108.}$
$\sqrt{.\frac{408}{3}}$ ———— $\sqrt{.108}$ ———— $\sqrt{.\frac{204}{3}}$ ———— $\sqrt{.54.}$

$\sqrt{.272}$ ——|—— $\sqrt{.216}$ ——|—— $\sqrt{.136}$ ——|—— $\sqrt{.108.}$
$-\sqrt{.68}$ ———— $\sqrt{.54}$ ———— $\sqrt{.136}$ ———— $\sqrt{.108.}$

$\sqrt{.68}$ ——|—— $\sqrt{.54.}$

First I multiplie $\sqrt{.\frac{136}{3}}$ by $\sqrt{.6.}$ and there commeth
$\sqrt{.\frac{816}{3}}$ that is. $\sqrt{.272}$. Again I doe multiplie $.6.$ or $\sqrt{.36}$
by $\sqrt{.6.}$ and it maketh $\sqrt{.216.}$ Then I multiplie $\sqrt{.\frac{68}{3}}$
by. $\sqrt{.6.}$ & it giueth. $\sqrt{.\frac{408}{3}}$ whiche is. $\sqrt{.136.}$ Fourthly
$\sqrt{.18.}$ multiplied by. $\sqrt{.6.}$ dooeth make. $\sqrt{.108.}$ All
whiche I set doune with their conueniente signes.

After that I multiplie. $\sqrt{.\frac{136}{3}}$ by. $\sqrt{.3.}$ and it yeldeth
$\sqrt{.\frac{408}{3}}$ that is. $\sqrt{.136.}$ whiche I sette doune with his
signe ——|—— . Then $\sqrt{.36}$ by. $3.$ maketh $\sqrt{.108.}$ Third=
ly. $\sqrt{.\frac{68}{3}}$ by $\sqrt{3.}$ doeth giue. $\sqrt{.68.}$ and last of all, $\sqrt{.18.}$
multiplied by. $\sqrt{.3.}$ bryngeth forthe. $\sqrt{.54.}$

When all these be placed conueniently, I doe con=
sider that ——|—— $\sqrt{.136.}$ and ———— $\sqrt{.136.}$ maie bee
bothe cancelled, bicause the one doeth abate the other.
And likewaies, ——|—— $\sqrt{.108.}$ and ———— $\sqrt{.108.}$ eche
abate other : so that thei must bothe be reiected.

Then I see, that $\sqrt{.68.}$ beyng abated out of $\sqrt{.272}$
there will remain. $\sqrt{.68.}$ And in like. $\sqrt{.54.}$ beyng a=
bated out of. $\sqrt{.216.}$ doeth leaue. $\sqrt{.54.}$ So that the
whole multiplicatiõ doth make iustly $\sqrt{68}$ ——|—— $\sqrt{54}$

Rr.i. whiche

The Arte

whiche is the firste diuidende. And so is that diuision approued good.

An other example.

Master. Yet for you exercise, you shall haue some examples moare of diuision.

$\sqrt{}$.456. ——— .$\sqrt{}$.72. is sette to bee diuided by $\sqrt{}$.18 —+— $\sqrt{}$.6.

Scholar. That diuisor must I multiplie by his contrarie, whiche is the *Residualle.* $\sqrt{}$.18 —— $\sqrt{}$.6. And so, as you maie sone perceiue, there will rise. 18. ——— 6. that is 12. whiche must be kepte for the newe diuisor.

$$\sqrt{}.18. \quad —+— \quad \sqrt{}.6.$$
$$\sqrt{}.18. \quad ——— \quad \sqrt{}.6.$$
$$18 \quad ——— \quad .6.$$
$$\text{That is, } 12.$$

Then shall I multiplie the former diuidende, that is $\sqrt{}$.456 —— $\sqrt{}$.72 by the same *residualle* $\sqrt{}$.18 —— $\sqrt{}$6

$$\sqrt{}. \; 456 \quad ——— \quad \sqrt{}. \; 72.$$
$$\sqrt{}. \quad 18 \quad ——— \quad \sqrt{}. \; 6.$$
$$\sqrt{}.8208 \quad ——— \quad \sqrt{}.1296.$$
$$\sqrt{}.432. \quad ——— \quad \sqrt{}.2736.$$

$\sqrt{}$.8208 —+— $\sqrt{}$.432 ——— $\sqrt{}$.2736 ——— $\sqrt{}$.1296.

And there will rise of that multiplication, as here by example appereth $\sqrt{}$.8208 —+— $\sqrt{}$.432 ——— $\sqrt{}$.2736 ——— 1296. whiche nōber I shall diuide by. 12. that was founde for the newe diuisor. And then will the quotiente bee. $\sqrt{}$.57 —+— $\sqrt{}$.3 ——— $\sqrt{}$.19. ——— $\sqrt{}$.9. As here in woorke doeth appeare.

$\sqrt{}$.8208 —+— $\sqrt{}$.432 ——— $\sqrt{}$.2736 ——— $\sqrt{}$.1296 ($\sqrt{}$.57 —+— $\sqrt{}$.3 ——— $\sqrt{}$ 19 ——— $\sqrt{}$ 9
$\sqrt{}$.144. $\sqrt{}$ 144 $\sqrt{}$.144 $\sqrt{}$.144.

Where I haue set. $\sqrt{}$.144. for. 12. seyng thei be all one : but that. $\sqrt{}$.144. is moare apte for this woorke. And I haue repeated it as often tymes, as the diuisor should be remoued.

The proofe.

But now to trie this woorke, whether it bee well wroughte, I shall multiplie this *quotiente* by the firste diuisor, & then ought the firste diuidende to amounte.

As

of Surde nombers.

As here in example, you see wroughte.

√. 57. ——— √.3. ——— √. 19. ——— √.9.
√. 18. ——— √.6.

√.1026 ——— √.54 ——— √.342 ——— √.162.
√. 342 ——— √.18 ——— √.114 ——— √. 54.

√.1026 ——— √.18. ——— √.114 ——— √.162.

Where ——— √.54. doeth cancell ——— √.54.
and is cancelled by it.

So ——— √.342. and ——— √.342. exclude one
an other, and therefoze muſt bee bothe reiected. And
then remaineth onely,

√.1026 ——— √.18. ——— √.114 ——— √.162.

whiche nombers J dooe well examine : and kinde that
√.114. beyng abated out of. √.1026. there will re=
maine. √456. Againe if ——— √.18. be ſubtracted
out of ——— √.162. there will reſte ——— √.72. And
ſo is that whole multiplicatiõ onely. √.456 ——— √72
agreable to the firſte diuidende. Wherby it is mani=
feſte, that the foamer diuiſion was good.

Maſter. How can you woozke this example?
Where. 24. is ſet to be diuided by. 3. ——— √.8.

*The thirde
example.*

Scholar. J muſt ſtill obſerue the generalle rule.
And multiplie bothe thoſe nombers, by the contrarie
of the diuiſoz, that is, by the *reſidualle.* 3 ——— √.8. And

24.
3 ——— √.8
———————
72 ——— √.4608

of the firſte multiplication of it,
with the diuidende. 24, there ri=
ſeth. 72 ——— √.4608. Of the
ſeconde multiplica=
tion, where the *Binomialle* is multiplied
by the *Reſidualle,* that is his contrary, the
totalle will be. 9 ——— 8. that is by. 1.
And therfoze ſeyng. 1. doeth nother mul=
tiplie noz diuide, the foamer nomber.

3. ——— √.8
3. ——— √.8
9. ——— .8.
That is. 1.

That is. 72 ——— .√.4608. is the *quotiente,* when
24. is diuided by 3. ——— √.8.

Kr.ii.   Foz

The Arte

The proofe.

Foʒ pʒofe whereof, I multiplie 72 ——— √.4608 that is the *quotiente*, by. 3. —+— √.8. And there riſeth 216 —+— √.41472 ——— √.41472 ——— √.36864. whereof. 2. nombers differyng but by —+— & —— muſte bothe bee reiected, as nombers ſuperfluouſe.

$$
\begin{array}{l}
72 \text{ ——— } \sqrt{.}4608. \\
3 \text{ —+— } \sqrt{.}8. \\
\hline
216 \text{ ——— } \sqrt{.}41472 \\
\sqrt{.}41472 \text{ ——— } \sqrt{.}36864 \\
\hline
216 \text{ ——— } .192.
\end{array}
$$

That is. 24.

Then. 36864. is a *ſquare* nomber, and hath. 192 foʒ his roote. Wherfoʒe the whole nomber is, 216———192 that is (as it is manifeſte inough) 24. And ſo is the whole wooʒke pʒoued good.

The fourthe example.

Maſter. You ſhall haue one example moare, and then will I make an ende of diuiſion.

When √.6570. —+—. √.254. is pʒopounded to bee diuided by √.54 ——— √.6. I would knowe the *quotiente.*

Scholar. I ſee the newe diuiſoʒ will be. 54———6. that is. 48.

And then foʒ to finde a diuidende conueniente, I ſhall multiplie the firſte diuidende, by the contra= rie of the firſte diuiſoʒ, that is by √.54 —+— √6 And there will riſe, as you ſee. √. 354780.

$$
\begin{array}{l}
\sqrt{.} \ 6570 \text{ —+— } \sqrt{.}254. \\
\sqrt{.} \ \ \ 54 \text{ —+— } \sqrt{.} \ 6. \\
\hline
\sqrt{.}354780 \text{ —+— } \sqrt{.}13716. \\
\sqrt{.} \ 39420 \text{ —+— } \sqrt{.} \ 1524.
\end{array}
$$

—+— √.13716 —+— √.39420. —+— √.1524.

That diuidende muſt be diuided by. 48. oʒ moare ap= tly by. √.2304. And the *quotiente* will bee.

$$
\sqrt{.}153\tfrac{189}{192} \text{ —+— } \sqrt{.}5\tfrac{183}{192} \text{ —+— } \sqrt{.}17\tfrac{63}{576} \text{ —+— } \sqrt{.}\tfrac{127}{192}.
$$

As here appeareth in wooʒke.

$$
\sqrt{.}354780 \text{ —+— } \sqrt{.}13716 \text{ —+— } \sqrt{.}39420 + \sqrt{.}1524 \ (\sqrt{.}153\tfrac{189}{192} + \sqrt{.}5\tfrac{183}{192} + \sqrt{.}17\tfrac{63}{576} + \sqrt{.}\tfrac{127}{192}
$$

$$
\sqrt{.}2304. \qquad \sqrt{.}2304 \qquad \sqrt{.}2304 \qquad \sqrt{.}2304.
$$

The proofe.

And that this wooʒke is good, I will pʒoue it by multiplication.

of Surde nombers.

multiplication. As the example folowyng dooeth de=
clare. Where by the firſte multiplication there com=
meth. 8. nombers, that is. 4. with. ──── . and. 4.
with ──── .

$$\sqrt{.\tfrac{29565}{192}} + \sqrt{.\tfrac{1143}{192}} + \sqrt{.\tfrac{3285}{192}} + \sqrt{.\tfrac{127}{192}}$$

$$\sqrt{.54.} \quad ──── \quad \sqrt{.6.}$$

$$\sqrt{.\tfrac{1596510}{192}} + \sqrt{.\tfrac{61722}{192}} + \sqrt{.\tfrac{177390}{192}} + \sqrt{.\tfrac{6858}{192}}$$

$$──── \sqrt{.\tfrac{177390}{192}} \quad \sqrt{.\tfrac{6858}{192}} \quad \sqrt{.\tfrac{19710}{192}} \quad \sqrt{.\tfrac{762}{192}}$$

$$\sqrt{.\tfrac{1596510}{192}} + \sqrt{.\tfrac{1722}{192}} ──── \sqrt{.\tfrac{19710}{192}} ──── \sqrt{.\tfrac{762}{192}}$$

$$\sqrt{.6570.} + \sqrt{.254.}$$

And bicauſe the firſte nōber with ──── , is equalle
to the thirde with ──── , therfoze thei bothe muſt be
reiected. Again in as moche as the ſeconde nōber with
──── is equalle to the fourthe nomber with ──── ,
thei bothe ſhall bee cancelled. And then remaineth. 2.
nombers with ──── , and other. 2. with ──── .

So if you abate the thirde ──── out of the firſte
──── , the *quotiente* will be. $\sqrt{.6570.}$

Likewaies if you abate the fourthe ──── out of
the ſeconde ──── , the *quotiente* will yelde. $\sqrt{.254.}$

And thei bothe will make the firſte diuidende. $\sqrt{.}$
6570. Wherby the foзmer diuiſiō is appзoued good.
Maſter. This ſhall ſuffice foз diuiſion.

Of extraction of rootes.

He nexte wooзke is extraction of rootes :
whiche you maie very eaſilie wooзke, by
puttyng the ſigne of the roote, that you
deſire, befoзe the whole nomber. As if
you would haue the ſquare roote of $\sqrt{.10}$
──── $\sqrt{.5}$. this is it $\sqrt[\sqrt{}]{.10}$ ──── $\sqrt{.5}$. The *Cubike*
roote of the ſame nōber is. $\sqrt[\sqrt{\sqrt{}}]{}.\sqrt{.10}$ ──── $\sqrt{.5}$. And
the *zenzizenzike* roote of it is $\sqrt[\sqrt{}]{}.\sqrt{.10}.$ ──── $\sqrt{.5}.$
But if you will haue the *Square* roote of. 10 ──── $\sqrt{.5}$

Kr.iii.   it

The Arte

it is. √.10 ——— √.5. And his *Cubike* roote is. ⋀⋀.10
——— √.5. Likewaies his *zenzizenzike* roote is
⋁⋁.10 ——— √.5.

So of. ⋀⋀.18 ——— 2. the *Square* roote is √. ⋀⋀.18
——— 2. The *Cubike* roote is. ⋀⋀. ⋀⋀.18 ——— 2.
And the *zenzizenzike* roote is. ⋁⋁. ⋀⋀. ——— 2.

Scholar. Hereby I perceiue that the later parte of
the côpofition, is not varied at all, but onely the firfte
parte taketh vnto it the figne of the roote. And that
figne is referred to the whole compounde nomber.

Mafter. Thefe rootes therefoze bee called *vniuer=*
falle rootes, bicaufe thei are the rootes, not of the feue=
ralle partes of the compounde nôber, but of the whole
compounde nomber. And that is the difference, be=
twene the common *Surde* nombers, and *vniuerfalle roo=*
tes, Foz if √.24 ——— √.144. be fette foz a common
Surde nomber, then doeth it betoken, that I muft take
2. rootes, that is. √.24. and √.144. and ioyne theim
together. But if it ftande foz an *vniuerfalle roote*,it re=
pzefenteth the roote of this whole nomber. 24 ———
√.144. whiche is. 6. foz the whole *Square* is. 36.

Scholar. I perceiue it well. Foz. √.144. beeyng
12, that. 12. with. 24. dooeth make. 36. And therefoze
muft the *vniuerfalle roote* of. 24 ——— √.144. bee. 6.
And fo √.24 ——— √.144. is iuft. 6.

But if. √.24 ——— √.144. doe ftande foz a com=
mon *Surde* nomber compounde : then is it made of. 2.
rootes, that is √.24. whiche is almofte. 5. and √.144
beyng. 12. And fo the whole compounde roote, in that
fozte is almofte. 17. And is nighe. 3. tymes fo moche
as the fame nomber, beyng an *vniuerfalle roote*.

Mafter. Bicaufe you maie perceiue it the better,
I will put an example in *Square* nombers, made like
Surdes. As this. √.81 ——— √.361 if it be an *vniuerfalle*
roote, then it is equalle to 10. Foz I muft take firft the
roote of the lafte nomber, whiche is. 19. And adde it
with

of Surde nombers.

with. 81. wherby there amounteth. 100. whose roote
is. 10. But if it stand after the common sorte of *Surde*
nombers, it betokeneth the roote of. 81. and the roote
of. 361. (that is. 9. and. 19) to bee added together. And
so thei make. 28. which is farre aboue. 10.

But farther now, if it stande for a common *Surde*
nomber : And I would haue the *Square* roote of it, then
is that. √ √ .81 —┼— √.361. And betokeneth the
Square roote of the square roote of. 81. and the Square
roote of. 361. added together, that is the square roote
of. 28. But moste generally and moste aptly, it beto=
keneth the roote of the *vniuersalle roote* of. 81. & √.361.

Scholar. Now I perceiue that in Addition, and
Subtraction of *Surdes*, the last nombers that did result
of that woorke, were *vniuersalle rootes.*

Master. You saie truthe. But harke what mea=
neth that hastie knockyng at the doore?

Scholar. It is a messenger.

Master. What is the message? tel me in mine eare

Yea sir is that the mater? Then is there noe reme=
die, but that I must neglect all studies, and teaching,
for to withstande those daungers. My fortune is not
so good, to haue quiete tyme to teache.

Scholar. But my fortune and my fellowes, is
moche woorse, that your vnquietnes, so hindereth our
knowledge. I praie God amende it.

Master. I am inforced to make an eande of this
mater : But yet will I promise you, that whiche you
shall chalenge of me, when you see me at better laiser :
That I will teache you the whole arte of *vniuersalle
rootes.* And the extraction of rootes in all *Square Surdes* :
with the demonstration of theim, and all the former
woorkes.

If I mighte haue been quietly permitted, to reste
but a litle while löger, I had determined not to haue
ceased, till I had ended all these thinges at large. But
                                                              now

The Arte

now farewell. And applie your studie diligently in
this that you haue learned. And if I maie gette any
quietnesse reasonable, I will not forget to performe
my promise with an augementation.

Scholar. My harte is so oppressed with pensifenes,
by this sodaine vnquietnesse, that I can not expresse
my grief. But I will praie, with all theim that
loue honeste knowledge, that God of his
mercie, will sone ende your troubles,
and graunte you soch reste, as
your trauell doeth merite.
And al that loue lear=
nyng : saie ther
to. Amen.
Master. Amen,
and Amen.

❡ Imprinted at London,
by Jhon Kyngston.

Anno domini. 1557.

www.ingramcontent.com/pod-product-compliance
Lightning Source LLC
Chambersburg PA
CBHW071355170526
45165CB00001B/59